Petroleum Microbiology

In the oil and gas industry, technologies have been developed to address microbial-related issues such as oil field souring, microbiologically influenced corrosion, bio-fouling, and targeted measures for risk assessment and mitigation. Microorganisms have also benefited the oil sector through microbial-enhanced oil recovery and bio-remediation of petroleum-contaminated environments. However, during the current transitional phase in the oil and gas industry, the role of the microbiome within the current infrastructure and its potential impact on future systems remains an open question. *Petroleum Microbiology: The Role of Microorganisms in the Transition to Net Zero Energy* explores technological advances in applied microbiology in the oil and gas sector that can be utilized in its transition to renewable energy systems.

- Provides insights on the potential of applying microbiological techniques in oil systems to pave the way to achieving net-zero energy.
- Presents the major industrial problems caused by microbes and their beneficial activities from both fundamental and applied perspectives.
- Covers such technologies as next-generation sequencing, sampling, and diagnostics.
- Offers a solid foundation on the importance of microbes to key aspects of the energy industry.
- Seeks to answer the question: what role will microorganisms play in the evolution of energy systems?

Featuring chapters from interdisciplinary experts spanning academia and industry, this is an excellent reference for microbial ecologists, molecular biologists, operators, engineers, chemists, and academics involved in the oil and gas sector, working toward energy transition.

Microbes, Materials, and the Engineered Environment

Series Editor
Torben Lund Skovhus and Richard B. Eckert

Microbial Bioinformatics in the Oil and Gas Industry: Applications to Reservoirs and Processes
Kenneth Wunch, Marko Stipaničev, and Max Frenzel

Failure Analysis of Microbiologically Influenced Corrosion
Torben Lund Skovhus and Richard B. Eckert

Petroleum Microbiology: The Role of Microorganisms in the Transition to Net Zero Energy
Biwen Annie An Stepec, Kenneth Wunch, and Torben Lund Skovhus

For more information about this series, please visit: https://www.routledge.com/ Microbes-Materials-and-the-Engineered-Environment/book-series/CRCMMEE

Petroleum Microbiology
The Role of Microorganisms in the Transition to Net Zero Energy

Edited by
Biwen Annie An Stepec
Kenneth Wunch
Torben Lund Skovhus

CRC Press
Taylor & Francis Group
Boca Raton London New York

CRC Press is an imprint of the
Taylor & Francis Group, an **informa** business

Designed cover image: ©Shutterstock images

First edition published 2024
by CRC Press
2385 NW Executive Center Drive, Suite 320, Boca Raton FL 33431

and by CRC Press
4 Park Square, Milton Park, Abingdon, Oxon, OX14 4RN

CRC Press is an imprint of Taylor & Francis Group, LLC

© 2024 Biwen Annie An Stepec, Kenneth Wunch, and Torben Lund Skovhus

Library of Congress Cataloging-in-Publication Data
Names: An Stepec, Biwen Annie, editor. | Wunch, Kenneth, editor. | Skovhus, Torben Lund, editor.
Title: Petroleum microbiology: the role of microorganisms in the transition to net zero energy / edited by Biwen Annie An Stepec, Kenneth Wunch, and Torben Lund Skovhus. Description: First edition. | Boca Raton, FL : CRC Press, 2024. | Series: Microbes, materials, and the engineered environment | Includes bibliographical references and index. | Summary: "In the oil and gas industry, technologies have been developed to address microbial-related issues such as oil field souring, microbiologically influenced corrosion, biofouling, and targeted measures for risk assessment and mitigation. Microorganisms have also benefited the oil sector through microbial enhanced oil recovery and bioremediation of petroleum-contaminated environments. However, during the current transitional phase in the oil and gas industry, the role of the microbiome within the current infrastructure and its potential impact on future systems remains an open question. Petroleum Microbiology: The Role of Microorganisms in the Transition to Net Zero Energy explores technological advances in applied microbiology in the oil and gas sector that can be utilized in its transition to renewable energy systems. Features: Provides insights on the potential of applying microbiological techniques in oil systems to pave the way to achieving net-zero energy. Presents the major industrial problems caused by microbes and their beneficial activities from both fundamental and applied perspectives. Covers such technologies as next-generation sequencing, sampling, and diagnostics. Offers a solid foundation on the importance of microbes to key aspects of the energy industry. Seeks to answer the question: what role will microorganisms play in the evolution of energy systems? Featuring chapters from interdisciplinary experts spanning academia and industry, this is an excellent reference for microbial ecologists, molecular biologists, operators, engineers, chemists, and academics involved in the oil and gas sector, working toward energy transition"-- Provided by publisher.
Identifiers: LCCN 2023048703 (print) | LCCN 2023048704 (ebook) | ISBN 9781032262055 (hbk) | ISBN 9781032269566 (pbk) | ISBN 9781003287056 (ebk)
Subjects: LCSH: Petroleum--Microbiology. | Microbial biotechnology. | Industrial microbiology.
Classification: LCC QR53.5.P48 P483 2024 (print) | LCC QR53.5.P48 (ebook) | DDC 547/.83--dc23/eng/20240129
LC record available at https://lccn.loc.gov/2023048703
LC ebook record available at https://lccn.loc.gov/2023048704

ISBN: 978-1-032-26205-5 (hbk)
ISBN: 978-1-032-26956-6 (pbk)
ISBN: 978-1-003-28705-6 (ebk)

DOI: 10.1201/9781003287056

Typeset in Times
by SPi Technologies India Pvt Ltd (Straive)

Contents

SECTION IV   *Subsurface Reservoir Microbiome and Hydrocarbon Degradation*

SECTION V   *Microbial Based Emerging Technologies in Energy Systems*

SECTION VI Future Perspectives on Microorganisms in the Energy Transition

Foreword

The awareness that microbial activity may impact operations in the oil and gas industry began roughly 100 ago when Richard Gaines, in 1910,[1] reported on the influence of bacterial activity on iron and steel corrosion and Bastin et al., in 1926,[2] reported the presence of sulfate-reducing bacteria as a potential source of H_2S in oil field waters. These two seminal papers unlocked a new area of scientific interest in how microbes impact our materials and industrial processes.

Since these early days, our understanding of the specific microbial reactions that drive these risks, the species associated with them, and relevant environmental factors have greatly increased. In addition, as we gained more knowledge about the underlying microbiological fundamentals, we have also developed the tools the oil and gas industry relies on for its risk assessments and mitigation strategies to protect the integrity of our assets, the health and safety of our people, and the environment.

Beyond unraveling the fundamentals, we have also greatly benefited from advances in biotechnology from a wide range of disciplines that spurred significant progress in our ability to manage microbial risks, for instance, biofilm research. Whether it is in the oil and gas context or in other areas such as shipping, medicine, or hygiene, unwanted microbes are usually present as biofilm and any new insights into their formation and manipulation in one discipline are potentially pertinent across different industries. Similarly, the development and transfer of affordable "Next Generation Sequencing" (NGS), catalyzed by the human genome project, machine learning approaches and the different "Omics" technologies (transcriptomics, metabolomics, and proteomics) greatly boost our ability to understand and monitor the presence and activity of "risk carrying" organisms.

It is this theme of transferability of knowledge and expertise that the authors of this book highlight as they point out the wealth of relevant transferable knowledge and expertise that already exists in the field of petroleum microbiology and how it can be harnessed to anticipate and manage microbial risks associated with the new processes and technologies that will facilitate the energy transition. The potential impact of microbial activity during H_2 storage is just one key example to showcase this. Additionally, we are encouraged to recognize that the energy industry's move to lower-carbon energy sources is closely related to an increased commitment to minimize environmental impacts and improved sustainability during operations. It is these ambitions which are shared across different industries that open up an even wider range of opportunities for continued cross-disciplinary fertilization.

To realize the rapid pace of transition needed to meet the world's climate goals, it undoubtedly will require contributions across many disciplines. Toward this, the field of petroleum microbiology has a lot to offer as it provides a large body of relevant and transferable know-how, which will be needed to manage potential microbial risks and inspire innovation as we transition to a "Net Zero" future.

1. Gaines R. 1910. Bacterial activity as a corrosive influence in the soil. *Journal of Industrial and Engineering Chemistry*, 2, 128–130.
2. Bastin E, Greer F, Merrit C, Moulton G, 1926. The Presence of Sulphate Reducing Bacteria in Oil Field Waters. *Science*, 63 (1618), 21–24.

Dr. Reinhard Paul Dirmeier
Principal Scientist – BP, USA

Preface

The rapid energy transition from petroleum to renewable energies represents a profound and transformative shift in our global energy landscape. According to the International Energy Agency's World Energy Outlook 2022, the proportion of fossil fuels in the global energy market is expected to be around 60% by 2050. Whereas the rise of energy production from renewable sources is expected to account for the rest. An intricate interplay between environmental, economic, and geopolitical challenges is intertwined with this energy transition. Renewable energies, such as solar, wind, hydro, CO_2 storage, hydrogen, and geothermal power, have evolved as sustainable options to drive this energy transition. However, the vast knowledge generated from the fossil fuel sector is extremely valuable for the development of renewable systems as they may face similar challenges as before.

One of the most versatile aspects of this energy transition is centered around the role of microorganisms. Microorganisms, both discovered and unexplored, play a significant role in our energy system. Not only are they extremely diverse, but they are also capable of adapting to a wide range of extreme environments. For decades, the study of petroleum microbiology has offered insights on the potential impacts of microorganisms at an industrial level, including microbiologically influenced corrosion (MIC), hydrocarbon biodegradation, reservoir souring, and microbially enhanced oil recovery (MEOR). Whereas in the renewable sector, the discussions around microorganisms are mostly surrounding biofuel production, biogas generation, microbial fuel cells, biomaterials, hydrogen production, and carbon capture and storage (CCS). At first glimpse, there are no immediate common grounds between the two sectors. But at a closer look, microorganisms that have the same metabolic properties may be undesired on one side while valuable for the other. For example, microorganisms capable of direct electron uptake from metal surfaces cause microbiologically influenced corrosion, but their metal-cell interaction proved to be extremely valuable for driving the development of microbial fuel cells. How we can effectively transfer the knowledge generated from the field of petroleum microbiology to the emerging renewable sector will be the main focus of this book.

As editors of this comprehensive book, we are honored to present a collection of cutting-edge research and expert insights that illuminate the intricate relationship between microorganisms and energy systems. The journey through these pages is a testament to the dedication and passion of the authors, offering a comprehensive perspective on how microorganisms shape our energy landscape. **Section I** of the book, authored by a team of experts, provides an in-depth look at the evolution of petroleum microbiology and its pivotal role in the energy transition. This section serves as a foundation upon which the subsequent chapters build, offering readers a broad understanding of the field's significance. **Section II – Microbial Ecology of Energy Systems** explores the impact of microbial biofilms on subsurface energy systems and the potential for managing microorganisms to improve the environmental footprint of oil and gas operations. These chapters, authored by leading researchers, highlight the potential for sustainable energy practices through microbial intervention. **Section III – Microbiologically Influenced Corrosion (MIC) and Souring** delves into the complex world of microorganism-induced souring and corrosion in energy systems. Authors provide valuable insights into the effects of high salinity practices on sulfidogenesis and using novel bioinformatic approaches to analyze MIC. The research presented here is essential for safeguarding our energy infrastructure. **Section IV – Subsurface Reservoir Microbiome and Hydrocarbon Degradation** takes us deep into the reservoirs, exploring the ecological interactions of microbial co-occurrence in oil degradation. This section unveils the intricacies of hydrocarbon metabolism and the role of microorganisms in this critical process. **Section V – Microbial-based Emerging Technologies in Energy Systems** showcases innovative solutions for managing and mitigating MIC. These chapters provide a glimpse into the future of microbial management and sustainability. **Section VI – Future Perspectives on Microorganisms in the Energy Transition** offers reflections on the road ahead. As we contemplate the future of energy, these authors guide us through the possibilities and challenges that lie ahead.

The editors wish to thank all of the authors and reviewers who contributed their time, knowledge, and expertise to this book, making it an invaluable resource for many years to come. The editors hope this book will stimulate further research, discussions, and developments in the field of oilfield bioinformatics and its importance to the oil and gas industry.

Biwen Annie An Stepec
Norwegian Research Centre (NORCE)
Norway

Kenneth Wunch
Lanxess
United States

Torben Lund Skovhus
VIA University College
Denmark

Editors

Biwen Annie An Stepec (formerly An), PhD, holds the position of Senior Researcher at the Norwegian Research Centre (NORCE) in Bergen, Norway. Her research portfolio encompasses diverse topics, including microbial processes and risks during hydrogen underground storage, microbiologically influenced corrosion, and microbial processes in extreme environments, particularly high salinity conditions. She earned a PhD in environmental microbiology at the University of Calgary (Canada), followed by a postdoc position at the Bundesanstalt für Materialforschung und -prüfung (Germany) working on understanding the intricacies between materials and microorganism during microbiologically influenced corrosion. During this time, she was a guest lecturer at the Humboldt University of Berlin, where she designed and lectured a course on the topic of "electroactive" microorganisms and their role within our society. Currently, Dr. An Stepec actively participates in numerous national and international projects, including HyLife-Microbial risks associated with hydrogen underground storage in Europe (awarded by Clean Energy Transition Partnership) aimed at unraveling the pivotal roles played by microorganisms in subsurface energy storage systems. As a young scientist, Dr. An Stepec has published numerous papers on the topic of environmental and applied microbiology. She is currently the Roadmap Leader for RM04-Hydrogen Storage at Hydrogen Europe Research, Work Group leader for COST Action Euro-MIC Network (CA20130), and part of the technical committee at the International Symposium on Applied Microbiology and Molecular Biology in Oil Systems (ISMOS).

Kenneth Wunch, PhD, holds the position of Energy Technology Fellow at Lanxess in Houston responsible for business development, technology transfer, and shaping the innovation pipeline and strategy for global oil and gas applications. He earned a PhD in environmental microbiology at Tulane University, followed by an Exxon-funded postdoc working on bioremediation of the *Valdez* spill and Prince William Sound. He then accepted a professorship at the Texas Research Institute for Bioenvironmental Studies and became Director of the SHSU Disease Vector Program for the Air Force Border Health and Environmental Threats Initiative. Dr. Wunch moved into petroleum microbiology at Baker Hughes with responsibilities in oilfield production chemistry and development and application of technologies associated with oilfield microbiology, microbially influenced corrosion (MIC), sulfide control, and corrosion inhibition. He later moved to BP as a production microbiologist with responsibilities in development and implementation of R&D strategies for reservoir souring and MIC before his current role at Lanxess. Dr. Wunch is the author of more than 40 publications and patents and has chaired several committees, including SPE/NACE Deepwater Field Life Corrosion Prevention, Detection, Control and Remediation, ISMOS, NACE Control of Problematic Microorganisms in the Oil and Gas Industry, and the Energy Institute and Reservoir Microbiology Forum.

Torben Lund Skovhus, PhD, is Docent and Project Manager at VIA University College in the Research Centre for Built Environment, Climate, Water Technology and Digitalisation. He graduated from Aarhus University, Denmark, in 2002 with a master's degree (cand.scient.) in biology. In 2005, he earned a PhD from the Department of Microbiology, Aarhus University. In 2005, Torben was employed at Danish Technological Institute (DTI) in the Centre for Chemistry and Water Technology, where he was responsible for the consultancy activities for the oil and gas industry around the North Sea. Torben was heading DTI Microbiology Laboratory while he was developing several consultancy and business activities with the oil and gas industry. He founded DTI Oil and Gas in both Denmark and Norway, where he was team and business development leader for five years. Thereafter, Torben worked as project manager at DNV GL (Det Norske Veritas) in the field of corrosion management in both Bergen and Esbjerg. Torben is currently chair of AMPP SC-22 on Biodeterioration and ISMOS TSC, an organization he cofounded in 2006. He is an international scientific reviewer and the author of 150+ technical and scientific papers and book chapters related to industrial microbiology, applied biotechnology, corrosion management, oilfield microbiology, water treatment and safety, reservoir souring, and biocorrosion. He is scientific/technical reviewer with over 30 international journals in the same fields. He is co-editor of eight books and was honored with the NACE Technical Achievement Award in 2020 for outstanding research and outreach on the MIC in the Energy Sector. He was awarded the AMPP Fellow Honor title March 2024 for his extraordinary and outstanding contributions to the scientific or technical body of knowledge in the field of corrosion.

Contributors

Biwen Annie An Stepec
Energy & Technology
Norwegian Research Centre (NORCE)
Bergen, Norway

Charles D. Armstrong
Research & Innovation Group
Syensqo Oil and Gas Solutions
The Woodlands, TX, United States

Damon Brown
Group 10 Engineering Ltd
Calgary, Canada

Tanmay Chaturvedi
AAU Energy
Aalborg University
Aalborg, Denmark

Beate Christgen
School of Natural and Environmental
    Sciences
Newcastle University
Newcastle upon Tyne, United Kingdom

Renato M. De Paula
Business Development Group
Syensqo Oil and Gas Solutions
The Woodlands, TX, United States

Reinhard Paul Dirmeier
BP Biosciences Center
BP
San Diego, CA, United States

Nicole Dopffel
Energy & Technology
Norwegian Research Centre (NORCE)
Bergen, Norway

Richard B. Eckert
Microbial Corrosion Consulting, LLC
Commerce Township, MI, United States

Mary Eid
Eni US Operating Co., Inc.
Anchorage, AK, United States

Sarah E. Gasda
Energy & Technology
Norwegian Research Centre (NORCE)
Bergen, Norway

Neil Gray
School of Natural and Environmental
    Sciences (SNES)
Newcastle University
Newcastle upon Tyne, United Kingdom

Ian Head
Faculty of Science, Agriculture &
    Engineering
Newcastle University
Newcastle upon Tyne, United Kingdom

Clara Di Iorio
Eni S.p.A.
San Donato Milanese, Italy

Arul Jayaraman
Artie Mc Ferrin Department of
    Chemical Engineering
Texas A&M University
College Station, TX, United States

Cory Klemashevich
Artie Mc Ferrin Department of
    Chemical Engineering
Texas A&M University
College Station, TX, United States

Andrea Koerdt
Division 4.1 Biodeterioration and
    Reference Organisms
Bundesanstalt für Materialforschung
    und -prüfung (BAM)
Berlin, Germany

Susmitha Purnima Kotu
DNV Energy Systems
Houston, TX, United States
and
Artie Mc Ferrin Department of
    Chemical Engineering
Texas A&M University
College Station, TX, United States

Na Liu
Department of Physics and
    Technology
University of Bergen
Bergen, Norway

M. Sam Mannan
Mary Kay O'Connor Process Safety
    Center
Texas A&M University
College Station, TX, United States

Markus Pichler
RAG Austria AG
Vienna, Austria

Luciano Procópio
Laboratory of Biocorrosion and
    Biodegradation – LABIO
National Institute of Technology
Rio de Janeiro, Brazil

Julia R. de Rezende
The Lyell Centre
Heriot-Watt University
Edinburgh, United Kingdom

Jerzy Samojluk
Faculty of Geology, Geophysics and
    Environmental Protection
AGH University of Science and
    Technology
Kraków, Poland

Angela Sherry
Department of Applied Sciences
Northumbria University
Newcastle upon Tyne, United Kingdom

Mohammed Sindi
Research & Analytical Services
    Department
Saudi Aramco
Dhahran, Saudi Arabia

Torben Lund Skovhus
Research Centre for Built Environment,
    Energy, Water and Climate
VIA University College
Horsens, Denmark

Jakob Lykke Stein
AAU Energy
Aalborg University
Aalborg, Denmark

Carla N. Thomas
Product Stewardship Group
Syensqo Oil and Gas Solutions
The Woodlands, TX, United States

Mette H. Thomsen
AAU Energy
Aalborg University
Aalborg, Denmark

Raymond J. Turner
Department of Biological Sciences
University of Calgary
Calgary, Canada

Kenneth Wunch
Lanxess
New Orleans, LA, United States

Fang Yang
Centene Corporation
United States
and
Texas A&M University
College Station, TX, United States

Xiangyang Zhu
Research & Analytical Services
    Department
Saudi Aramco
Dharan, Saudi Arabia

Section I

Introduction

1 Petroleum Microbiology's Metamorphosis
Expert Insights on the Energy Transition

Biwen Annie An Stepec
Norwegian Research Centre, Bergen, Norway

Kenneth Wunch
Lanxess, New Orleans, LA, United States

Torben Lund Skovhus
VIA University College, Horsens, Denmark

Julia R. de Rezende
Heriot-Watt University, Edinburgh, United Kingdom

Markus Pichler
RAG Austria AG, Vienna, Austria

Susmitha Purnima Kotu
DNV, Houston, TX, United States

Sarah E. Gasda
Norwegian Research Centre, Bergen, Norway

Nicole Dopffel
Norwegian Research Centre, Bergen, Norway

DOI: 10.1201/9781003287056-2

3

In the dynamic landscape of today's energy sector, the shift from fossil fuels to sustainable energy sources has become a focal point of exploration and innovation. Yet, microorganisms continue to be a dominant player amid this transition period. In a series of enlightening interviews, we engage with experts who stand at the forefront of this research, offering valuable perspectives on the pivotal role of microorganisms in the ongoing energy transition. These conversations provide a unique window into the intricate relationships between microbiology and the energy transition, shedding light on how these tiny organisms hold the potential to drive significant change in our quest for a more sustainable energy future.

Please tell us a bit about your professional background and current line of work.

I am an environmental microbiologist with expertise in the microbiology of subsurface and engineered environments. I am particularly interested in industrial issues associated with microbial activity (e.g., souring, MIC), microbial control, and applications such as biogas production.

From your experience, how has the field of petroleum microbiology research changed over the past 10 years?

The use of molecular microbiology and easy access to DNA sequencing have certainly opened many doors to understanding the complex microbial community associated with petroleum microbiology questions. This has led to many answers and, interestingly, to further questions as well. I believe we have become more aware of the need for interdisciplinary collaboration as well.

In recent years, we have seen a massive shift within the energy landscape. To what extent do you think microbiology would play a role during this transition phase?

It should play an important role if we want to avoid expensive issues in the future. We already know of the importance of subsurface microbial communities and their potential to cause significant changes in engineered environments. By regularly monitoring from "time zero", well-informed mitigating decisions can be made in a timely manner.

Dr. Julia R. de Rezende
Assistant professor
Heriot-Watt University, UK

What are some research topics you believe are important to transfer from the oil and gas industry to the renewable sector?

Microbial ecology of subsurface environments, souring, and MIC. This should be supported by interdisciplinary research with (geo)chemists, engineers, modelers, and materials scientists, and should involve field monitoring and lab experiments. Genomics and metagenomics have a huge potential to unravel knowledge that can lead to groundbreaking solutions, and ecophysiological experiments will help test hypotheses generated by omics techniques and simulate scenarios applicable to the field.

With your expert opinions, how can microbial populations positively and negatively impact green energy as the energy landscape is quickly changing?

As seen for decades in the oil industry, microbial activity can have a detrimental effect on operations with the production of unwanted gases that can contaminate gas reserves, contribute to climate change, and put workers at risk. Biocorrosion can also lead to increased costs in infrastructure maintenance. However, we could also harness the activity of microbes for the production of energy-rich gases or other compounds that can themselves be energy sources with a lower impact on the environment.

Would microbial control remain an important aspect of renewables, and why?

For sure. Microbes are incredibly resilient and can make a living in unexpected environments and surfaces, reducing the lifespan of materials, infrastructure, or operations. Monitoring and early, strategic control are essential.

What would you like to see in the next 10 years in the field of petroleum/energy microbiology?

Interdisciplinary collaboration to tackle these complex, real-life industrial issues, as well as support for research that leads to significant understanding of underlying mechanisms and microbial interactions, which can really lead to meaningful, effective solutions and field-applicable technological advancement.

Please tell us a bit about your professional background and current line of work.

My background is in environmental microbiology with a special focus on subsurface and anaerobic microbiology. I am currently focusing on the effects of microbes in energy systems.

Dr. Nicole Dopffel
Senior Researcher
Norwegian Research Centre
(NORCE), Norway

From your experience, how has the field of petroleum microbiology research changed over the past 10 years?

I have not been active for 10 years in this field (so I guess I am still young), but I have already experienced the changes of several oil price crises and now the energy crisis. Prices for energy are either in the sky or at the bottom. The effects on research are always strong. When I started, enhanced oil recovery (EOR) was a very active topic including microbially enhanced oil recovery (MEOR). This is now over, and EOR is not attractive from a research standpoint. I also see that MIC got more and more attention over the past few years. I noticed a strong change when Dennis Enning's paper on EMIC was published. I had the feeling that suddenly the research exploded. Now real petroleum microbiology is difficult and as mentioned oil-related research is not very interesting. But I am not sad about this.

In recent years, we have seen a massive shift within the energy landscape. To what extent do you think microbiology would play a role during this transition phase?

Everything in the subsurface is affected by microbes. So, microbiology will play a vital role also in the future energy systems. Even a more important role as CO_2 and H_2 storage will be key pillars of the energy system and they need to work. Here there are still big question marks when it comes to microbiology. Also, corrosion of wind pillars and wind turbines is a major research area which has been ignored up until now.

What are some research topics you believe are important to transfer from the oil and gas industry to the renewable sector?

- Effects on H_2 underground storage
- Effects on CO_2 storage
- Biomethanation processes
- Biohydrogen processes
- Geothermal installations and microbial problems

What are some key technological advances you see that are important to transfer from petroleum to renewable energy?

I would say all of them. Reservoir microbiology is not specifically linked to oil but to everything which is inside a reservoir.

Would microbial control remain an important aspect of renewables and why?

Sure! Just speaking of hydrogen storage, the question of how to protect the tasty hydrogen from microbial consumption will probably be a research topic for many years. It has not even started yet.

What would you like to see in the next 10 years in the field of petroleum/energy microbiology?

I would love to have a stronger focus on microbial communities and not only single strains. Communities are the drivers in the subsurface but so very hard to research. I also would love if old-school cultivation work will still be done. With the new DNA methods, sometimes it is forgotten that you actually need enrichments and strains growing in the lab to understand your genomic data. Here I see a lot of cool developments in culturing difficult microbes and communities.

Please tell us a bit about your professional background and current line of work.

I am a reservoir engineer with more than 10 years of experience in storage of different gases in the subsurface. This includes, but is not limited to, the storage of hydrogen in porous depleted oil and gas reservoirs. Within my company RAG Austria AG, I am currently charged with supporting research on the subsurface aspects of hydrogen storage and the identification and development of future hydrogen storage projects.

Markus Pichler
Reservoir Engineer Subsurface Storage Development
RAG Austria AG, Austria

From your experience, how has the field of petroleum microbiology research changed over the past 10 years?

There is definitely a change from "those beasts are annoying" to how they can be utilized. It is still a fact that the industry needs to control and deal with microbial life in the subsurface in order to prevent negative effects on our operation. However, projects like Bio-EOR and Underground Sun Conversion are good initiatives in not only worrying about, but also utilizing microbes for the benefit of the industry.

In recent years, we have seen a massive shift within the energy landscape. To what extent do you think microbiology would play a role during this transition phase?

The most important topics are microbial fouling and microbial-induced corrosion, which will stay with the industry as long as we are injecting foreign fluids into the subsurface. Not only in hydrogen storage but also in geothermal applications do microbial-induced changes play an important role when developing a future monitoring and incident-preventing system. The growing awareness for these topics is not only reflected by numerous publications that have been produced in the past years but also by the fact that many oil and gas companies are now establishing their own microbial divisions and are actively spending money on research.

What are some research topics you believe are important to transfer from the oil and gas industry to the renewable sector?

I do not really see that big a shift, to be honest. We are and will still be an energy industry generating and distributing energy to our customers. So, our basic know-how will go in full into this new field of application. It is my understanding that all disciplines that have been supporting the energy industry for the past decades have their role to play in a future energy system and their know-how will be needed. In detail there are changes that need to be addressed, but to give a technical answer, the basic equations stay the same.

With your expert opinions, how can microbial populations positively and negatively impact green energy as the energy landscape is quickly changing?

If we only focus on how microbes can harm and prevent new technologies from arising, the perception will grow that change is impossible and companies as well as investors will no longer be interested. Don't get me wrong. It is important to list all the possible ways in which microbes could prevent a project from happening; however, it depends if this list is presented with an alarmist attitude or one where researchers actually can contribute to a solution. As I see it, the communication that should be done is rather simple. The aim must be to understand the processes that might happen for example in hydrogen storage. If the understanding is there, it can be verified by field tests and solutions can and will be found that will lead toward a positive outcome. Finally, with technologies like renewable methane, and microbial remediation of contaminated sites, there are already examples of how microbes can support a renewable energy future and who knows if there isn't a bug out there that can bind CO_2 in huge quantities.

What would you like to see in the next 10 years in the field of petroleum/energy microbiology?

Pilots and commercial projects but also a huge increase in public engagement. The transformation we are seeing is still a niche of our industry and although there is one press release after another common people are not yet really informed nor engaged. Lab experiments are great and dearly needed to explain the basics that we are seeing in the field. But nothing beats a long-term field experiment. Only by verifying lab experiments in the field can we really build confidence in emerging technologies. Also, in a public engagement it is something completely different if you show lab experiments compared to the actual application of the technology. Especially if you deliver this technology to the public so they can experience it and see the benefit they are gaining for themselves. If we manage the step from lab to field together with the public, then in my opinion the biggest showstopper for a future renewable energy system is out of our way.

Please tell us a bit about your professional background and current line of work.

I have a PhD in chemical engineering with a focus on microbiology. During my PhD, I studied the physiology of co-culture biofilms, particularly the ones implicated in microbial corrosion. I also investigated the metage-nome and metabolome associated with microbial corrosion from oilfield-produced water. After my PhD, I started working at DNV (an independent, energy-consulting company) managing microbiol-ogy projects covering topics like microbial corro-sion, biofilms, bioremediation, biofouling, etc. We work with oil and gas companies needing technical support in microbiology.

Dr. Susmitha Purnima Kotu
Senior Engineer
DNV, USA

From your experience, how has the field of petroleum microbiology research changed over the past 10 years?

It was 10 years ago that I started my PhD and was exposed to the field of petroleum microbiology for the first time. The biggest shift has been the widespread adoption of molecular microbiological methods for understanding the microbial community and the role of microorganisms. I see more professionals aware of the appropriate methods for sampling and preservation when conducting corrosion failure analysis. The most interesting of all these is the focus in research shifting from solely trying to solve microbial challenges in petroleum microbiology to also investigating the uses of microorganisms to help with energy transition. A couple of examples include a) using microorganisms for carbon capture and conversion to useful products such as jet fuel and beverage bottles (work done by LanzaTech) and b) using depleted oil and gas reservoirs to make hydrogen (work done by Cemvita Factory).

In recent years, we have seen a massive shift within the energy landscape, to what extent do you think microbiology would play a role during this transition phase?

Many energy companies have pledged net-zero emissions by 2050. This involves efficient, safe, and sustainable use of oil and gas assets ensuring no leaks or failures. Hence, the biggest role of microbiology in this energy transition is to reliably diag-nose the microbial threats and optimize the mitigation treatments.

What are some research topics you believe are important to transfer from the oil and gas industry to the renewable sector?

I think the concepts of microbial communities, microbial metabolism, appropriate sampling and preservation, and reliable understanding of microbial threats are critical.

An integration in the innovation of digital technologies and molecular biology is important during this transition. This can be particularly useful for identifying micro-bial biomarkers and applying the knowledge of these to microbial threats in oil and gas systems. This can also be helpful for meta-analysis of microbiological metabolic processes and identifying specific metabolic pathways of interest and tweaking these metabolic pathways to produce low-carbon products.

With your expert opinions, how can microbial populations positively and negatively impact green energy as the energy landscape is quickly changing?

Despite the increase in the use of renewables, fossil fuels will still contribute to 49% of the global energy mix (per DNV Energy Outlook 2022). This means that the negative impacts of microbiological populations in existing oil and gas operations cannot be ignored and should be accounted for appropriately. The most interesting aspect of energy transition is the emerging field of using microbiology and synthetic biology for energy transition to produce biofuels and bioproducts that have lower carbon emissions than the traditionally used methods.

Would microbial control remain an important aspect of renewables, and why?

Microbial control for renewables is an emerging field with a lot of unknown unknowns. The widespread adoption of renewables may highlight these microbial challenges soon. Being aware of some of these microbial challenges and implementing effective microbial control is important.

What would you like to see in the next 10 years in the field of petroleum/energy microbiology?

I would like to see a deeper understanding of microbiological challenges in the solar, wind, and hydrogen industries. To be more specific, an understanding of the impact of soil microbiology and any microbial threats associated with solar farms, the impact of microorganisms and potential for microbial corrosion in the flooded wind turbine foundations and offshore wind, and the unintended microbial consequences of underground hydrogen storage.

Please tell us a bit about your professional background and current line of work.

I have a Ph.D. in Environmental Microbiology with a Post-Doc on bioremediation of the Exxon Valdez spill in Alaska. I had a brief tenure in Academia before being recruited into the oil and gas industry and working with an operator (BP), service company (Baker Hughes) and now a manufacturer (Lanxess). Currently, I hold the position of Energy Technology Fellow at Lanxess Microbial Control in Houston responsible for business development, technology transfer, and shaping the innovation pipeline and strategy for global oil and gas applications.

Dr. Kenneth Wunch
Energy Technology Fellow
Lanxess, USA

From your experience, how has the field of petroleum microbiology research changed over the past 10 years?

The most dramatic change has been the application of molecular tools in the industry. A decade ago, evaluating the microbial contamination in a system involved serial dilution culturing with "bug bottles" to quantify sulfate-reducers, acid-producers, or general heterotrophic bacteria present. Clearly, this left gaping holes in understanding population dynamics. Application of qPCR, metagenomics, and bioinformatics now allows us to thoroughly investigate oil and gas systems to determine what organisms are present; what is their involvement in corrosion, souring, or biofouling; and how they are being introduced into the system. Ironically, the chemistries developed to control these populations have changed little over the past 10 years mainly due to regulatory costs. However, their application has evolved due to the adoption of modern molecular techniques.

In recent years, we have seen a massive shift within the energy landscape. To what extent do you think microbiology would play a role during this transition phase?

I think we are on the precipice of a shift in the energy landscape away from fossil fuels but have not yet reached it. Europe is leading the way, but it will take years before North America and Australia adopt this transition and decades before countries like India, Russia, and China do so. In the renewable technologies that are becoming commercially viable (solar, wind, geothermal, hydro, tidal), microbiology is more problematic than beneficial. However, economically harnessing the biomass potential of algae to produce biofuels and the development of microbial fuel cells will most likely lead the way in beneficial contributions from microbes.

What are some research topics you believe are important to transfer from the oil and gas industry to the renewable sector?

The continued advancement of how microbial communities contaminate energy systems and their potential impacts, including corrosion, souring, biofouling, and other biogeochemical transformations.

Also, the energy industry has made huge strides in understanding how and where microbial communities "infect" systems. I think it is imperative to take these learnings into account when designing and developing new assets for renewable energy.

With your expert opinions, how can microbial populations positively and negatively impact green energy as the energy landscape is quickly changing? Would microbial control remain an important aspect of renewables, and why?

I don't have practical experience in how microbial populations positively influence green energy so I will combine the subsequent question about microbial control into my answer. Outside of solar, microbial processes have the potential to negatively impact the efficiency of current commercial green energy technologies. Equipment in aquatic environments is susceptible to macro biofouling, which is initiated by the attachment of biofilms to the infrastructure surface. This biofouling can have deleterious effects on hydrodynamics in tidal and hydro technologies along with impacting mooring lines or power cables in offshore wind turbines. Biofouling also impacts geothermal processes as microbial contamination can reduce the efficiency of heat exchange. However, the green energy technology that has the potential to be most impacted by microbial populations is hydrogen storage. As green hydrogen is generated from low-carbon power often found in remote locations, it must be stored and transported to population centers. Current solutions for storage are focused on depleted salt caverns and oil and gas reservoirs. These environments may be convenient for geologically storing large quantities of hydrogen but are also ideal for the growth of halophilic, hydrogenotrophs that metabolize hydrogen as a source of energy. The resultant problems are ones very familiar to oil and gas energy, including souring (production of sulfide), corrosion, and biofouling.

What would you like to see in the next 10 years in the field of petroleum/energy microbiology?

A more structured and comprehensive strategy for risk management in traditional and novel energy development. This would include:

- **Modeling** – determine operational risk of microbial contamination.
- **Preventive Barriers** – development of primary barriers (chemical, operational, mechanical, etc.) to mitigate the risks modeled.
- **Barrier Assurance** – confirmation that preventive barriers are working as intended by developing and routinely measuring KPIs (key performance indicators). Failure of preventive barriers requires an immediate operational response.
- **Reactive Barriers** – manage preventive barrier failure by operational design or other chemical or physical barriers.

Please tell us a bit about your professional background and current line of work.

I have a background in environmental engineering, specializing in computational methods and modeling of multiphase flow in porous media. My research is centered around solving engineering challenges within subsurface energy resources, in particular for geological CO_2 storage applications. I currently lead a research group of computational geoscientists who develop new models and simulation tools that can be used for a variety of energy applications within petroleum, CCS, subsurface energy storage, and wind. I also head up several large research initiatives, including one of the three national centers for petroleum research, Centre for Sustainable Subsurface Resources (CSSR). Among the many things we research in CSSR is the influence of microbiological activity on underground hydrogen storage.

As an applied researcher within geosciences, I interact with many different specialists in geology, geochemistry, petrophysics, and geomicrobiology. My professional interest is to build so-called multi-physics models that accurately reflect the complex interplay of multiple physical-chemical-biological-thermal processes in deep geological systems. The purpose of these models is to help practitioners understand and manage the energy resources in their portfolio. The multidisciplinary nature of developing complicated models is as challenging as it is rewarding.

Dr. Sarah E. Gasda
Research Director,
Energy & Technology Division,
NORCE, Norway

From your experience, how has the field of petroleum microbiology research changed over the past 10 years?

I have only been peripherally involved in petroleum microbiology research. About five years ago, I had a small part in a research project on MEOR where laboratory and modeling studies tried to understand how microbes preferentially clog flow paths, thus encouraging better reservoir sweep. That was the last such project in our project portfolio. Industry interest dried up due to shifting priorities. It costs a lot to perform experiments to screen and characterize microbes for use in MEOR. The money is not there anymore to support this line of research.

In recent years, we have seen a massive shift within the energy landscape. To what extent do you think microbiology would play a role during this transition phase?

I think microbiology will play a role, even if it doesn't always seem that way now. I will give an example from my experience with CO_2 storage research. During my PhD coursework, I was connected to the Princeton Environmental Institute where I was exposed to the existence and complexity of microorganisms in the subsurface. But for the first 10 years of my research career within CO_2 storage (prior to 2010), microbiology was not a topic that anyone talked about. One reason for this was a single study by a well-respected microbiologist at Princeton that concluded there were no environmental impacts of CO_2 storage on microbiological communities living thousands of meters underground. Since that time, microbiology has become popular again in CCS, this time to understand how we can use microbes to remediate leakage.

What are some research topics you believe are important to transfer from the oil and gas industry to the renewable sector?

The laboratory and experimental capabilities that have been built up over the past decades to study petroleum resources are extremely sophisticated. The image resolution we can achieve now allows us to see fluid interfaces moving from pore to pore in almost real time, which is really exciting. Analysis of these data has dramatically improved our understanding of the fundamental nature of fluid flow in porous media, including how fluids are trapped, remobilized, transferred to new phases, and react with minerals. We have achieved many key insights that are applicable to other types of porous media well beyond oil and gas. It is important to continue this direction of research in porous media science as we move into the renewable sector.

What are some key technological advances you see that are important to transfer from petroleum to renewable energy?

There are many. I think the most important is digitalization. The oil and gas industry has been driver of many important advancements in digital technology. For example, the push toward remote operations and automation of platforms helps increase safety and reduce costs. These advancements can have wide-reaching repercussions in other sectors, which is positive. Another example is simulation technology, where the size of reservoirs, increasing complexity of thermal-mechanical-hydraulic-chemical processes (and microbiology), and the level of geological uncertainty have the danger of pushing CPU time through the roof. Simulation experts have to be very smart in finding ways to simplify models without sacrificing accuracy. There's good progress, but more needs to be done.

What would you like to see in the next 10 years in the field of petroleum/energy microbiology?

In our underground hydrogen storage projects, we see a clear need for more inter-disciplinary research at the intersection between fluid flow and transport in porous media and microbiology. I want to see more experiments where the microbes are not so well fed and where the conditions are more realistic, even varying in time and space. I also want to see models developed that really detail the close coupling between the microbial-chemical processes and the flow and transport of fluids in porous rocks. These are very challenging topics, but the most important thing is to work together to design good experiments, and even invent new ways of studying and modeling these complex systems. I also want more field pilots to test our under-standing, acquire data, and reveal the gaps so we can fine-tune our research moving forward.

Please tell us a bit about your professional background and current line of work.

I'm currently Docent and Project Manager at VIA University College in the Research Center for Built Environment, Energy, Water, and Climate (Horsens, Denmark). I graduated from Aarhus University, Denmark (2002), with a master's degree (cand. scient.) in biology. In 2005, I earned a PhD from the Department of Microbiology, Aarhus University. The same year I was employed at the Danish Technological Institute (DTI) in the Centre for Chemistry and Water Technology, where I was responsible for the consultancy activities for the oil and gas industry around the North Sea and later also worldwide. While heading DTI Microbiology Laboratory I was also developing several consultancy and business activities with the oil and gas industry. I founded DTI Oil and Gas in both Denmark and Norway, where I was the team and business development leader for five years. Thereafter, I worked as a project manager at DNV (Det Norske Veritas) in the field of corrosion management in both Bergen and Esbjerg.

I'm the current chair of AMPP SC-22 on Biodeterioration and ISMOS TSC, an organization I co-founded in 2006 with Dr. Corinne Whitby. I'm an international scientific reviewer and the author of 150+ technical and scientific papers and book chapters related to industrial microbiology, applied biotechnology, corrosion management, oilfield microbiology, water treatment and safety, reservoir souring, and biocorrosion. I spent quite some time editing and reviewing book proposals in my field of expertise and I have co-edited the books *Applied Microbiology and Molecular Biology in Oilfield Systems* (Springer, 2011); *3rd International Symposium on Applied Microbiology and Molecular Biology in Oil Systems* (Elsevier, 2013); *Applications of Molecular Microbiological Methods* (Caister Academic Press, 2014); *Microbiologically Influenced Corrosion in the Upstream Oil and Gas Industry* (CRC Press, 2017); *Microbiological Sensors for the Drinking Water Industry* (IWA Publishing, 2018); *Oilfield Microbiology* (CRC Press, 2019); and *Failure Analysis of Microbiologically Influenced Corrosion* (CRC Press, 2021). And also, the current book you are sitting with right now. While I remember: I was honored with the NACE Technical Achievement Award in 2020 for outstanding research on MIC in the energy sector. That was a great endorsement of my current work. I think I'll leave it with this and tell the story of the origin and development of Euro-MIC for another time.

Dr. Torben Lund Skovhus
Docent and Project Manager
VIA University College, Denmark

From your experience, how has the field of petroleum microbiology research changed over the past 10 years?

I took a quick look at the book Dr. Corinne Whitby and I edited and published in 2011 with Springer called *Applied Microbiology and Molecular Biology in Oilfield Systems* to get inspired. In fact, the problem in the industry remains the same (MIC, souring, and biofouling), but our toolbox has increased with more sophisticated molecular methods that are now also reaching the commercial market (e.g., long-read nanopore DNA sequencing). In the ISMOS community, we have published numerous books and special journal issues from the beginning, enriching knowledge for industry professionals and academics alike. One major accomplishment I want to highlight is our effort to transform all the excellent scientific work from the ISMOS events (and from other communities) into industry-relevant recommended practices, guidelines, and standard documents. Just to mention a few milestones are the following: DNVGL-RP-G101, NACE-TM0212, NACE-TM0106, ASTM-D8412-21, and the recently published Energy Institute document *Selection, Applicability, and Use of Molecular Microbiological Methods (MMM) in the Oil and Gas Industry.*

Further standard documents are on the way from AMPP SC-22 and ISO, which is very encouraging for implementing the correct use and applicability of the latest methods to investigate petroleum microbiology and MIC.

In recent years, we have seen a massive shift within the energy landscape. To what extent do you think microbiology would play a role during this transition phase?

For clarification, it might be worth splitting this question into two parts: 1) the positive role microorganisms could play in the energy transition away from coal, oil, and gas to a more sustainable energy landscape, and 2) the negative impact microorganisms will have on the new infrastructure that comes with the energy transition.

For the first part, I'm rather optimistic as we will find great use of microorganisms in large-scale fermentation plants for e-fuel production based partly on genetic engineering and synthetic biology. This is already ongoing as we speak and will increase in focus over the coming decade. For the second part, the learning from the current research is that microorganisms will always find a way to assist degrading (man made) materials in the environment, if they get something out of the effort – so our new energy infrastructures will also suffer from MIC, biofouling, etc. Here we need to transform our current knowledge from primarily the oil and gas industry to the new industries and their materials. An example of this knowledge transition is the ongoing Euro-MIC (euro-mic.org) project where training of the coming generation of researchers and industry professionals is a key element via training programs, conferences, developing standards, and hosting industry workshops.

What are some research topics you believe are important to transfer from the oil and gas industry to the renewable sector?

I'll highlight three areas where I see an important transfer of knowledge and skills. 1) production chemistry testing, 2) monitoring and testing via a multiple lines of evidence (MLOE) approach, and 3) the future use of AI in monitoring the threat of corrosion in general and MIC in particular.

First, we need to transfer all the existing knowledge on how to prevent corrosion by the use of production chemistry. Several good standards and guidance documents have been produced in this field recently. Second, a MLOE approach for diagnosing MIC is required and is now being implemented worldwide in operating companies based on solid work over the past decades. Finally, we see an increase in the application of AI in material selection, prediction of corrosion failures, and optimization of mitigation approaches. Procedures and algorithms will be important to transfer to the new energy landscape for building a more sustainable future.

Would microbial control remain an important aspect of renewables and why?

Sure. Wherever we have water, nutrients, favorable growth conditions for microorganisms, and a surface to colonize, we will find the potential for MIC to happen. Not that it always will – but often it is the case. This will eventually lead to material degradation. Take for example the Power-to-X (PtX) revolution. PtX covers technologies that produce fuels, chemicals, and materials based on green hydrogen made by electrolysis. Electrolysis is a process where electrical energy from, for example, wind turbines and solar cells is used to split water into hydrogen and oxygen. The hydrogen can then be used directly as an energy source. It can also be refined into methanol and e-jet fuels using CO_2, just as it can be refined into ammonia using nitrogen. During the PtX process, a vast amount of ultra-clean water is needed. When water is part of the process – so are the microorganisms and the threat of MIC will be present.

What would you like to see in the next 10 years in the field of petroleum/energy microbiology?

In my opinion, we have most of the tools and technologies we need to assess, mitigate, and monitor corrosion and MIC. The task is to use it wisely and come up with smarter and more updated ways of regulating the industry, so we avoid leaks, spills, and other disasters. Standardization is picking up slowly – but there is still a good way to go. For many operating companies, production chemicals are applied as insurance. We have done something – we hope it works. Few invest the effort in establishing robust corrosion management systems to ensure that the treatment effectively accomplishes its intended purpose of corrosion control, which should be a focus area over the next decade. Finally, we need much more updated training courses for industry professionals, students, and established academics. Today, they have very limited access to the latest knowledge through MIC training courses worldwide, and there is a great demand for such courses. This is the main reason we recently established the MIC E-learning Academy (www.mic-learning.com), which features free tutorials, an online expert community, and certified online training material. What I'll very much like to see is more students and industry professionals getting upgraded on their knowledge in the field of petroleum microbiology and MIC in the decades to come.

Section II

Microbial Ecology of Energy Systems

2 Impact of Microbial Biofilms on Subsurface Energy Systems

From Oil and Gas to Renewable Energy

Na Liu
University of Bergen, Bergen, Norway

Nicole Dopffel and Biwen Annie An Stepec
Norwegian Research Centre, Bergen, Norway

2.1 BIOFILM FORMATION IN SUBSURFACE ENVIRONMENTS

Biofilms are complex biological systems, which are very commonly formed by microbial communities adhered to a surface by producing a matrix of extracellular polymeric substances (EPS), proteins, and extracellular DNA (1, 2). The biofilm growth cycle generally includes the following stages (Figure 2.1): (i) initial adhesion of planktonic cells to a surface, (ii) irreversible adhesion by EPS generation, (iii) early structural development, (iv) maturation of the biofilm, (v) detachment of cells from the biofilm matrix, and (vi) regrowth in old and new places (3). Biofilm substrate provides a surface for the attachment and growth of microorganisms, immobilizing biomass on their surfaces and within their internal pores (4). The EPS can also shield the embedded microbial cells from environmental virulence, stress factors, and stripping (3, 5), such as changes in temperature, pH, and nutrient availability or contact with biocides, enabling them to survive for extended periods of time (6).

Biofilm formation in subsurface environments is a common phenomenon that occurs in soil, groundwater, and geological formations, such as oil and gas reservoirs (7–10). Their accumulation in the pore can cause clogging, leading to significant changes in reservoir conductivity and physical/chemical properties of porous media (11, 12). Meanwhile, biofilms are also utilized in environmental restoration and protection applications in porous media, including remediation of toxic metals (13), oil spills (14), and microplastics (15). Therefore, subsurface biofilms can have several important ecological and biogeochemical roles that can significantly influence the subsurface ecosystem. Understanding the fundamental mechanisms that contribute to biofilm formation, as well as estimating the potential microbial risks, is important

DOI: 10.1201/9781003287056-4

21

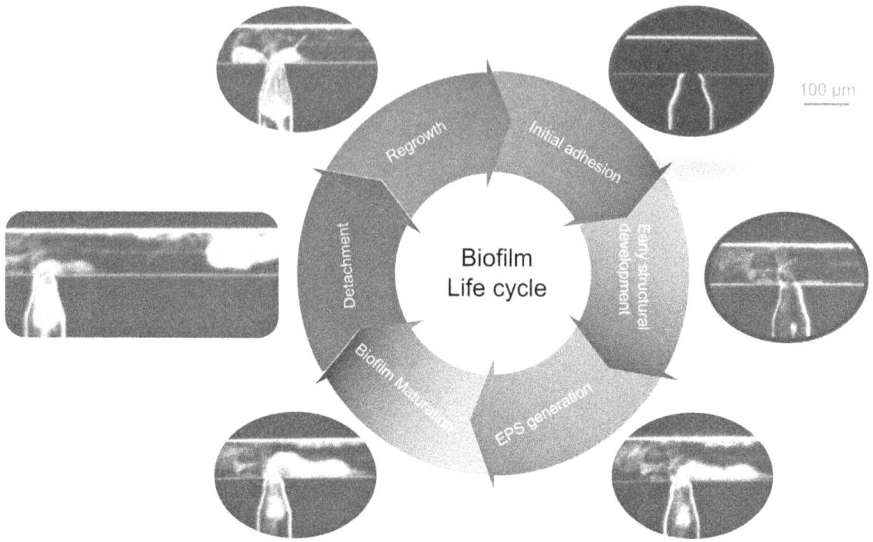

FIGURE 2.1 The life cycle of biofilm formation. The bacterium used in the study was *Thalassospira strain* A216101, a facultative anaerobic, nitrate-reducing bacteria (NRB), capable of growing under both aerobic and anaerobic conditions. Flow direction was from left to right.

for developing effective strategies to manage subsurface ecosystems and mitigate the impacts of biofilm formation on industrial processes (16, 17).

Hydrocarbon reservoirs are one of the most important industrial-relevant subsurface environments, as they are a primary source of the world's energy supply. Through decades of research on petroleum and reservoir microbiology, significant knowledge and large data sets have been accumulated on the potential microbial risks and impacts. As our society undergoes the important energy transition from using fossil fuel to renewable energy, these learnings on subsurface microbiology become important and need to be taken into consideration when developing new energy systems in the underground. Geologic CO_2 sequestration is seen as one of the most feasible industrial solutions to significantly reduce CO_2 emissions and mitigate climate change from human activities. Microorganisms can influence the fate and behavior of CO_2 in the subsurface by directly utilizing CO_2 as a carbon source for their metabolism or changing the storage capacity and stability of CO_2 by altering the geochemical conditions. Hydrogen storage in subsurface reservoirs plays a pivotal role in establishing a robust renewable energy system by storing excess hydrogen produced from renewable sources, such as wind and solar power. This stored hydrogen can be used as an energy carrier for direct use in heavy industry (like steel, shipping, or transport) or for fuel cell applications to produce electricity during times when renewable energy sources are not available. In this chapter, we summarize the positive and negative impacts of biofilm formation on the oil/gas industry and estimate the potential problems during underground hydrogen storage. Furthermore, the purpose of this work is to provide an overview of biofilm impacts on subsurface applications to engineers, operators, and others, with the goal of enhancing their understanding of potential microbial risks and aiding in the successful large-scale implementation of gas storage in the future.

2.2 IMPACTS OF BIOFILM FORMATION IN HYDROCARBON RESERVOIRS

Currently, fossil fuels remain the key source of energy worldwide, representing about 80–90% of the global energy production. Among them, oil and gas contribute significantly, representing approximately 60% of the overall energy mix (18). With the continuous rise of worldwide energy demands and the difficulty in discovering new hydrocarbon fields, efficient utilization of existing resources has been and still is an important research area for many fields. Despite constant development of novel enhanced oil recovery (EOR) technologies, about 25–55% of oil is still left behind as residual oil in the reservoirs (19, 20). In recent years, with the increasing request for green and environmentally sound technologies, the focus shifted to microbial-enhanced oil recovery (MEOR), which relies on the activity of microbial cells and their metabolites to increase the oil/gas recovery. Promising results on improving oil recovery performance by MEOR have been reported based on various experimental studies (8, 10, 19, 21), but significant large-scale field use has not been achieved. There are several mechanisms involved that microorganisms may contribute to tertiary oil recovery, including interfacial tension reduction, emulsification, gas production, wettability alteration, and selective bioplugging (22, 23). Here we specifically focus on the impact of biofilms formation on MEOR technologies in hydrocarbon reservoirs.

In the oil and gas industry, biofilms can form in any area where water is present, including pipelines, tanks, and reservoirs (24). Of these, the presence of biofilms in reservoirs can have both positive and negative effects on operations (Figure 2.2). Biofilms positively induce wettability alteration, and selective bioplugging has been explored as a viable technique for in situ bioremediation and MEOR (25, 26). Conversely, biofilm formation is correlated to fouling, and pore clogging, which can lead to production loss, injectivity reduction, formation damage, and increased operational costs (27, 28). Maintaining the control of microbial activity and biofilm growth is a crucial factor for successful operation in hydrocarbon reservoirs across various industrial applications.

2.2.1 Biofilms and Microbiologically Influenced Corrosion

Several microorganisms including sulfate-reducing microbes, iron-oxidizing microbes, iron-reducing microbes, and acid-producing microbes are known to mobilize and/or metabolize corrosive compounds. These microorganisms are either indigenous to the reservoir or introduced during operations, i.e., waterflooding (28, 29). As defined by the Association for Materials Protection and Performance, microbiologically influenced corrosion (MIC) is caused by corrosive microbial activity in biofilms on corrosion-susceptible materials (30). Biofilm formation was reported to greatly accelerate the corrosion process by concentrating corrosive compounds,

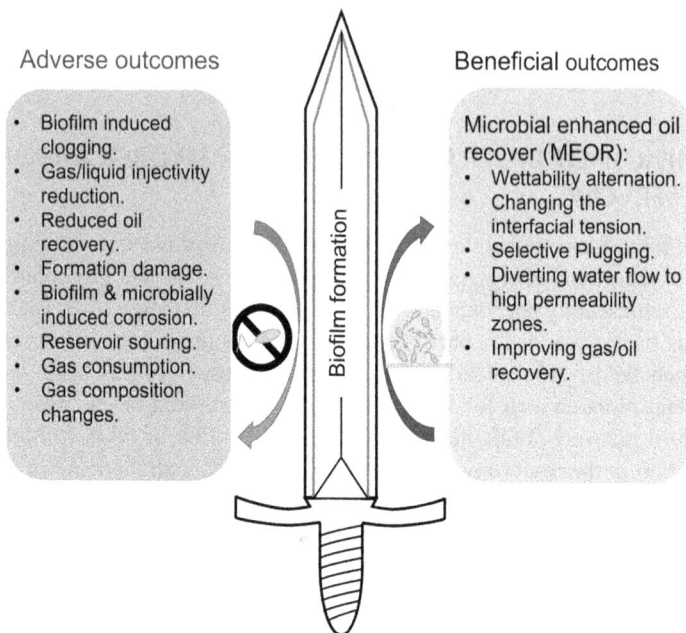

Adverse outcomes

• Biofilm induced clogging.
• Gas/liquid injectivity reduction.
• Reduced oil recovery.
• Formation damage.
• Biofilm & microbially induced corrosion.
• Reservoir souring.
• Gas consumption.
• Gas composition changes.

Biofilm formation

Beneficial outcomes

Microbial enhanced oil recover (MEOR):
• Wettability alternation.
• Changing the interfacial tension.
• Selective Plugging.
• Diverting water flow to high permeability zones.
• Improving gas/oil recovery.

FIGURE 2.2   Biofilm accumulation in porous media: a double-edged sword for its applications.

facilitating electron transfer, changing the electrochemical properties, and reducing effectiveness of corrosion-mitigation methods (28, 31).

Sulfate-reducing bacteria (SRB) are commonly found across several industrial environments, i.e., hydrocarbon reservoirs, and identified as one of the major culprits of MIC. They prefer temperatures around 38 °C and near-neutral pH conditions (32) but can also be active at high salinities (current known limit is 24% (33)) and temperatures even above 100 °C (34). Some estimates (27) suggest that the activities of SRB are responsible for corrosion in 77% of oil-producing wells. Molecular hydrogen present in the environment and on steel surfaces serves as the electron donor for sulfate reduction, resulting in the production of hydrogen sulfide. In terms of electrochemical reactions, they are shown as below (35, 36):

Cathodic Reactions:

$$2H^+ + 2e^- \rightarrow H_2$$

$$SO_4^{2-} + H_2 \rightarrow S^{2-} + 4H_2O$$

Anodic reactions:

$$H_2O \rightarrow H^+ + OH^-$$

$$\text{Metal} \rightarrow \text{Metal}^{2+} + 2e^-$$

$$\text{Metal}^{2+} + S^{2-} \rightarrow \text{MetalS}$$

Biofilms can promote non-uniform bacterial colonization on metallic substrates, leading to the formation of differential aeration cells in the presence of aerobic respiration. Thicker bacterial colonies, with higher respiration activity and lower oxygen concentration, create anodic areas, while thinner colonies create cathodic areas, thereby facilitating corrosion (31). More severe corrosion was reported to take place in a multispecies biofilm as the interactions between different species may induce a cascade of biochemical reactions (37). In some cases, certain groups of microorganisms, such as nitrate-reducing bacteria (NRB), have an adverse effect on SRB via nutrient competition and production of nitrite. However, it's worth noting that the activities of some nitrate reducers are characterized to be corrosive (38). While bacterial biofilms are conventionally associated with corrosion, a few research studies also discussed the protective role of certain bacterial biofilms in preventing corrosion of metals (31). A biofilm matrix can act as a transport barrier, hindering the penetration of corrosive agents such as oxygen, chloride, and other harmful substances and reducing their contact with the metal surface. In addition, competition between microorganisms further reduces risks of MIC through production of antimicrobial agents and bio-competitive exclusion (39, 40).

2.2.2 BIOFILMS-INDUCED CLOGGING AND POTENTIAL USE FOR CO_2 STORAGE

The clogging effect induced by biofilm accumulation in pores might have negative impacts on many industrial applications, resulting in increased costs associated with remediation and prevention. For instance, unspecific biofilm growth in the near-wellbore areas can cause formation damage and reduced injectivity, significantly reducing oil recovery efficiency (41, 42). Studies have demonstrated that several chemical additives, such as organic compounds, during water injection can serve as a growth substrate for microorganisms, causing biofilm formation and serious injectivity reduction (43). However, selective bioplugging has emerged as a promising technique in several applications, including in situ bioremediation (44), soil injection (45), waste treatment (46), groundwater recharge (47), and MEOR (7–10).

Selective bioplugging aims at the selective accumulation of biofilms in high-permeability zones of the reservoir, which can divert injection fluids toward lower-permeability oil-filled zones, facilitating improved oil recovery (see Figure 2.3a). This bioplugging strategy has been shown to be efficient for improving water flood efficiency and oil recovery based on various experimental studies. The *Enterobacter* sp. CJF-002 was found to selectively grow and form clogging in the high permeable zones of the reservoir, resulting in an increase of oil production and concomitant reduction in water cut (48). *Bacillus licheniformis* TT33 was also observed to form biomass in highly permeable zones in a sand pack column, thereby increasing the sweep force of the injected water (49). The formation of biofilms is heavily influenced by the surrounding environment, including factors such as shear stress,

(a)

Microbes flow with
water phase.

Biofilm formation when
bacteria adhere to surfaces.

Biofilm diverts water flow.

| ↘ streamline | ⬭ sand grains | ⸰ microorganisms | ⸰ biofilm | ● trapped oil |

(b)

Flow velocity increases:
1.66 mm/s   2.50 mm/s   3.33 mm/s   4.17 mm/s

Nutrient concentration increases:
0.1 mM   5.0 mM   10.0 mM   20.0 mM

Scale bar = 100 μm

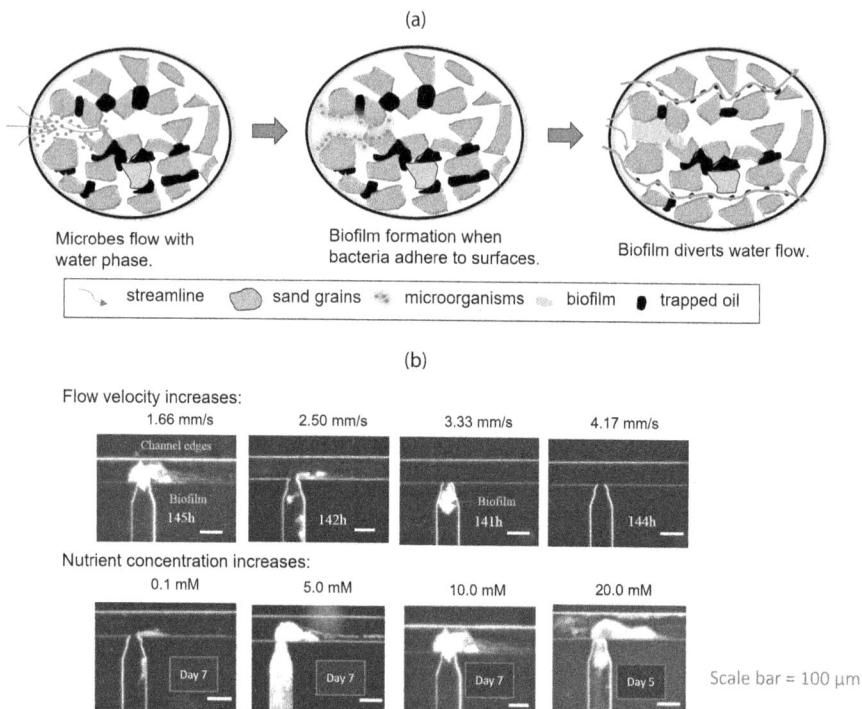

FIGURE 2.3 (a) Scheme of the selective bioplugging method to improve water flood efficiency and oil recovery. (b) Effects of flow rate and nutrient concentrations on biofilm formation in a microchannel: optical images of biofilm growth in microchannels at various nutrient concentrations and velocities. *The figure is edited with permission from N. Liu et al. DOI: 10.1007/s10295-019-02161-x.*

nutrient availability, temperature, and pH (50, 51). The biofilm growth and detachment rates can both increase with injection velocity, as the increased mass transfer facilitates the supply of nutrients for bacterial growth, while the increased shear force can cause detachment (11, 52).

The concept of biofilm-induced bio-clogging is also proposed to enhance CO_2 sequestration and storage by reducing upward leakage of the injected CO_2 through fractures and faults in cap-rock or near injection wells (53). Supercritical CO_2 (SC-CO_2) can highly reduce the number of viable cells of subsurface microorganisms in planktonic cultures but affects much less cells within a biofilm matrix due to the protection from EPS (54). Engineered biofilm barriers for mitigating gas leakage have been proven to be resilient to the effects of high-pressure SC-CO_2 (53). In addition, biofilms can induce the mineralization of carbonate minerals (microbially induced calcite precipitation – MICP) (55). Especially effective are urease-producing bacteria, which can catalyze the hydrolysis of urea, promoting $CaCO_3$ precipitation, and reducing porosity and permeability of fractures and faults in the reservoir. EPS and suspended biomass can provide nucleation sites for this precipitation process.

This concept has shown great potential to close cement fractures, reduce near-wellbore permeability, and remediate CO_2-related corrosion (56, 57). Additionally, it might reduce undesirable migration of CO_2 or other fluids (58). Overall, this makes MICP a promising solution to reduce the environmental risk from leakage of the sequestered CO_2.

2.2.3   SURFACE EFFECTS OF BIOFILMS – WETTABILITY ALTERATION

Wettability is one of the key parameters in determining the recovery factor of an oil/gas production process by governing the waterflooding efficiency, relative permeability, and capillary pressure in porous media (59). Microbial-induced wettability alteration is regarded as one of the most efficient mechanisms that can lead to increased oil recovery during MEOR processes (21, 60, 61). While the precise underlying mechanisms of wettability alteration by microbes are still being investigated, biosurfactant adsorption, bacterial adhesion, and biofilm formation are widely believed to be the primary processes responsible (62). Biofilms can cause wettability changes due to the produced EPS on the surface altering its physical and chemical properties. EPS can act as a hydrogel, reducing the contact angle between the surface and liquids and thus making it more hydrophilic or hydrophobic. From these perspectives, the wetting behavior of the surface could be changed markedly by biofilm formation. Based on the experimental results presented in the literature, wettability can be altered toward both more water-wet and more oil-wet conditions depending on the initial wetting condition, surface properties, types of microorganisms, and metabolites involved in the process (63). For instance, carbonate reservoirs, which are commonly found in the oil and gas industry, are typically neutral or preferential oil wet, which requires more water to recover trapped oil compared to a water-wet condition. To increase the efficiency of oil recovery, numerous attempts have been made to change the wettability to be more hydrophilic (59, 64).

Additionally, surface wettability can also strongly affect bacterial initial attachment. However, there are clear differences described in the literature when it comes to the favoring wettability status for bacterial adhesion. Some studies demonstrated that enhancing the hydrophilic properties of the surfaces causes an increase in cell adhesion (65, 66), while some strains seem to prefer to attach to the substrate with hydrophobic properties (67). Lee et al. (68) observed that the maximum cell adhesion appeared at a water contact angle of 55° for all types of tested cells (Chinese hamster ovary, fibroblast, and endothelial cells), while Yuan et al. (69) showed that moderate hydrophobicity with a water contact angle of about 90° produced the highest level of bacterial adhesion. Our previous observations of a *Thalassospira* strain to hydrophobic and hydrophilic surfaces (see Figure 2.4) show that *Thalassospira* prefers to attach to a moderate hydrophobic surface compared to a strong hydrophilic surface. This shows that preferred adhesion seems to be strain dependent. Understanding and manipulating surface wettability can help control bacterial attachment and biofilm formation, leading to improved operational efficiency, reduced maintenance costs, and enhanced product quality in various industrial applications. Surface modifications, coatings, and treatments can be employed to alter the wettability of surfaces and discourage bacterial attachment (70).

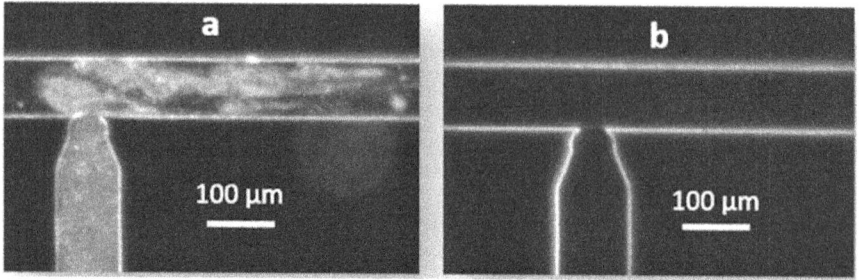

FIGURE 2.4 Biofilms initial attachment at different surface wettability in T-shape glass microchannels: (a) a hydrophobic surface, contact angle = 105~110° (water in air); (b), a hydrophilic surface, contact angle = 0° (water in air). Before bacterial inoculation, both microchannels were cleaned in the same procedure. The microchannels were injected with the same concentration of bacterial solution for a period of 15 days.

2.3 POTENTIAL EFFECTS ON UNDERGROUND H_2 STORAGE

The use of fossil fuels results in the emission of greenhouse gases in the atmosphere, which causes climate change and global warming. The transition from fossil fuels to renewable energy sources is a critical step in mitigating climate change and building a sustainable, reliable, and secure energy supply system. Renewable energy sources, such as wind, solar, and hydropower, face a significant challenge due to the imbalance between their supply and demand, which are greatly influenced by seasonal fluctuations in atmospheric conditions, such as varying intensity of sunlight and wind force (71). The generation of H_2 gas through water electrolysis (known as green H_2) during times of electricity oversupply and then storing it to be used during periods of high energy demand is considered a crucial element in addressing supply–demand imbalances (59–61). Underground H_2 storage (UHS) in geological formations, including salt caverns, depleted oil and gas fields and saline aquifers, is one of the options for large-scale H_2 storage and has been suggested to be suitable for mid- to long-term energy demand fluctuations (72, 73). However, the presence of large volumes of H_2 means introducing high amounts of electron donors for many different microbial metabolisms and therefore creates favorable conditions for respiration of indigenous microorganisms in the receiving reservoirs. The potential biogeochemical activity followed by H_2 injection calls for caution and revision of conventional storage practices (74, 75) (Figure 2.5).

The stored H_2 gas can be microbially converted to CH_4, H_2S, or CH_3COOH through metabolism reactions of methanogenesis, sulfate reduction, and acetogenesis (76). In recent pore-scale experiments with a halophilic SRB, *Desulfohalobium retbaense* DSM 5692, growing with H_2 gas, four microbial effects were observed: microbial-induced clogging, H_2 loss from bacterial consumption, wettability alteration, and increased residual trapping of the H_2 phase. Microbial-induced wettability alteration appears to be the only positive effect of microbial growth in an H_2 storage reservoir. Microbial growth changed the surface wettability from a water-wet to a neutral-wet state, leading to minimal interfacial surface areas and favorable gas recovery due to reduction in capillary pressure (60, 61). This in turn leads to decreased disconnected gas phase and an increase in H_2 relative permeability during imbibition.

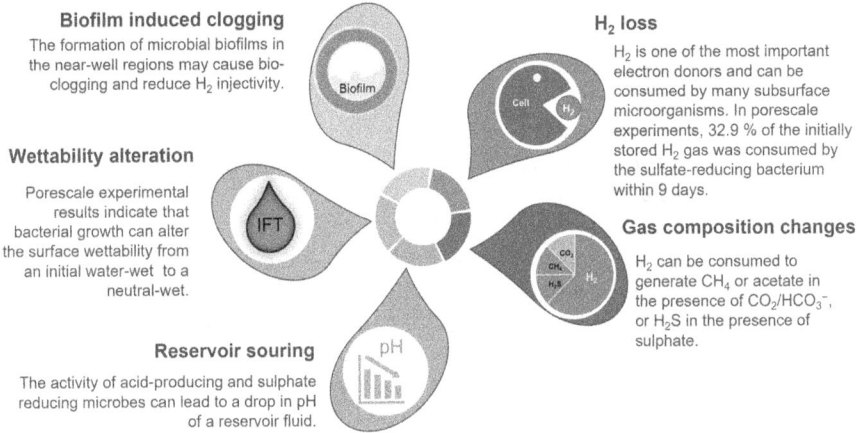

Biofilm induced clogging
The formation of microbial biofilms in the near-well regions may cause bio-clogging and reduce H_2 injectivity.

Wettability alteration
Porescale experimental results indicate that bacterial growth can alter the surface wettability from an initial water-wet to a neutral-wet.

Reservoir souring
The activity of acid-producing and sulphate reducing microbes can lead to a drop in pH of a reservoir fluid.

H_2 loss
H_2 is one of the most important electron donors and can be consumed by many subsurface microorganisms. In porescale experiments, 32.9 % of the initially stored H_2 gas was consumed by the sulfate-reducing bacterium within 9 days.

Gas composition changes
H_2 can be consumed to generate CH_4 or acetate in the presence of CO_2/HCO_3^-, or H_2S in the presence of sulphate.

FIGURE 2.5 Experimental results of microbial effects during underground H_2 storage: biofilms in the pore throats were observed to block the flow of H_2 gas and reduced gas injectivity occurred; the gas-liquid interfacial area reduced to near-minimal values when the H_2 gas was exposed to the bacterial solution for one day; rapid bacterial consumption was observed (75). The experiments were conducted in a silicon-wafer micromodel with a pore pattern from natural sandstone for direct observations of the microbial-induced sulfate reduction at 35 bar and 37 °C. *Desulfohalobium retbaense* DSM 5692, a halophilic sulfate-reducing bacterium, was cultured with H_2 gas for nine days.

The formation of biofilms led to bio-clogging, which was primarily observed in the inlet regions. This phenomenon resulted in a reduction in H_2 injectivity and altered the subsurface transport properties, as illustrated in Figure 2.5. Despite its negative impact on hydrogen injectivity, the effect of bio-clogging during H_2 storage is not entirely detrimental. In fact, it was discovered that biofilms formed in areas of high hydrogen saturation can impede the vertical rise of H_2 gas and facilitate more uniform radial gas penetration into the reservoir, which makes biofilm-induced clogging beneficial for UHS (77). Overall, in all major H_2 storage field trials to this date no injectivity issues were reported indicating that bio-clogging and biofilm formation might not be an immediate occurring issue (78, 79).

Significant gas loss from microbial hydrogen consumption with sulfate reduction was observed not only in our pore-scale experiment using a single bacterial strain but also in a recent test using the original brine from a gas reservoir (80). The production of hazardous gases, such as H_2S from sulfate reduction metabolism, would lead to contamination of the stored gas or even storage souring. Microbial consumption can break the continuity of the gas phase in pores and generate isolated gas bubbles, which will further speed up the microbial consumption process and increase the flow resistance, resulting in a low gas recovery.

In this regard, there is every reason to expect microbial risks during UHS, in which extremely favorable conditions are created for the existence and growth of the population of certain types of microorganisms (62). The microbial-induced gas loss and contamination must be taken into account when planning underground hydrogen storage at sites, particularly for storage in aquifers and depleted gas reservoirs due to the abundance of sulfate-reducing microorganisms and sulfate in the formation water (71).

2.4   MITIGATION AND TREATMENT OF BIOFILMS

The biofilm matrix can behave as a viscous fluid to protect the embedded cells against environmental threats such as mechanical shear, temperature changes, and chemical toxins. The high resistance and low permeability of biofilms significantly increase the difficulty for anti-biofilm trials (6, 81, 82). There are several strategies that can be used to mitigate the effects of biofilms in the oil/gas industry.

1. Biocide treatment: Biocides can be applied to control microbial growth and prevent biofilm formation. Biocides can be added to production fluids, injection water, or other process streams to inhibit microbial activity. However, biocide injection is very often limited in its efficiency due to the reduced diffusive transport of effective components across the low permeable biofilm matrix (83, 84). It has been reported that biofilm tolerance to biocides is about 10–1,000 times higher compared with planktonic bacteria (81, 82).
2. Physical cleaning: Regular cleaning and maintenance can help prevent the accumulation of biofilms on surfaces. High-pressure water jetting, pigging, and other mechanical methods can be used to remove biofilms and other deposits. This is of course only possible in pipelines and accessible infrastructure.
3. Corrosion inhibitors: Corrosion inhibitors can be added to production fluids to prevent or slow down the corrosion of metal surfaces. Some inhibitors can also help prevent biofilm formation by altering the surface properties of materials. Smooth surfaces, adequate drainage, and effective sealing can help minimize the adhesion of microorganisms and the accumulation of deposits.
4. Alternate treatment methods: Osmotic pressure has been experimentally proved to be an alternative method for anti-biofilm trials by creating a high osmotic imbalance between the biofilm and the external environment to allow an influx of biocides-rich water into the biofilm (16). The application and effectiveness in the field still need to be investigated.

Regular monitoring of microbial activity and biofilm formation can help identify potential issues early and allow for timely intervention. Advanced monitoring technologies, such as microbial DNA analysis and real-time sensors, can provide more accurate and timely information on microbial activity.

2.5   CONCLUSIONS AND RECOMMENDATIONS FOR FUTURE BIOFILM RESEARCH

Biofilms play an essential role in subsurface ecosystems. Human activities such as oil and gas extraction and injection in subsurface reservoirs have led to various environmental impacts. Biofilm-induced clogging and corrosion can cause injectivity reduction, production loss, and increased operational costs. Meanwhile, the presence of biofilms has spurred the development of biofilm-enhanced technologies, which harness the utility of biofilm communities to form barriers to flow and mass transport in subsurface environments. Selective bioplugging technology has been researched as a strategy for enhancing oil recovery and mitigating gas (i.e., CO_2 and H_2) upward movement and leakage in reservoirs and aquifers. Wettability alternation induced by microbial can reduce capillary pressure and improve waterflooding efficiency in porous media, thereby enhancing oil recovery. The experiences on biofilm control from the oil and gas industry provide the incentive for undertaking a unique research and development effort aimed at examining the problems and opportunities of biofilms within other subsurface systems like underground CO_2 and H_2 storage.

CONFLICT OF INTEREST

The authors declare that the research was conducted in the absence of any commercial or financial relationships that could be construed as a potential conflict of interest.

REFERENCES

1. J. W. Costerton, Z. Lewandowski, D. E. Caldwell, D. R. Korber and H. M. Lappin-Scott: Microbial biofilms. *Annual Review of Microbiology*, 49, 34 (1995).
2. T. C. Zhang and P. L. Bishop: Density, porosity, and pore structure of biofilms. *Water Research*, 28(11), 11 (1994).
3. N. Liu, T. Skauge, D. Landa-Marbán, B. Hovland, B. Thorbjørnsen, F. A. Radu, B. F. Vik, T. Baumann and G. Bødtker: Microfluidic study of effects of flow velocity and nutrient concentration on biofilm accumulation and adhesive strength in the flowing and no-flowing microchannels. *Journal of Industrial Microbiology and Biotechnology*, 46(6), 855–868 (2019) doi:10.1007/s10295-019-02161-x
4. S. Al-Amshawee, M. Yunus and A. Azoddein: A novel microbial biofilm carrier for wastewater remediation. In: *IOP Conference Series: Materials Science and Engineering*. IOP Publishing (2020). doi:10.1088/1757-899X/736/7/072006
5. Y. Zhang, A. Xu, X. Lv, Q. Wang, C. Feng and J. Lin: Non-invasive measurement, mathematical simulation and in situ detection of biofilm evolution in porous media: A review. *Applied Sciences*, 11(4), 1391 (2021).
6. H.-C. Flemming and J. Wingender: The biofilm matrix. *Nature Reviews Microbiology*, 8(9), 623–633 (2010).
7. P. Hosseininoosheri, H. R. Lashgari and K. Sepehrnoori: A novel method to model and characterize in-situ bio-surfactant production in microbial enhanced oil recovery. *Fuel*, 183, 501–511 (2016) doi:10.1016/j.fuel.2016.06.035
8. M. Karimi, M. Mahmoodi, A. Niazi, Y. Al-Wahaibi and S. Ayatollahi: Investigating wettability alteration during MEOR process, a micro/macro scale analysis. *Colloids and Surfaces B: Biointerfaces*, 95, 129–136 (2012) doi:10.1016/j.colsurfb.2012.02.035
9. H. Khajepour, M. Mahmoodi, D. Biria and S. Ayatollahi: Investigation of wettability alteration through relative permeability measurement during MEOR process: A micromodel study. *Journal of Petroleum Science and Engineering*, 120, 10–17 (2014) doi:10.1016/j.petrol.2014.05.022
10. A. Rabiei, M. Sharifinik, A. Niazi, A. Hashemi and S. Ayatollahi: Core flooding tests to investigate the effects of IFT reduction and wettability alteration on oil recovery during MEOR process in an Iranian oil reservoir. *Applied Microbiology and Biotechnology*, 97(13), 5979–5991 (2013) doi:10.1007/s00253-013-4863-4

11. N. Weiss, K. E. Obied, J. Kalkman, R. G. Lammertink and T. G. van Leeuwen: Measurement of biofilm growth and local hydrodynamics using optical coherence tomography. *Biomedical Optics Express*, 7(9), 3508–3518 (2016) doi:10.1364/BOE.7.003508

12. H. Wang, J. Xin, X. Zheng, Y. Fang, M. Zhao and T. Zheng: Effect of biofilms on the clogging mechanisms of suspended particles in porous media during artificial recharge. *Journal of Hydrology*, 619, 129342 (2023).

13. Z. Rahman and V. P. Singh: Bioremediation of toxic heavy metals (THMs) contaminated sites: Concepts, applications and challenges. *Environmental Science and Pollution Research*, 27, 27563–27581 (2020).

14. M. Omarova, L. T. Swientoniewski, I. K. Mkam Tsengam, D. A. Blake, V. John, A. McCormick, G. D. Bothun, S. R. Raghavan and A. Bose: Biofilm formation by hydrocarbon-degrading marine bacteria and its effects on oil dispersion. *ACS Sustainable Chemistry & Engineering*, 7(17), 14490–14499 (2019).

15. J. Wang, X. Guo and J. Xue: Biofilm-developed microplastics as vectors of pollutants in aquatic environments. *Environmental Science & Technology*, 55(19), 12780–12790 (2021).

16. N. Liu, N. Dopffel, B. Hovland, E. Alagic, B. F. Vik and G. Bødtker: High osmotic stress initiates expansion and detachment of Thalassospira sp. biofilms in glass microchannels. *Journal of Environmental Chemical Engineering*, 8(6), 104525 (2020) doi: 10.1016/j.jece.2020.104525

17. R. Hao, C. Meng and J. Li: Impact of operating condition on the denitrifying bacterial community structure in a 3DBER-SAD reactor. *Journal of Industrial Microbiology and Biotechnology*, 44(1), 9–21 (2017) doi:10.1007/s10295-016-1853-4

18. W. Graus, M. Roglieri, P. Jaworski, L. Alberio and E. Worrell: The promise of carbon capture and storage: Evaluating the capture-readiness of new EU fossil fuel power plants. *Climate Policy*, 11(1), 789–812 (2011) doi:10.3763/cpol.2008.0615

19. B. Shibulal, S. N. Al-Bahry, Y. M. Al-Wahaibi, A. E. Elshafie, A. S. Al-Bemani and S. J. Joshi: Microbial enhanced heavy oil recovery by the aid of inhabitant spore-forming bacteria: An insight review. *ScientificWorldJournal*, 2014, 309159 (2014) doi:10.1155/2014/309159

20. R. Sen: Biotechnology in petroleum recovery: The microbial EOR. *Progress in Energy and Combustion Science*, 34(6), 714–724 (2008) doi:10.1016/j.pecs.2008.05.001

21. H. Khajepour, M. Mahmoodi, D. Biria and S. Ayatollahi: Investigation of wettability alteration through relative permeability measurement during MEOR process: A micromodel study. *Journal of Petroleum Science and Engineering*, 120, 10–17 (2014) doi:10.1016/j.petrol.2014.05.022

22. I. Lazar, I. G. Petrisor and T. F. Yen: Microbial enhanced oil recovery (MEOR). *Petroleum Science and Technology*, 25(11), 1353–1366 (2007) doi:10.1080/10916460701287714

23. L. R. Brown: Microbial enhanced oil recovery (MEOR). *Current Opinion in Microbiology*, 13(3), 316–320 (2010) doi:10.1016/j.mib.2010.01.011

24. T. L. Skovhus, R. B. Eckert and E. Rodrigues: Management and control of microbiologically influenced corrosion (MIC) in the oil and gas industry—Overview and a North Sea case study. *Journal of biotechnology*, 256, 31–45 (2017).

25. C. Nikolova and T. Gutierrez: Use of microorganisms in the recovery of oil from recalcitrant oil reservoirs: Current state of knowledge, technological advances and future perspectives. *Frontiers in Microbiology*, 10 (2020) doi:10.3389/fmicb.2019.02996

26. J. Patel, S. Borgohain, M. Kumar, V. Rangarajan, P. Somasundaran and R. Sen: Recent developments in microbial enhanced oil recovery. *Renewable and Sustainable Energy Reviews*, 52, 1539–1558 (2015) doi:10.1016/j.rser.2015.07.135

27. S. Al-Jaroudi, A. Ul-Hamid and M. Al-Gahtani: Failure of crude oil pipeline due to microbiologically induced corrosion. *Corrosion Engineering, Science and Technology*, 46(4), 568–579 (2011).

28. A. R. Al-Shamari, A. W. Al-Mithin, O. Olabisi and A. Mathew: Developing a metric for microbilogically influenced corrosion (MIC) in oilfield water handling systems. In: *Paper presented at CORROSION 2013*. Orlando, FL: OnePetro (2013).

29. B. A. An, Y. Shen and G. Voordouw: Control of sulfide production in high salinity Bakken shale oil reservoirs by halophilic bacteria reducing nitrate to nitrite. *Frontiers in Microbiology*, 8, 1164 (2017).

30. NACE TM0212-2012: *Detection, Testing, and Evaluation of Microbiologically Influenced Corrosion on Internal Surfaces of Pipelines*. Houston, TX: NACE (2012).

31. R. B. Eckert: Emphasis on biofilms can improve mitigation of microbiologically influenced corrosion in oil and gas industry. *Corrosion Engineering, Science and Technology*, 50(3), 163–168 (2015) doi:10.1179/1743278214Y.0000000248

32. W. Schwartz: Postgate, J. R., The sulfate-reducing bacteria (2nd Edition) X + 208 S., 20 Abb., 4 Tab. University Press, Cambridge 1983. US $ 39.50. *Journal of Basic Microbiology*, 25(3), 202–202 (1985) doi:10.1002/jobm.3620250311

33. B. Ollivier, C. E. Hatchikian, G. Prensier, J. Guezennec and J.-L. Garcia: Desulfohalobium retbaense gen. nov., sp. nov., a halophilic sulfate-reducing bacterium from sediments of a hypersaline lake in Senegal. *International Journal of Systematic and Evolutionary Microbiology*, 41(1), 74–81 (1991) doi:10.1099/00207713-41-1-74

34. B. B. Jørgensen, M. F. Isaksen and H. W. Jannasch: Bacterial sulfate reduction above 100°C in deep-sea hydrothermal vent sediments. *Science*, 258(5089), 1756–1757 (1992) doi:10.1126/science.258.5089.1756

35. H. A. Videla and L. K. Herrera: Microbiologically influenced corrosion: Looking to the future. *International Microbiology*, 8(3), 169 (2005).

36. P. Elumalai, M. S. AlSalhi, S. Mehariya, O. P. Karthikeyan, S. Devanesan, P. Parthipan and A. Rajasekar: Bacterial community analysis of biofilm on API 5LX carbon steel in an oil reservoir environment. *Bioprocess and Biosystems Engineering*, 44(2), 355–368 (2021) doi:10.1007/s00449-020-02447-w

37. J. S. Lee, J. M. McBeth, R. I. Ray, B. J. Little and D. Emerson: Iron cycling at corroding carbon steel surfaces. *Biofouling*, 29(10), 1243–1252 (2013).

38. S. Lahme, D. Enning, C. M. Callbeck, D. Menendez Vega, T. P. Curtis, I. M. Head and C. R. J. Hubert: Metabolites of an oil field sulfide-oxidizing, nitrate-reducing Sulfurimonas sp. cause severe corrosion. *Applied and Environmental Microbiology*, 85(3) (2019) doi:10.1128/aem.01891-18

39. L. Gieg, T. Jack and J. Foght: Biological souring and mitigation in oil reservoirs. *Applied microbiology and biotechnology*, 92, 263–282 (2011) doi:10.1007/s00253-011-3542-6

40. N. Kip and J. A. van Veen: The dual role of microbes in corrosion. *The ISME Journal*, 9(3), 542–551 (2015) doi:10.1038/ismej.2014.169

41. C. C. Ezeuko, A. Sen and I. D. Gates: Modelling biofilm-induced formation damage and biocide treatment in subsurface geosystems. *Microbial Biotechnology*, 6(1), 53–66 (2013) doi:10.1111/1751-7915.12002

42. B. Yuan and D. A. Wood: Chapter One - Overview of formation damage during improved and enhanced oil recovery. In: *Formation Damage during Improved Oil Recovery*. Ed. B. Yuan and D. A. Wood. Gulf Professional Publishing (2018) doi:10.1016/B978-0-12-813782-6.00001-4

43. C. D. Hsi, D. S. Dudzik, R. H. Lane, J. W. Buettner and R. D. Neira: Formation injectivity damage due to produced water reinjection. In: *SPE Formation Damage Control Symposium*. Society of Petroleum Engineers, Lafayette, Louisiana (1994) doi:10.2118/27395-MS

44. S. Joshi, S. Goyal, A. Mukherjee and M. S. Reddy: Microbial healing of cracks in concrete: A review. *Journal of Industrial Microbiology and Biotechnology*, 44(11), 1511–1525 (2017) doi:10.1007/s10295-017-1978-0

45. G. K. Oka and G. F. Pinder: Multiscale model for assessing effect of bacterial growth on intrinsic permeability of soil: Model description. *Transport in Porous Media*, 119(2), 267–284 (2017) doi:10.1007/s11242-017-0870-8

46. C. M. Manuel, O. C. Nunes and L. F. Melo: Dynamics of drinking water biofilm in flow/non-flow conditions. *Water Research*, 41(3), 551–562 (2007) doi:10.1016/j.watres.2006.11.007

47. A. Franco-Rivera, J. Paniagua-Michel and J. Zamora-Castro: Characterization and performance of constructed nitrifying biofilms during nitrogen bioremediation of a wastewater effluent. *Journal of Industrial Microbiology and Biotechnology*, 34(4), 279–287 (2007) doi:10.1007/s10295-006-0196-y

48. K. Fujiwara, Y. Sugai, N. Yazawa, K. Ohno, C. X. Hong and H. Enomoto: Biotechnological approach for development of microbial enhanced oil recovery technique. *Petroleum Biotechnology: Developments and Perspectives*, 151, 405–445 (2004).

49. H. Suthar, K. Hingurao, A. Desai and A. Nerurkar: Selective plugging strategy based microbial enhanced oil recovery using Bacillus licheniformis TT33. *Journal of Microbiology and Biotechnology*, 19(10), 8 (2009) doi:10.4014/jmb.0904.04043

50. R. Hao, C. Meng and J. Li: Impact of operating condition on the denitrifying bacterial community structure in a 3DBER-SAD reactor. *Journal of Industrial Microbiology and Biotechnology*, 44(1), 9–21 (2017) doi:10.1007/s10295-016-1853-4

51. J. H. Lee, J. B. Kaplan and W. Y. Lee: Microfluidic devices for studying growth and detachment of Staphylococcus epidermidis biofilms. *Biomed Microdevices*, 10(4), 489–498 (2008) doi:10.1007/s10544-007-9157-0

52. C. David, K. Bühler and A. Schmid: Stabilization of single species Synechocystis biofilms by cultivation under segmented flow. *Journal of Industrial Microbiology and Biotechnology*, 42(7), 1083–1089 (2015) doi:10.1007/s10295-015-1626-5

53. A. C. Mitchell, A. J. Phillips, R. Hiebert, R. Gerlach, L. H. Spangler and A. B. Cunningham: Biofilm enhanced geologic sequestration of supercritical CO_2. *International Journal of Greenhouse Gas Control*, 3(1), 90–99 (2009) doi:10.1016/j.ijggc.2008.05.002

54. M. McHugh and V. Krukonis: *Supercritical Fluid Extraction: Principles and Practice*. Stoneham, MA: Elsevier (2013).

55. N. Liu, M. Haugen, B. Benali, D. Landa-Marbán and M. A. Fernø: Pore-scale spatiotemporal dynamics of microbial-induced calcium carbonate growth and distribution in porous media. *International Journal of Greenhouse Gas Control*, 125, 103885 (2023) doi:10.1016/j.ijggc.2023.103885

56. A. B. Cunningham, E. Lauchnor, J. Eldring, R. Esposito, A. C. Mitchell, R. Gerlach, A. J. Phillips, A. Ebigbo and L. H. Spangler: Abandoned well CO_2 leakage mitigation using biologically induced mineralization: Current progress and future directions. *Greenhouse Gases: Science and Technology*, 3(1), 40–49 (2013) doi:10.1002/ghg.1331

57. A. J. Phillips, E. Lauchnor, J. Eldring, R. Esposito, A. C. Mitchell, R. Gerlach, A. B. Cunningham and L. H. Spangler: Potential CO_2 leakage reduction through biofilm-induced calcium carbonate precipitation. *Environmental Science & Technology*, 47(1), 142–149 (2013) doi:10.1021/es301294q

58. D. Landa-Marbán, S. Tveit, K. Kumar and S. E. Gasda: Practical approaches to study microbially induced calcite precipitation at the field scale. *International Journal of Greenhouse Gas Control*, 106, 103256 (2021) doi:10.1016/j.ijggc.2021.103256

59. B. Pan, T. Ni, W. Zhu, Y. Yang, Y. Ju, L. Zhang, S. Chen, J. Gu, Y. Li and S. Iglauer: Mini review on wettability in the methane–liquid–rock system at reservoir conditions:

Implications for gas recovery and geo-storage. *Energy & Fuels*, 36(8), 4268–4275 (2022) doi:10.1021/acs.energyfuels.2c00308

60. Q. Lin, B. Bijeljic, S. Berg, R. Pini, M. J. Blunt and S. Krevor: Minimal surfaces in porous media: Pore-scale imaging of multiphase flow in an altered-wettability Bentheimer sandstone. *Physical Review E*, 99(6), 063105 (2019) doi:10.1103/PhysRevE.99.063105

61. Q. Lin, B. Bijeljic, S. Foroughi, S. Berg and M. J. Blunt: Pore-scale imaging of displacement patterns in an altered-wettability carbonate. *Chemical Engineering Science*, 235, 116464 (2021) doi:10.1016/j.ces.2021.116464

62. H. Alkan, M. Szabries, N. Dopffel, F. Koegler, R.-P. Baumann, A. Borovina and M. Amro: Investigation of spontaneous imbibition induced by wettability alteration as a recovery mechanism in microbial enhanced oil recovery. *Journal of Petroleum Science and Engineering*, 182, 106163 (2019) doi:10.1016/j.petrol.2019.06.027

63. D. Tiab and E. C. Donaldson: *Petrophysics: Theory and Practice of Measuring Reservoir Rock and Fluid Transport Properties*. Waltham, MA: Gulf Professional Publishing (2015).

64. Y. Yao, M. Wei and W. Kang: A review of wettability alteration using surfactants in carbonate reservoirs. *Advances in Colloid and Interface Science*, 294, 102477 (2021).

65. L. Jin Ho and L. Hai Bang: A wettability gradient as a tool to study protein adsorption and cell adhesion on polymer surfaces. *Journal of Biomaterials Science, Polymer Edition*, 4(5), 467–481 (1993) doi:10.1163/156856293X00131

66. G. Altankov and T. Groth: Reorganization of substratum-bound fibronectin on hydrophilic and hydrophobic materials is related to biocompatibility. *Journal of Materials Science: Materials in Medicine*, 5(9), 732–737 (1994) doi:10.1007/BF00120366

67. M. Hannig: Transmission electron microscopy of early plaque formation on dental materials in vivo. *European Journal of Oral Sciences*, 107(1), 55–64 (1999) doi:10.1046/j.0909-8836.1999.eos107109.x

68. J. H. Lee, G. Khang, J. W. Lee and H. B. Lee: Interaction of different types of cells on polymer surfaces with wettability gradient. *Journal of Colloid and Interface Science*, 205(2), 323–330 (1998) doi:10.1006/jcis.1998.5688

69. Y. Yuan, M. P. Hays, P. R. Hardwidge and J. Kim: Surface characteristics influencing bacterial adhesion to polymeric substrates. *RSC Advances*, 7(23), 14254–14261 (2017) doi:10.1039/C7RA01571B

70. P. Narayana and P. Srihari: A review on surface modifications and coatings on implants to prevent biofilm. *Regenerative Engineering and Translational Medicine*, 6(3), 330–346 (2020) doi:10.1007/s40883-019-00116-3

71. D. Zivar, S. Kumar and J. Foroozesh: Underground hydrogen storage: A comprehensive review. *International Journal of Hydrogen Energy*, 46(45), 23436–23462 (2021) doi:10.1016/j.ijhydene.2020.08.138

72. J. Simon, A.M. Ferriz, L.C. Correas, HyUnder – Hydrogen Underground Storage at Large Scale: Case Study Spain. *Energy Procedia*, 73, 136–144 (2015), ISSN 1876-6102, doi:10.1016/j.egypro.2015.07.661

73. N. Heinemann, J. Alcalde, J. M. Miocic, S. J. T. Hangx, J. Kallmeyer, C. Ostertag-Henning, A. Hassanpouryouzband, E. M. Thaysen, G. J. Strobel, C. Schmidt-Hattenberger, K. Edlmann, M. Wilkinson, M. Bentham, R. Stuart Haszeldine, R. Carbonell and A. Rudloff: Enabling large-scale hydrogen storage in porous media – The scientific challenges. *Energy & Environmental Science*, 14(2), 853–864 (2021) doi:10.1039/D0EE03536J

74. M. Panfilov: Underground storage of hydrogen: In situ self-organisation and methane generation. *Transport in Porous Media*, 85(3), 841–865 (2010) doi:10.1007/s11242-010-9595-7

75. N. Liu, A. R. Kovscek, M. A. Fernø and N. Dopffel: Pore-scale study of microbial hydrogen consumption and wettability alteration during underground hydrogen storage. *Frontiers in Energy Research*, 11 (2023) doi:10.3389/fenrg.2023.1124621

76. N. Dopffel, S. Jansen and J. Gerritse: Microbial side effects of underground hydrogen storage - Knowledge gaps, risks and opportunities for successful implementation. *International Journal of Hydrogen Energy*, 46(12), 8594–8606 (2021) doi:10.1016/j.ijhydene.2020.12.058

77. N. Eddaoui, M. Panfilov, L. Ganzer and B. Hagemann: Impact of pore clogging by bacteria on underground hydrogen storage. *Transport in Porous Media*, 139(1), 89–108 (2021) doi:10.1007/s11242-021-01647-6

78. A. Pérez, E. Pérez, S. Dupraz, & J. Bolcich: Patagonia wind-hydrogen project: Underground storage and methanation. In *21st world hydrogen energy conference*, Zaragoza, Spain. 13–16th June, 2016.

79. S. Bauer and R. Austria: *Underground Sun Storage*. Final Report, Vienna, Austria, 1369 (2017).

80. A. B. Dohrmann and M. Krüger: Microbial H(2) consumption by a formation fluid from a natural gas field at high-pressure conditions relevant for underground H(2) storage. *Environmental Science & Technology*, 57(2), 1092–1102 (2023) doi:10.1021/acs.est.2c07303

81. P. Gilbert, T. Maira-Litran, A. J. McBain, A. H. Rickard and F. W. Whyte: The physiology and collective recalcitrance of microbial biofilm communities. *Advances in Microbial Physiology*, 46, 202–256 (2002).

82. M. R. Parsek and C. Fuqua: Biofilms 2003: Emerging themes and challenges in studies of surface-associated microbial life. *Journal of Bacteriology*, 186(14), 4427–4440 (2004) doi:10.1128/jb.186.14.4427-4440.2004

83. P. S. Stewart: Mechanisms of antibiotic resistance in bacterial biofilms. *International Journal of Medical Microbiology*, 292(2), 107–113 (2002) doi:10.1078/1438-4221-00196

84. P. S. Stewart and J. W. Costerton: Antibiotic resistance of bacteria in biofilms. *Lancet*, 358(9276), 135–138 (2001) doi:10.1016/s0140-6736(01)05321-1

3 Microbial Control and Sustainability

Can Managing Microorganisms Improve the Environmental Footprint of Oil and Gas Operations?

Renato M. De Paula, Charles D. Armstrong, and Carla N. Thomas
Syensqo Oil and Gas Solutions, The Woodlands, TX, United States

3.1 INTRODUCTION

In the past few years, topics related to climate change, sustainability, and renewable energies have been at the forefront of public discussion, activism, and political agendas throughout the world. Climate change is now an accepted fact, albeit its cause and the anthropogenic contribution to it remain debatable. Nonetheless, within the scientific community, it is widely accepted that the generation of carbon dioxide by human-induced activities has had a dramatic impact on climate change. A recent study in 2021 showed that 99% of scientific papers indicate that humans cause climate change.[1] Combustion of fossil fuels (oil, natural gas, coal) to generate energy is among the human activities most credited for climate change, and the oil and gas industry has been implicated as a contributor to global warming. As populations grow and access to modern life amenities increases, energy consumption must also increase to sustain economic activities and human well-being. Carbon dioxide emissions from fossil fuels have risen seven times since the 1950s, mostly driven by transportation, heating, and electricity.[2]

The conundrum of high-energy demand versus the impact on the planet's climate has led to constant calls for energy transition, from burning fossil fuels to more sustainable energy sources, with less detrimental impact on the planet. In 2020, 79% of the global required energy came from fossil fuels and only 21% from non-fossil fuels (wind, biofuels, hydroelectricity, and other renewable sources).[3]

While public consensus and policy-making decisions continue to evolve, many signs indicate that energy transition, or at least a more balanced energy portfolio, will indeed occur in the next decade. Larger corporations, including major petroleum

DOI: 10.1201/9781003287056-5

37

producers, have made pledges to either completely decarbonize their operations or make significant contributions to minimize the negative impact of burning fossil fuels and seek alternative sources of energy.[4,5] Nonetheless, it is widely understood that this transition must be progressive, incremental, and without abrupt depletion of existing energy supplies. Energy deficiency can have a profound negative effect on global economies, which can lead to long-lasting unemployment, depletion of resources, and erosion of national security.

It is unclear how long it will take for the energy transition to occur, but many oil and gas producers are exploring ways to minimize the environmental footprint of their operations. For example, in the past years, many producers that operate shale-producing wells have shifted from using fresh water to using flow-back/produced water or co-mingled water for hydraulically fracking new wells.[6] When the produced water is not suitable for reuse in activities related to oil exploration, efforts to reutilize the water in other industries, such as agriculture, have been documented.[7] Moreover, major oil producers have established tighter guidelines for operations around the globe to minimize the risks of oil spills, production of toxic gases, and discharge of chemicals offshore.

In this chapter, focus will be given to microbial control during the production of oil and gas, highlighting steps where proper control of microorganisms can lead to a positive environmental footprint. The maintenance of the integrity and safety of oil and gas operations can help to minimize the negative impact on the environment, while energy transition occurs, and new sources of renewable energy are further developed.

3.2   ASSURANCE OF CONTAINMENT: CONTROL OF MICROBIAL-INFLUENCED CORROSION (MIC)

Corrosion of transmission pipelines and oilfield equipment is one of the major causes of oil spills and poses a significant challenge to decreasing environmental pollution by the petroleum industry. Although microbial corrosion can occur in any part of a producing facility, failures occurring in topside vessels can be easily identified and contained, thereby avoiding major leaks and spills. On the other hand, leaks in transmission lines often in remote locations can result in significant environmental impact before they can be restrained. The impact of oil and gas spills on the environment can have long-lasting effects. The damage can persist for years after the spills. In most cases, environmental recovery is achieved between 2 and 10 years, although it may take several decades if salt marshes and mangrove swamps are affected.[8] In 2015, a failure attributed to MIC released over 100,000 tons of methane from a well casing in a storage field in Aliso Canyon, CA, USA. This event highlighted the environmental impact of poorly controlled microorganisms in gas production.[9] Not only are the releases of crude oil and gases (natural gas, carbon dioxide, hydrogen sulfide) detrimental, but the efforts to control and remediate the spills can also cause large-scale ecological damage.[10]

In the United States, the costs associated with corrosion are estimated to be about $170 billion a year, with the oil and gas industry incurring more than half of these costs.[11] Due to the complex nature of MIC and constant discovery of new mechanisms and organisms involved in the process, it is unclear how much MIC contributes to the overall corrosion problems faced by the petroleum industry. Some estimations suggest that MIC causes up to 20% of the corrosion failures worldwide.[12]

From an environmental standpoint, transportation of hydrocarbon production by pipelines has a lower carbon footprint than fuel-burning carriers such as trucks, barges, tankers, and railroad tank cars. Pipelines can be engineered to avoid highly sensitive areas; they can be buried and, if properly maintained, they can safely transport fluids with a lower risk for spills. As natural gas production grows in the United States, the demand for new pipelines has increased.[13] Nonetheless, offshore pipelines carry more risk for leaks and environmental impact than onshore pipelines, despite significant improvements in technologies for pipeline material and monitoring systems. As increases in pipeline networks are foreseen as a means to reduce carbon footprint in oil and gas production, more focus will be placed on the integrity of these pipeline networks, and control of MIC will certainly be a considerable topic.

Microbial corrosion in pipelines is the result of accumulation of stagnant water in areas of low flow along the pipeline topography. In these areas, anaerobic microorganisms can grow and form biofilms on the internal metal surface of the pipe, leading to pitting corrosion. The pits can continue to grow through the metal wall until they reach the external wall of the pipe, compromising the integrity of the line. As most transmission pipelines are under pressure, a single pit or a few pitting features can result in the outburst of the pipe.

Microbial-influenced corrosion is a complex process that relies on different mechanisms involving many microorganisms, and details about these mechanisms are beyond the scope of this chapter. A considerable amount of information and data are available in the literature,[14–16] but two mechanisms seem to be prevalent in oilfield systems. The classical chemical MIC mechanism involves the production of corrosive byproducts (e.g., hydrogen sulfide and organic acids) from the metabolism of anaerobic organisms. These byproducts can react with the metallurgy, leading to transfer of electrons from the surface material.[17] Another more recently identified mechanism, named electronic MIC, relies on the direct transport of electrons from the material to the microbial cells via specific proteins present in the cell wall/membrane.[18,19] Thermophilic and mesophilic sulfate-reducing bacteria (e.g., *Halanaerobium sp, Thermovirga sp, Geotoga sp*) and methanogenic archaea (e.g., *Methanococcus sp, Methanocalculus sp*), among, other genera, are often identified in genetic analysis of corroded pipes where MIC is suspected.[20,21]

Control of MIC in pipelines is particularly challenging due to the lack of proper mechanisms for monitoring.[22] Long onshore pipelines are usually equipped with corrosion coupons/probes along the line, which can be used to monitor the accumulation of biofilms, if the coupons are correctly installed in the water phase. For offshore pipelines, corrosion coupons, if present, are only located at the producing platform (s) and at the onshore processing facility. Nonetheless, routine assessment of coupons is not easily performed due to accessibility, lack of personnel, and difficulties in preserving the samples for proper analysis and a comprehensive approach to data gathering.[23]

Control of MIC in pipelines and topside equipment relies heavily on the use of biocides. Pipelines can also be cleaned by mechanical means using automated brushes (commonly known as pigging), although that practice is not widely adopted in all pipeline networks due to differences in pipeline diameters and lack of pigging launching and receiving ends. Ideally, a combination of pigging and chemical treatment should be used to render better results; maintaining the lines clean of solids and debris and avoiding the formation of biofilms in the areas of low or stagnant flow is paramount. In topside equipment (wellheads, separation vessels, tanks), mechanical cleaning is rare and only done during scheduled shutdown of operations.

Although numerous biocide products are commercially available for treatment and control of MIC in production systems and pipelines, they are mostly composed of 1) electrophilic biocides such as glutaraldehyde and tetrakis hydromethyl phosphonium sulfate (THPS), 2) surfactants and lytic biocides (e.g., quaternary ammonium compounds (Quats)), and 3) blends of these molecules among themselves or with specialty non-biocidal polymers. Based on extensive laboratory and field data, the combination of surfactants and non-surface-active polymers with either glutaraldehyde or THPS significantly improves the performance of biocide products by increasing their penetration into biofilms and minimizing the corrosion process (Figure 3.1).[24–26]

It is generally accepted that a combination of mechanical (pigging) and chemical cleaning is the best approach to control solid deposition and formation of biofilms that can lead to MIC in pipelines. Nonetheless, pigging runs are not easily performed and are very expensive. Some estimates indicate that the cost of mechanically cleaning pipelines varies around \$35,000/mile.[27] By extrapolation, to clean all the pipelines throughout the United States only once, the cost would be about \$50 billion. Thus, the majority of control of microorganisms in pipelines relies on chemical use of biocides.

In that regard, the use of proper biocidal products, and most importantly the regimen of application and optimization of treatment, plays one of the most crucial roles

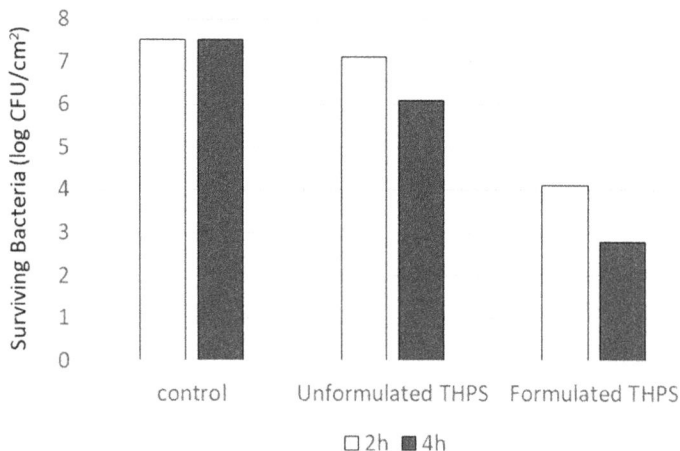

FIGURE 3.1 Microbial kill of a 10-day-old *Desulfovibrio vulgaris* biofilm. Microbial reduction was measured by swabbing the biofilms from the surface of coupons followed by culturing in Modified Postgate B medium. Formulated THPS contains a sulfonate polymer.

in controlling biofilms and avoiding leaks. Even highly efficacious biocides will not completely remove or mitigate biofilms in a single treatment. Multiple applications, at consistent intervals, are necessary to produce a net reduction of sessile microorganisms over time. Microbial rebound and biofilm regrowth are common after chemical stress.[28,29] Thus, the application of a biocide should follow a strict schedule that considers the amount of biofilm removed on a single dosage and the time for the biofilm to regrow. Additional applications of biocides prior to full regrowth will lead to general reduction over time and/or maintenance of the existing microbial population at low risk for corrosion until a pigging run is finally put in place.

Thus, although MIC is a significant threat to the safe operations of oil and gas production, there are significant opportunities for improvement to control the detrimental effects of MIC on pipeline integrity. The development of better methods for monitoring of microbial corrosion in long pipelines, improvements in mechanical cleaning that can reduce the costs of pigging and increase the deployment of the technology, and a more systematic and well-controlled application of biocides can result in better control of biofilms. That will ultimately lead to a lower risk of MIC in transmission lines and reduce the chances for leaks and spills of hydrocarbon in the environment.

3.3 BIOGENIC METHANE RELEASE FROM RESERVOIRS

Production of natural gas from reservoirs is fundamental to hydrocarbon production. Natural gas, also called fossil gas, is a mixture of gaseous hydrocarbons (under atmospheric conditions) containing up to four carbon atoms. Methane is the highest component of natural gas, followed by ethane, butane, and propane. Pentane (C5) and hexane (C6) are also components of natural gas but, at room temperature, they are highly volatile liquids. Other gases such as carbon dioxide, nitrogen, hydrogen sulfide, and helium are also present in natural gas, albeit at trace levels.

Natural gas is primarily used for heating and cooling houses. According to the US Energy Information Administration (EIA), natural gas accounted for 44% of residential energy consumption in 2019.[30] From an environmental impact standpoint, natural gas is a much cleaner energy source for heating than coal burning. Natural gas emits about 50% less carbon dioxide than coal while generating the same amount of energy. Nonetheless, it is important to note that the main component of natural gas, methane, is a very potent greenhouse gas. One molecule of methane has 25 times more heat absorption than carbon dioxide.[31] As of 2020, methane represented 11% of all greenhouse gases in the atmosphere.[32]

When methane is released into the atmosphere, it traps heat, which can contribute to climate change. Although the persistence of methane in the atmosphere is relatively short (~12 years) compared to that of other greenhouse gases (e.g., nitrous oxide has a persistence of 114 years), it is more efficient at trapping heat than other greenhouse gases. Increased emissions of methane have been documented to be associated with increases in stratospheric moisture that could increase future global warming and hinder ozone stratospheric recovery.[33]

The indirect effects of methane on ecosystems, due to the depletion of the ozone layer, have been well documented. Reduced ozone levels result in increased exposure to harmful UVB radiation at the Earth's surface. Environmental effects include those impacting the physiological and developmental processes of terrestrial and aquatic plants and their ability to function normally as producers in ecosystems. UVB radiation has also been documented to have deleterious effects on early developmental stages of fish, crustaceans, amphibians, and other marine animals. Decreases in reproductive success and impaired larval development are the most severe effects.[34]

Regarding the negative impact on human health, methane is an odorless gas that can displace oxygen in confined areas, which makes it highly hazardous as an asphyxiant. Exposure to methane can cause loss of consciousness, drowsiness, suffocation, acute pulmonary toxicity, and death at high concentrations. OSHA has no permissible exposure limit (PEL) for methane; however, the NIOSH maximum recommended safe methane concentration for workers during an eight-hour period is 1,000 ppm (0.1%).

Methane plays an important role in the degradation of the ozone layer when transformed into water in the stratosphere. Indirect adverse effects from the release of methane into the atmosphere include the impact of exposure to UVB radiation caused by the depletion of the ozone layer. UVB radiation has been shown to cause non-melanoma skin cancer and is associated with the development of malignant melanoma.[35]

Methanogenic biodegradation is the leading process responsible for the degradation of hydrocarbons in crude oil reservoirs. This process entails the transformation of hydrocarbon substrates by syntrophic bacteria to produce degradants, that are subsequently utilized by methanogens to produce methane.[36,37] The role of Methanogenic Archaea, or methanogens, has drawn significant attention in the past years, primarily because of their role in producing methane from carbon dioxide and volatile fatty acids, and hydrogen.[38]

Methanogenic archaea produce methane through the reduction of carbon dioxide, hydrogen, acetate, formate, methanol, methyl sulfides, and methylamines under strictly anaerobic conditions. Common electron acceptors are nitrate, sulfate, and iron. The key enzyme in methanogenesis is the methyl-coenzyme M reductase (*Mcr*) complex, which catalyzes the final step of reduction of methyl-coenzyme M to methane.[39] Methanogen accumulation in the environment can have a profound impact on climate change. A study has suggested that methanogenesis played a significant role in the Permian extinction event, about 250 million years ago. This was reported to be caused by Methanosarcina's ability to use acetate for methane production, which allowed proliferation of these organisms in marine sediments, leading to an increase in atmospheric methane and elimination of more than 90% of species in the planet.[40]

While it is impractical to control the activity of methanogens in the reservoir, oil and gas operators are increasingly facing scrutiny concerning methane leaks during production. Numerous cases of methane leaks have been reported in the media, including noteworthy cases in West Texas[41] and in the Gulf of Mexico.[42] These methane leaks have increased public awareness and oversight from environmental agencies. New technologies have been developed to monitor methane leaks, including methane sensors, infrared cameras, and aerial and satellite imagery, to detect methane accumulation in the atmosphere.

Moreover, new policies have been implemented to curb methane release from oil and gas production. On May 12, 2016, the US Environmental Protection Agency

(EPA) issued a rule outlining measures to curb the emission of methane from oil and gas resources with a goal of cutting methane emissions in this sector by 40–45% from 2012 levels, by 2025. Approximately one-third of the methane emissions that are due to human activities have been reported to be attributable to oil production and the production, processing, transmission, and storage of natural gas.[30]

3.4   RESERVOIR SOURING

Reservoir souring refers to the process whereby sweet oil reservoirs (i.e., ones containing no hydrogen sulfide (H_2S)) start producing sour fluids containing H_2S. Although reservoir souring often occurs when oil reservoirs are intentionally injected with (sea) water for secondary recovery, souring can also occur at the very early stages of production in unconventional wells, where large volumes of water are used for hydraulic fracturing and completion of the well.[43]

The production of H_2S from sulfate in waters was documented as a biological process as early as 1864.[44] Sulfate-reducing prokaryotes (sulfate-reducing bacteria and sulfate-reducing archaea) are now generally accepted as the primary microorganisms that elicit the reduction of sulfur compounds (sulfate, thiosulfate, etc.) to H_2S, thereby causing the souring of oilfield reservoirs.[45] Reservoir souring not only has the potential to impact oilfield operational costs, due to the need for mitigation of iron sulfide scaling and corrosion of equipment, but can also be deleterious to human health and the environment.

Hydrogen sulfide is a colorless, flammable gas that is extremely hazardous, with a noticeable rotten egg odor. The production of H_2S in sour reservoirs can deteriorate air quality, thereby posing potential occupational safety risks to oilfield workers. The primary route of exposure is inhalation leading to rapid absorption by the lungs. While the acute toxicity of this gas has been well documented, there is paucity of information on the long-term effects on human health.

Several regulatory agencies, including the US Occupational Health and Safety Administration (OSHA), have established H_2S exposure limits to ensure the safety of workers and the public. The PEL set by OSHA is 20 ppm not to be exceeded at any time and a 50 ppm limit at maximum peak not to be exceeded during any 10-minute work period. Other exposure limits include a NIOSH-recommended airborne exposure limit (REL) of 10 ppm, which should not be exceeded during a 10-minute work period.[46]

In water, H_2S dissociates to form the following ionic species in equilibrium:

$$H_2S <=> H^+ + HS^-$$

$$HS^- <=> H^+ + S^{2-}$$

The toxicity of hydrogen sulfide to aquatic organisms is typically expressed in terms of molecular H_2S concentration or total sulfide in water. The toxicity has been demonstrated to be pH- and temperature-dependent, with the neutral H_2S molecule being the most toxic form.[47] In a multi-species study, macroinvertebrates exposed to sulfide exhibited increased hypoxic and anoxic effects.[48] In particular, crustaceans were documented to have a lower tolerance to sulfides compared with other benthic invertebrates. The 96-hour lethal concentration 50 (LC50) values for hydrogen sulfide to freshwater fish species have been documented to range from 20 to 50 μg/L and lower concentrations have been shown to increase the susceptibility of fish to diseases. Comparatively, shrimp and other marine species tend to be more tolerant to hydrogen sulfide than freshwater species.[49] Table 3.1 presents the acute toxicity of H_2S, expressed as un-ionized H_2S, to various aquatic organisms:

Sour fluids can have a significant impact on the integrity of the oil and gas assets, leading to corrosion failures by sulfide stress cracking.[50] Moreover, chemical scavenging of H_2S gas often uses triazine-based chemicals, which produce toxic byproducts that must be removed from the system and disposed elsewhere.

The mechanisms of reservoir souring and the aspects of the subsurface microbiology that lead to the phenomenon have been extensively studied, and excellent reviews on the subject are available in the literature.[41,51,52] However, the solutions to solve reservoir souring and minimize the negative effects of H_2S on operations and the environment remain elusive and are a considerable focus of research. Competitive exclusion by injection of nitrate is, by far, the most applied solution for reservoir souring, although results are often controversial and inconsistent.[53] Nitrate is fed into the reservoir to serve as an electron acceptor under anaerobic conditions. Nitrate stimulates the growth of nitrate-reducing organisms, which in turn compete with sulfate reducers for the carbon sources. Moreover, certain sulfate reducers can also reduce nitrate preferentially over sulfate, as this process is more energetically favorable. While the application of nitrate can have positive effects in certain fields,[54] a number of parameters can render the treatment ineffective, including the presence of organisms recalcitrant to nitrate[55] and the availability of specific carbon sources.[56]

A number of specific inhibitors of the sulfate reduction pathway have been studied in the past years as potential solutions to stop reservoir souring. These include perchlorate,[57] monofluoro phosphate,[58] molybdate,[59] and iodonium salts.[60] However, these molecules, while promising, have not been extensively tested in the field, and neither have they been evaluated and approved by environmental agencies for field application.

TABLE 3.1
Aquatic Toxicity of Hydrogen Sulfide

Organism and Exposure Duration		LC_{50} (μg/L)
Fish	*Coregonus clupeaformis* (48–96 hr)	2
	Carassius auratus (48–96 hr)	4
	Salmo trutta (48–96 hr)	7
	Pimephales promelas (48–96 hr)	710
	Oncorhynchus mykiss (5–29-day)	6–22
Freshwater crustaceans	*Branchiura sowerbyi* (96 hr)	19,500
Freshwater insects	*Chironomus* spp. (96 hr)	23,000–33,400
Freshwater mollusk	*Lymnaea luteola* (96 hr)	6,000
Marine mollusk	*Mytilus edulis* (48 hr EC_{50})	1.5
Marine echinoderm	*Strongylocentoatus purpuratus* (48 hr EC_{50})	3
Marine crustacean	*Palaemonetes pugio, Rhepoxynius abronius, Eohaustorius estuarius* (48–96 hr)	24–112

Biocides have also been evaluated for control of reservoir souring, although only a few products have been proven to withstand reservoir conditions, transverse the formation, remain active during the transition time, and effectively control souring in regions farther from the injection wells. THPS has been shown to control souring in a 60°C sandstone reservoir in a producing field in the North Sea, persisting for five months in the reservoir, and capable of being detected in producing wells.[61] Moreover, THPS has been shown to work synergistically with nitrate to control souring in high-temperature (up to 84°C) reservoirs.[62]

In conclusion, management of reservoir souring is an important area for improvement of oil and gas production. Minimizing the presence of toxic gases generated by microbial activity can substantially increase the sustainability of the operations and decrease the exposure of workers and the environment to these contaminants.

3.5   APPLICATION AND DISCHARGE OF BIOCIDES

Biocides are frequently used to control microbial issues in oilfield operations. Due to the large volumes of water accumulated during the production of hydrocarbons, the quantities of biocides applied on a daily basis in onshore and offshore production systems are significantly large. Biocides have an inherently toxic nature, as they are designed to penetrate biomass and kill a broad range of microbial cells that can be harmful to industrial operations, human health, and the environment if not properly controlled. Nonetheless, understanding the nature of specific chemicals, their fate when discharged, and the guidelines imposed by government agencies can help operators to make educated decisions about how to handle these toxic substances. Thus, proper management of biocide applications during production of hydrocarbons can significantly contribute to increasing the sustainability of oil and gas operations.

Biocides are highly regulated products. Registration of biocides is required in many countries and regions, such as the United States, Canada, Great Britain, and the European Union. Even in lightly regulated countries, the use of biocides must be approved by government officials and follow local guidelines. In each of these jurisdictions, biocide labels are placed on products to direct the handling and proper use and specify the acceptable range of dosage of these chemicals. Compliance of the label requirements is enforced and non-compliance can be legally prosecuted.

In the United States, all biocidal products come with labels on the product that are designed to manage any potential risks from the use of the biocide. Labels are legally enforceable and both state and federal agencies will enforce these requirements. Prior to granting a label for a biocide product, the EPA requires extensive scientific data on the potential health and environmental effects that might occur from use of the product. As such, the label will detail the requirements on who may use that product, where, under what conditions, and how much may be used.[63] In Canada, all biocidal products undergo premarket approval through the Pest Management Regulatory Agency (PMRA).[64] Similar to the EPA, PMRA requires scientific evidence when evaluating a product for authorization and labeling with the ultimate goal of protecting human and environmental health. In 2012, the European Parliament established Regulation (EU) No 528/2012, also known as Biocidal Products Regulation (BPR), to regulate biocidal products and protect human, animal, and environmental health.[65] This regulation establishes an authorized list of substances in biocidal product formulations and the registration process to bring these types of products to market. As of January 1, 2021, Great Britain (GB) was no longer part of the European Union but continued to use the BPR to regulate its biocidal process. The GB BPR now regulates the use of biocide products in Great Britain. While much of the law is the same between the EU BPR and the GB BPR, some differences have arisen. For instance, GB will no longer participate in the decision-making processes with the EU and does not have further access to technical information submitted to the EU GPR after the finalization of Brexit.[66]

It is crucial to consider the label requirements when selecting a biocide for a particular application and area. Consideration about human, animal, and environmental health must also be evaluated before selecting a biocide to be used in the field and which can be eventually discharged. Topics such as biodegradability, bioaccumulation, environmental persistence and mobility, and inherent toxicity must all be considered. Clearly, most operators initially consider other parameters such as the economics and cost-benefit of individual biocidal products, but the environmental impact and public scrutiny have increasingly pushed the operators to consider the toxicity of the chemicals they use in their fields more than the cost of individual products.

The biodegradation of biocides depends on many different factors such as pH, salinity, heat, O_2 availability, and light/UV exposure. While oxidizing biocides are consumed rather quickly, usually within a matter of minutes or hours, some can generate dangerous decomposition products such as organochlorides, chloroform, and acids.[67] Additionally, these compounds can be extremely caustic (e.g., hypochlorite) or acidic (e.g., PAA), which can corrode the injection equipment and lead to chemical leaks if not properly managed.

Electrophilic biocides (e.g., Glutaraldehyde and THPS) undergo various modes of degradation, which varies from a few minutes to weeks, depending on the concentration utilized. According to the safety data sheet (SDS), 2,2-dibromo-3-nitrilopropion amide (DBNPA) has been classified as "not persistent" and found to degrade under normal environmental conditions.[68] Glutaraldehyde and THPS are readily biodegradable.[69,70] Glutaraldehyde meets the OECD test for Ready Biodegradability[71] at 73% after nine days. THPS also meets the OECD test at 70% after 21 days in aerobic conditions and 60% after 30 days in anaerobic conditions. UV light reduces the half-life of THPS by 0.4 days through indirect photolysis. Thus, THPS is more suited for down-hole applications, where the absence of oxygen and light renders the chemistry more persistent. Contrary to glutaraldehyde and THPS, Tris(hydromethyl)nitromethane (THNM), a formaldehyde releaser, is not readily biodegradable, with a biodegradation rate of only 13.4% over 28 days.[72,73]

Lytic biocides are usually surfactants with long organic chains that show varying degrees of biodegradation. For example, didecyldimethylammonium chloride (DDAC)[74] and benzalkonium chlorides (BKC/ADBAC)[75] both meet the biodegradability criteria as laid down in Regulation (EC) No. 648/2004[76] for detergents and are both considered readily biodegradable: 93.3% within 28 days and 95.5% within 28 days, respectively. On the other hand, tributyl tetradecyl phosphonium chloride (TTPC) does not fulfill the criteria for ready biodegradability or aerobic biodegradability.[77,78]

The biodegradation of chemical compounds should be considered when formulations are designed for applications in environmentally sensitive areas. Ideally, the antimicrobial molecule and the formulation that contains it should be persistent enough to complete the requirements of the application but not remain in the environment for extended periods of time.

Another key aspect of the environmental footprint of biocides is related to bioaccumulation. Since the release of the book *Silent Spring* by Rachel Carson in 1962,[79] which describes how DDT (dichlorodiphenyltrichloroethane) entered the food chain and accumulated in fatty tissues of humans and animals, public concern over the bioaccumulation of pesticides in the environment has been greatly justified. Most of the biocides used in the oil and gas industry, however, tend to not bioaccumulate, either due to their ability to breakdown into smaller, easier-to-metabolize products, or because they adsorbed onto surfaces which limits their mobility and accumulation. Additionally, most of these products are used in deep subterranean areas that are far removed from any potential places where they can cause much harm. Nonetheless, biocides are also applied in topside equipment and transmission pipelines and some may eventually be discharged into the environment, for instance into the ocean from offshore production systems. In those cases, any residual biocides either must be deactivated prior to discharge or diluted to levels that are deemed non-harmful when discharged. Thus, the treatment of surface waters, reuse of flow-back water, and potential communication with subsurface water supplies substantiates the need to understand the environmental fate and accumulation potential of any biocide.

Due to the fast degradation, oxidizer biocides are unlikely to bioaccumulate in the environment. Nonetheless, it is important to understand the byproducts of degradation, as they can be persistent and bioaccumulative.

Glutaraldehyde and THPS both have low bioaccumulation potentials and break down into substances that are readily absorbed and metabolized within the environment. DBNPA reacts quickly under both acidic and alkaline conditions and is not classified as persistent and with high bioaccumulation potential. However, similar to oxidizers, DBNPA decomposition products may pose some harm to the environment if not properly managed.

Although possessing high half-lives, surfactant-based biocides (ADBAC, DDAC, and TTPC) are minimally absorbed by animal tissues and do not bioaccumulate.[80] DDAC is further classified as having no substance considered to be persistent, bio accumulating, and toxic (PBT). This may be due to their limited mobility in the environment and their adsorption onto various surfaces.

The last aspect of the impact of biocides on the environment relates to ecological toxicity. An M-Factor (Multiplying Factor) is given to substances that have a Hazard Category of 1, an acute or chronic aquatic toxicity depending on the LC_{50} or the EC_{50}, with higher M-Factors for more toxic substances (CLP Regulation 1272/2008).[81] The Acute Aquatic Toxicity of a substance would be defined as that substance's ability to be injurious to aquatic organisms over a short-term exposure while the Chronic Aquatic Toxicity of a substance would be that substance's ability to cause adverse impacts to aquatic organisms in relation to the life cycle of the organism.

As shown in Table 3.1, the benzyl quats display the most toxicity to aquatic life and must be applied sparingly and carefully in situations where surface water

TABLE 3.2
M-Factors Describing the Aquatic Toxicity of Common Biocides Used in the O&G Business

	M-Factor		
Biocide	Acute	Chronic	Comments
THPS	1	Not Reported	Very toxic to aquatic life with long-lasting effects
Glutaraldehyde	<1	Not Reported	Moderately toxic to aquatic life
THMN	1	Not Reported	Highly toxic to aquatic organisms. Practically non-toxic to birds on a dietary basis
ADBAC	10	10	Highly toxic to aquatic life
DDAC	10	10	Highly toxic to fish
TTPC	10	10	Very toxic to aquatic life with long-lasting effects

ecosystems could be impacted. The remaining biocides reported in Table 3.2 show moderate to high toxicity in aquatic systems, but chronic toxicity has not been reported. This is most likely due to the high degradability and low bioaccumulation potential of these biocides.

The soil organic carbon-water partition constant (Koc) is the ratio of the mass of a chemical that is adsorbed in the soil per unit mass of organic carbon in the soil per the equilibrium chemical concentration in the soil. A high Koc indicates less mobility of a biocide through the soil, while a low Koc suggests a higher mobility of the biocide through the soil.

Kow is the Octinol-water partition constant (Kow) and is defined as the ratio of the hydrophobicity of a substance to its hydrophilicity. It can be a measure of how the biocide can be expected to be lost to hydrophobic reservoir fluids as well as an indicator of potential bioaccumulation. Low Kow (Table 3.3) indicates a preference for aqueous phases.[82] This can be interpreted as both increased solubility of the biocide in the aqueous reservoir fluids and a lower bioaccumulation in organic tissues of organisms.

TABLE 3.3
Mobility of Commonly Used Biocides through the Soil

Biocide	Log Kow	Log Koc
THPS	−9.77	1
Glutaraldehyde	−0.18	0
ADBAC	3.91	5.95
DDAC	4.66	5.69
TTPC	6.48	7.66

Data collected from EPI Suite; Kahrilas et al. (2015)

THPS has the lowest Kow of any of the biocides tested, suggesting that it is highly soluble and free to move through the reservoir as well as having a very low potential to bioaccumulate. Glutaraldehyde also shows high mobility coefficient. Nonetheless, the composition of the reservoir chemistry itself can affect the adsorption or migration of biocides through the formation. This suggests that a biocidal treatment using either of these biocides will lead to extensive dispersion in an area larger than the application area, so consideration must be given to the appropriate dosing of the product as it will eventually be diluted by migration. DBNPA has a higher mobility than either THPS or glutaraldehyde, although it is not expected to reach farther areas due to high degradability and low half-life. On the other hand, surfactant-based biocides have low mobility in the soil and are lost to adsorption by the reservoir. In fact, the experimental Koc for each of these is relatively high, with TTPC showing the highest Koc tested for all biocides. This explains the significant losses of benzyl quats in down-hole conditions. There appear to be two major adsorption mechanisms of cationic surfactants onto charged surfaces: Ion-exchange with existing adsorbed cations and hydrophobic (chain-to-chain) interaction mechanisms where the more hydrophobic the tail, the greater chance for the loss of these quaternary ammonium compounds upon the surface.[83] Additionally, as salt concentrations are increased, the solubility of these compounds increases as they compete with ion-binding sites.[84] This suggests that the higher the salt concentration of the reservoir fluids, the more soluble and, thus, mobile benzyl quats will move through the reservoir.

In conclusion, many different factors must be considered when selecting a biocide or formulation. There are not only legal ramifications to administering a biocide improperly but also health and environmental implications. Serious care must be taken in order to protect human, animal, and environmental health as well as the bottom line.

3.6   CONCLUSIONS

In the past decades, it has become unquestionable that burning of fossil fuels has had a dramatic influence on climate change. This has driven oil and gas operators to continuously seek improvements in their operations, while new technologies and sources of green, sustainable energy solutions continue to be developed and find space in the energy sector. Major oil-producing companies have pledged to decarbonize their operations, reduce energy consumption, and improve operations, even if that results in higher costs of operation. While no sole source of green energy can fulfill the demand of countries for energy, hydrocarbon combustion will still remain the main energy driver for a few decades ahead. Meanwhile, improvements that can reduce environmental risks, minimize the release of toxic gases, and make operations more efficient would have a positive impact on decarbonization processes.

Microorganisms may hold a very important role in this process, whether positively in the production of biofuels and hydrogen, or through better management of unwanted detrimental effects in industrial processes. As the Earth's master chemists, microorganisms may hold the keys to many of society's needs. Thus, understanding and being able to control and manage microorganisms are a crucial step in the final goal to develop sustainable energy while managing the negative effects on the planet.

REFERENCES

1. Lynas, M., Houlton, B.Z., Perry, S. (2021). Greater than 99% consensus on human caused climate change in the peer-reviewed scientific literature. *Environmental Research Letters*, 16(11), 114005. doi:10.1088/1748-9326/ac2966
2. Global Carbon Project. (2022). Global carbon budget. https://www.globalcarbonproject.org/carbonbudget/index.htm
3. Perera, F., Nadeau, K. (2022). Climate change, fossil-fuel pollution, and children's health. *The New England Journal of Medicine*, 386(24), 2303–2314. doi: 10.1056/NEJMra2117706. PMID: 35704482.
4. The 2019 Climate Pledge. https://www.theclimatepledge.com/us/en/Signatories
5. The Oil & Gas Climate Initiative. (n.d.). https://www.ogci.com/
6. Rassenfoss, S. (2011). From flowback to fracturing: Water recycling grows in the Marcellus Shale *Journal of Petroleum Technology* 63(07), 48–51.
7. Chen, C.-Y., Wang, S.-W., Kim, H., Pan, S.-Y., Fan, C., Lin, Y.J. (2021). Non-conventional water reuse in agriculture: A circular water economy, *Water Research*, 199, 117193.
8. Kingston, P.F. (2002). Long-term environmental impact of oil spills. *Spill Science & Technology Bulletin*, 7(1–2), 53–61. doi:10.1016/S1353-2561(02)00051-8
9. Conley, S., Franco, G., Faloona, I., Blake, D.R., Peischl, J., Ryerson, T.B. (2016). Methane emissions from the 2015 Aliso Canyon blowout in Los Angeles, CA. *Science*, 351, 1317–1320. doi:10.1126/science.aaf2348
10. Lusk, D., Gupta, M., Boinapally, K., Cao, Y. (2008). Armoured against corrosion. *Journal of Hydrologic Engineering*, 13, 115–118. doi:10.1061/(ASCE)1084-0699(2008)13:3(115)
11. Tuttle, R.N. (1987). Corrosion in oil and gas production. *Journal of Petroleum Technology*, 39, 756–762.
12. Shi, X, Xie, N., Gong, J. (2011). Recent progress in the research on microbially influenced corrosion: A Bird's eye view through the engineering lens. *Recent Patents on Corrosion Science*, 1, 118–131. doi:10.2174/1877610811101020118
13. Madigan, J. (2019). *Oil & Gas Pipeline Construction in the US* (IBISWorld Industry Report 23712, August 2019).
14. Little, B.J., Blackwood, D.J., Hinks, J., Lauro, F.M., Marsili, E., Okamoto, A., Rice, S.A., Wade, S.A., Flemming, H.-C. (2020). Microbially influenced corrosion—Any progress? *Corrosion Science*, 170, 108641, doi:10.1016/j.corsci.2020.108641
15. Kokilaramani, S., Al-Ansari, M.M., Rajasekar, A., Al-Khattaf, F.S., Hussain, A., Govarthanan, M. (2021). Microbial influenced corrosion of processing industry by recirculating waste water and its control measures - A review. *Chemosphere*, 265, 129075. doi:10.1016/j.chemosphere.2020.129075
16. Kip, N., van Veen, J. (2015). The dual role of microbes in corrosion. *The ISME Journal*, 9, 542–551. doi:10.1038/ismej.2014.169
17. Videla, H.A., Characklis, W.G. (1992). Biofouling and microbially influenced corrosion, *International Biodeterioration & Biodegradation*, 29(3–4), 195–212. doi:10.1016/0964-8305(92)90044-O
18. Venzlaff, H., Enning, D., Srinivasan, J., Mayrhofer, K.J.J., Hassel, A.W., Widdel, F., Stratmann, M. (2013). Accelerated cathodic reaction in microbial corrosion of iron due to direct electron uptake by sulfate-reducing bacteria. *Corrosion Science*, 66, 88–96. doi:10.1016/j.corsci.2012.09.006
19. Lahme, S., Mand, J., Longwell, J., Smith, R., Enning, D., Stams, A.J.M. (2021). Severe corrosion of carbon steel in oil field produced water can be linked to methanogenic archaea containing a special type of [NiFe] hydrogenase 2021. *Applied and Environmental Microbiology*, 87, 3. doi:10.1128/AEM.01819-20

20. Soler, A.J., Saavedra, A.U., Pagliaricci, M.C., Fernández, F.A., Morris, W., Vargas, W.A. (2021). "Identification and characterization of planktonic and sessile consortium associated with microbiologically influenced corrosion (MIC) in the oil and gas industry." *Paper presented at the CORROSION 2021*, Virtual, April 2021.

21. Leach, D.G., Wang, W., Yan, C., Mattis, D., MacLeod, R., Wei, W. "Molecular deep dive into oilfield microbiologically influenced corrosion: A detailed case study of MIC failure analysis in an unconventional asset." *Paper presented at the AMPP Annual Conference + Expo*, San Antonio, Texas, USA, March 2022.

22. De Paula, R., Keasler, V. (2017). MIC monitoring: Developments, tools, systematics, and feedback decision loops in offshore production systems. In *Microbiologically influenced corrosion in the upstream oil and gas industry*. doi:10.1201/9781315 157818-14

23. Abilio, A.A., Wolodko, J., Eckert, R.B., Skovhus, T.L. (2021). Review and gap analysis of MIC failure investigation methods in Alberta's oil and gas. In *Failure analysis of microbiologically influenced corrosion*. Edited by Richard B. Eckert and Torben L. Skovhus CRC Press, London, pp. 25–66.

24. Sharma, M., Liu, H., Chen, S. et al. (2018). Effect of selected biocides on microbiologically influenced corrosion caused by *Desulfovibrio ferrophilus* IS5. *Scientific Reports* 8, 16620. doi:10.1038/s41598-018-34789-7

25. Senthilmurugan, B., Radhakrishnan, J.S., Poulsen, M., Arana, V.H., Al-Qahtani, M., Jamsheer, A.F. (2019). Microbially induced corrosion in oilfield: Microbial quantification and optimization of biocide application. *Journal of Chemical Technology & Biotechnology*, 94, 2640–2650. doi:10.1002/jctb.6073

26. Gieg, L., Sargent, J., Bagaria, H., Place, T., Sharma, M., Shen, Y., Kiesman, D. "Synergistic effect of biocide and biodispersant to mitigate microbiologically influenced corrosion in crude oil transmission pipelines." *Paper presented at the CORROSION 2020*, NACE-2020-15090, June 2020.

27. https://theknowledgeburrow.com/how-much-does-a-pipeline-pig-cost/

28. Ezeuko, C.C., Sen, A., Gates, I.D. (2013). Modelling biofilm-induced formation damage and biocide treatment in subsurface geosystems. *Microbial Biotechnology*, 6(1), 53–66. doi: 10.1111/1751-7915.12002. Epub 2012 Nov 20. PMID: 23164434; PMCID: PMC3815385.

29. Bas, S., Kramer, M., Stopar, D. (2017). Biofilm surface density determines biocide effectiveness. *Frontiers in Microbiology*, 8,2443. doi:10.3389/fmicb.2017.02443. PMID: 29276508; PMCID: PMC5727120.

30. https://www.eia.gov/energyexplained/use-of-energy/homes.php

31. https://www.epa.gov/ghgemissions/overview-greenhouse-gases

32. https://www.epa.gov/ghgemissions/overview-greenhouse-gases#:~:text=In%202020% 2C%20methane%20(CH4,sources%20such%20as%20natural%20wetlands

33. Shindell, D. (2001). Climate and ozone response to increased stratospheric water vapor. *Geophysical Research Letters*, 28, 1551–1554.

34. United States Environmental Protection Agency, "Health and environmental effects of ozone layer depletion." [Online]. Available: https://www.epa.gov/ozone-layer-protection/ health-and-environmental-effects-ozone-layer-depletion. [Accessed 22 July 2022].

35. Kim, H.Y. (2014). Ultraviolet radiation-induced non-melanoma skin cancer: Regulation of DNA damage repair and inflammation. *Genes & Diseases*, 1(2), 188–198.

36. Jones, D.M. et al. (2008). Crude-oil biodegradation via methanogenesis in subsurface petroleum reservoirs. *Nature*, 451, 176–180.

37. Berdugo-Clavijo Carolina, G.L.M. (2014). Conversion of crude oil to methane by a microbial consortium enriched from oil reservoir production waters. *Frontiers in Microbiology*, 5, 1–10.

38. Zhou, Z., Zhang, C.-J., Liu, P.F. et al. (2022). Non-syntrophic methanogenic hydro-carbon degradation by an archaeal species. *Nature*, 601, 257–262. doi:10.1038/s41586-021-04235-2

39. Thauer, R., Kaster, A.K., Seedorf, H. et al. (2008). Methanogenic archaea: Ecologically relevant differences in energy conservation. *Nature Reviews Microbiology*, 6, 579–591. doi:10.1038/nrmicro1931

40. Rothman, D.H., Fournier, G.P., French, K.L., Alm, E.J., Boyle, E.A., Cao, C., Summons, R.E. (2014). Methanogenic burst in the end-Permian carbon cycle. *Proceedings of the National Academy of Sciences of the United States of America*, 111(15), 5462–5467. doi:10.1073/pnas.1318106111

41. https://www.adn.com/business-economy/energy/2022/07/28/massive-methane-leaks-in-texas-oil-and-gas-fields-speeds-climate-change-as-governments-fail-to-act/

42. https://www.houstonchronicle.com/business/energy/article/Massive-methane-leak-found-in-Gulf-of-Mexico-17452149.php#:~:text=The%2017%2Dday%20release%20over,environmental%20engineering%20at%20Rice%20University

43. Gieg, L.M., Jack, T.R., Foght, J.M. (2011). Biological souring and mitigation in oil reservoirs. *Applied Microbiology and Biotechnology*, 92(2), 263–282.

44. Meyer, L. (1864). Chemische Untersuchungen der Thermen zu Landeck in der Grafschaft Glatz. 91.

45. Johnson, R.J., Folwell, B.D., Wirekoh, A., Frenzel, M., Skovhus, T.L. (2017). Reservoir souring – Latest developments for application and mitigation. *Journal of Biotechnology*, 256, 57–67. doi:10.1016/j.jbiotec.2017.04.003

46. Occupational Safety and Health Administration, "Hydrogen sulfide." [Online]. Available: https://www.osha.gov/hydrogen-sulfide/hazards

47. Speece, R.E. (1983). Anaerobic biotechnology. *Environmental Science & Technology*, 17(9), 417A.

48. Bagarinao, T.U. (1993). Sulfide as a toxicant in aquatic habitats. *SEAFDEC Asian Aquaculture*, 15(3), 2–4.

49. Boyd, C.E. (2014). Hydrogen sulfide toxic, but manageable. *Global Aquaculture Advocate*, March/April 2014, 34–36.

50. Li, K., Zeng, Y., Luo, J.-L. (2021). Influence of H2S on the general corrosion and sulfide stress cracking of pipelines steels for supercritical CO2 transportation. *Corrosion Science*, 190, 109639, doi:10.1016/j.corsci.2021.109639

51. Gieg, L.M., Jack, T.R., Foght, J.M. (2011). Biological souring and mitigation in oil reservoirs. *Applied Microbiology and Biotechnology* 92(2), 263–282. doi: 10.1007/s00253-011-3542-6

52. Rajbongshi, A., Gogoi, S.B. (2021). A review on anaerobic microorganisms isolated from oil reservoirs. *World Journal of Microbiology and Biotechnology*, 37(7), 111. doi:10.1007/s11274-021-03080-9

53. Mitchell, A.F., Skjevrak, I., Waage, J. (2017). "A re-evaluation of reservoir souring patterns and effect of mitigation in a mature North Sea field." Publisher: Society of Petroleum Engineers (SPE). *Paper presented at the SPE International Conference on Oilfield Chemistry*, April 3–5, 2017. Paper Number: SPE-184587-MS; doi:10.2118/184587-MS

54. Bødtker, G., Thorstenson, T., Lillebø, B.L., Thorbjørnsen, B.E., Ulvøen, R.H., Sunde, E., Torsvik, T. (2008). The effect of long-term nitrate treatment on SRB activity, corrosion rate and bacterial community composition in offshore water injection systems. *Journal of Industrial Microbiology and Biotechnology*, 35(12), 1625–1636. doi:10.1007/s10295-008-0406-x

55. Kamarisima, K.H., Miyanaga, K., Tanji, Y. (2018). The presence of nitrate- and sulfate-reducing bacteria contributes to ineffectiveness souring control by nitrate injection. *International Biodeterioration & Biodegradation*, 129, 81–88. doi:10.1016/j.ibiod.2018.01.007

56. Navreet, S., Voordouw, J., Voordouw, G. (2017). The effectiveness of nitrate-mediated control of the oil field sulfur cycle depends on the toluene content of the oil. *Frontiers in Microbiology*, 8. https://www.frontiersin.org/articles/10.3389/fmicb.2017.00956

57. Engelbrektson, A.L., Cheng, Y., Hubbard, C.G., Jin, Yong T., Arora, B., Tom, L.M., Hu, P., Grauel, A.-L., Conrad, M.E., Andersen, G.L., Ajo-Franklin, J.B., Coates, J.D. (2018). Attenuating sulfidogenesis in a soured continuous flow column system with perchlorate treatment. *Frontiers in Microbiology*, 9. doi:10.3389/fmicb.2018.01575

58. Williamson, A.J., Carlson, H.K., Kuehl, J.V., Huang, L.L., Iavarone, A.T., Deutschbauer, A., Coates, J.D. (2018). Dissimilatory sulfate reduction under high pressure by *Desulfovibrio alaskensis* G20. *Frontiers in Microbiology*, 9, 1465. doi:10.3389/fmicb.2018.01465

59. Kögler, F., Hartmann, F.S.F., Schulze-Makuch, D., Herold, A., Alkan, H., Dopffel, N. (2021). Inhibition of microbial souring with molybdate and its application under reservoir conditions. *International Biodeterioration & Biodegradation*, 157, 105158. doi:10.1016/j.ibiod.2020.105158

60. Jones, A.M., Geissler, B., De Paula, R. (2018). "A comparison of chemistries intended to treat reservoir souring." *Paper presented at the CORROSION 2018*, Phoenix, Arizona, USA, April 2018.

61. MacLeod, N., Bryan, T., Buckley A.J., Talbot, R.E., Veale, M.A. (1990). A novel biocide for oilfield applications. In *SPE Conference*, Aberdeen.

62. Jurelevicius, D., Ramos, L., Abreu, F., Lins, U., de Sousa, M.P., dos Santos, V.V.C.M., Penna, M., Seldin, L. (2021). Long-term souring treatment using nitrate and biocides in high-temperature oil reservoirs. *Fuel*, 288, 119731. doi:10.1016/j.fuel.2020.119731

63. United States Environmental Protection Agency Website. https://www.epa.gov/pesticide-labels/introduction-pesticide-labels, June 15, 2022.

64. Pest Control Products Act (S.C. 2002, c.28). https://laws-lois.justice.gc.ca/eng/acts/P-9.01/index; June 29, 2022.

65. Regulation (EU) No 528/2012.

66. Health and Safety Executive Website. https://hse.gov.uk/biocides/brexit.htm; July 15, 2022.

67. The Chlorox Company, *Clorox ® Regular-Bleach, Safety Data Sheet*, 2015.

68. ICL, *Biobrom C-103, Safety Data Sheet*, 2016.

69. DOW, *Glutaraldehyde 50% Safety Data Sheet*, 2013.

70. Solvay, *Tolcide™ PS75 Safety Data Sheet*, 2020.

71. OECD (1992). Guideline for testing of chemicals, July 17, https://www.oecd.org

72. The Dow Chemical Company, *TRIS NITRO™ Antimicrobial, 50%, Safety Data Sheet*, 2008.

73. The Dow Chemical Company, *Tris(hydroxymethyl)nitromethane Product Safety Assessment*, 2014

74. Lonza, *Bardac™ 2250, Safety Data Sheet*, 2016.

75. Lonza, *Barquat™ MB-50, Safety Data Sheet*, 2015.

76. Regulation (EC) No 648/2004 of the European Parliament and the Council of 31 March 2004 on detergents.

77. The Dow Chemical Company, *TRIS NITRO™ Antimicrobial, 50%, Safety Data Sheet*, 2008.

78. Solvay, *Tolcide™ TP50, Safety Data Sheet*, 2019.

79. Carson, R. (1962). *Silent spring*. Houghton Mifflin Company, Boston, MA.

80. DeLeo, P.C., Huynh, C., Pattanayek, M., Schmid, K.C., Pechacek, N. (2020). Assessment of ecological hazards and environmental fate of disinfectant quaternary ammonium compounds. *Ecotoxicology and Environmental Safety*, 206, 111116. doi:10.1016/j.ecoenv.2020.111116

81. Regulation (EC) No 1272/2008 – Classification, labelling and packaging of substances and mixtures (CLP), March 19, 2021.

82. Kahrilas, G., Blotevogel, J., Stewart, P.S., Borch, T. (2015). Biocides in hydraulic fracturing fluids: A critical review of their usage, mobility, degradation, and toxicity. *Environmental Science & Technology*, 49, 16–22.

83. Ersoy, B., Çelik, M.S. (2003). Effect of hydrocarbon chain length on adsorption of cationic surfactants onto clinoptilolite. *Clays and Clay Minerals*, 51(2), 172–180.

84. Droge, S., Goss, K.U. (2012). Effect of sodium and calcium cations on the ion-exchange affinity of organic cations for soil organic matter. *Environmental Science & Technology*, 46(11), 5894–5901.

Section III

Microbiologically Influenced
Corrosion (MIC) and Souring

4 Effects of High Salinity PWRI Practice on Sulfidogenesis and Microbially Influenced Corrosion

Mohammed Sindi and Xiangyang Zhu
Saudi Aramco, Dhahran, Saudi Arabia

Beate Christgen
Newcastle University, Newcastle upon Tyne,
United Kingdom

Angela Sherry
Northumbria University, Newcastle upon Tyne,
United Kingdom

Neil Gray
School of Natural and Environmental Sciences
(SNES), Newcastle University, Newcastle upon Tyne,
United Kingdom

Ian Head
Newcastle University, Newcastle upon Tyne,
United Kingdom

4.1 INTRODUCTION: BACKGROUND AND DRIVING FORCES

Microbially influenced corrosion is a global issue influencing the premature failure of metallic infrastructure and accounting for 20% of all internal corrosion incidences (Zhu, *et al.*, 2003). Numerous microorganisms are implicated in MIC including sulphate-reducers, methanogens, nitrate-reducers, iron-oxidisers, iron-reducers, and acid-producers. The aim of this chapter was to explore the effects of high salinity (127 g/L TDS; 204 g/L TDS) in a range of temperature incubations (15°C, 30°C,

DOI: 10.1201/9781003287056-7

45°C, and 60°C) resulting from and simulating the mixing of synthetic injected sea-water (ISW) and synthetic PW on sulfidogenesis, and the underpinning fundamental processes associated with it such as sulphate reduction into sulphide, microbial community composition, and MIC. This wide range of incubating temperatures and high salinity ISW:PW mix represents plausible ISW, and PW thermal gradient encountered under different oil and gas industry operational settings including: A) On-shore Oil Production Facility Water Flooding Lines; B) On-shore Oil Production Facility ISW:PW Mixing Tanks; C) Drilling muds (i.e., mixed with a variety of water types including: ISW, PW, Utility Water, etc.) Mixing Tanks; D) Off-shore Platforms ISW:PW Mixing Tanks; and E) Oil Reservoir ISW:PW Mixing Zone.

Production Water for Reinjection practices are commonly utilised in the oil and gas industry. Additionally, seawater injection and PWRI lead to in-reservoir salinity gradients from seawater salinity (35 g/L TDS) at the injector, to the usually higher formation water salinity within the reservoir. Injection of cool seawater also leads to the formation of a temperature gradient with a significant impact on the growth and activity of microorganisms in petroleum reservoirs. A specific high salinity PWRI practice of 20% synthetic ISW mixing with 80% synthetic PW from the North Sea (NS) (127 g/L TDS) and the AG (204 g/L TDS) production water systems were selected for the following reasons:

A) As a result of synthetic (ISW:PW) mixing, the sulphate concentration will decrease from 21 mM in 100:0% ISW:PW(NS) to approximately 0.6 mM in 0:100% ISW:PW(NS), affecting the growth and activity of SRP.

B) At this (20:80%) ISW:PW (NS-127 g/L TDS) salinity from the previous work conducted by Sindi, *et al.* (2021), a shift from the SRP enrichments of *Desulfobacter* sp., *Desulfotignum* sp., and *Desulfobulubus* sp., observed under 100% ISW (NS) (42 g/L TDS), into members of endospore-formers *Peptococcaceae* family, and *Halanaerobium* sp., (sporulation, thiosulfate reduction, and/or fermentation metabolism) observed under the aforementioned high salinity 20:80% ISW:PW (NS) (127 g/L TDS) was evident.

C) These (20:80%) ISW:PW (NS – 127 g/L TDS; AG – 204 g/L TDS) mixes, therefore, represent the highest salinities with noticeable sulphate (mM) concentrations available for anaerobic sulphate respiration by SRP: (20:80% ISW:PW (NS) 5 mM sulphate; (20:80% ISW:PW (AG) 10 mM sulphate).

D) Considering the aforementioned reasons of (20:80% ISW:PW) being the highest in salinity mix with noticeable sulphate (mM) concentrations available for anaerobic sulphate respiration by SRP, and the observed microbial community successions at this higher salinity from the previous work conducted by Sindi, *et al.* (2021) (i.e., sulphate reduction – sporulation, thiosulfate reduction and/or fermentation), it was thought to be interesting to evaluate the MIC potential under such high salinity (ISW:PW) mixes and across a thermal gradient that represented plausible oil and gas industry operational settings.

Therefore, developing fundamental insights on the impact of high salinity PWRI mixing practices on microbial processes, microbial communities, and MIC will help inform knowledge-driven best field practices for PWRI, and other microbially implicated processes such as hydraulic fracturing and drilling, where also this high salinity PWRI or similar water mixing procedures will occur.

4.2 SUMMARY OF METHODS

4.2.1 INCUBATIONS

Microcosms were setup by inoculating the bicarbonate, trace elements, vitamin mixture basal mineral media (Widdel and Bak, 1992) with 5% w/v inoculum. Based on the selected NS and the AG mixing scenarios, all ISW:PW microcosms had sulphate as the main electron acceptor (21 mM, 10 mM, and 5 mM), and a mixture of volatile fatty acids (VFA: 10 mM acetate, 10 mM propionate, and 5 mM butyrate) used as electron donors, and the five basal salts (NaCl, $CaCl_2.2H_2O$, $MgCl_2.6H_2O$, KCl, and Na_2SO_4) calculated based on the average cationic and anionic compositions of Na, K, Cl, SO_4, and Mg from 52 producing wells in the NS (Bjørlykke, 1995), and five producing wells from the United Arab Emirates (U.A.E), the Kingdom of Saudi Arabia (KSA), and Kuwait (Bader, 2007). As a result, different ISW:PW salinity mixes (%) of 42 g/L TDS, 127 g/L TDS, and 204 g/L TDS were generated (Table 4.1). Each ISW:PW salinity mix (%) treatment condition microcosms were prepared in triplicates in 200 ml Wheaton bottles (190 ml final culture volume), (80 mm × 10 mm × 1 mm) of cleaned 0.2% carbon steel coupons in plastic fittings (Lahme, et al., 2019), closed with butyl rubber stoppers, crimped sealed with aluminium caps, and the headspace flushed with (CO_2/N_2) (20:80%). Three treatment conditions had no corrosion coupon additions in them and were: 100% ISW (NS) (42 g/L TDS), 20:80% ISW:PW (NS) (127 g/L TDS), and 20:80% ISW:PW (AG) (204 g/L TDS), and that the 100% ISW (NS) (42 g/L TDS) treatment condition was used as a positive control for sulfidogenesis (Table 4.1). All treatment conditions and controls microcosms were incubated horizontally for the corrosion coupons to be completely submerged in the ISW:PW solution for the corrosion process to initiate. Additionally, all microcosms were anaerobically incubated at 15°C, 30°C, 45°C, and 60°C, except

TABLE 4.1

Treatment Conditions and Controls in ISW:PW Anaerobic Microcosms across the Salinity and Temperature Gradients

Mixed Ratios of ISW:PW (%)	Incubating Temperature (°C)				Treatment Rationale	Replication	Final Sulphate (mM)	Final Salinity (g/L)
100% ISW (∅)	30				* A	3	21	42
20:80% ISW:PW (∅) (NS)	30				* A		4.6	127
20:80% ISW:PW (∅) (AG)	30				* A		9.8	204
20:80% ISW:PW (∇) (NS)	15	30	45	60	* B		4.6	127
20:80% ISW:PW (∇) (AG)	15	30	45	60	* B		9.8	204

* A (Sulfidogenesis only analysis)
* B (Salinity and temperature interactions (Full-scale analyses)
    (∇) (Corrosion coupons added), (∅) (Corrosion coupons not added); NS (The North Sea); and AG (The Arabian Gulf); ISW (Injected Sea water); and PW (produced water).

for the no coupon incubations which were incubated only at 30°C and resulting in 11 different treatment conditions.

Three no coupon ISW:PW anoxic microcosms were set up for sulfidogenesis analysis (Table 4.1). The sulfidogenesis positive control (100% ISW) (30°C), 20:80% ISW:PW (NS) (127 g/L TDS), and 20:80% ISW:PW (AG) (204 g/L TDS), without coupon incubations, were set up (Table 4.1). Additionally, the following eight anoxic ISW:PW microcosms incubated with corrosion coupons were set up: (20:80 ISW:PW) (127 g/L TDS) (15°C-NS), (20:80 ISW:PW) (127 g/L TDS) (30°C-NS), (20:80 ISW:PW) (127 g/L TDS) (45°C-NS), (20:80 ISW:PW) (127 g/L TDS) (60°C-NS), (20:80 ISW:PW) (204 g/L TDS) (15°C-AG), (20:80 ISW:PW) (204 g/L TDS) (30°C-AG), (20:80 ISW:PW) (204 g/L TDS) (45°C-AG), and (20:80 ISW:PW) (204 g/L TDS) (60°C-AG) (Table 4.1).

4.2.2 ANALYTICAL PROCEDURE

The ISW:PW treatment conditions were sampled for a period of approximately 7.5 months to 225 days). The concentration of sulphate was measured using an Ion Chromatography method; using column (AS14A), the autosampler (AS40), a Dionex Cd-25 Conductivity Detector Model 1S25 Na_2CO_3 (8mM)/$NaHCO_3$ (1 mM) as eluent, at 1ml/min flow rate, and 25 µL injection loop, nitrogen gas as the carrier at 9 psi, unicam 4851 integrator, and chromatograms visualised using the Chameleon Dionex Software. Aqueous sulphide was measured spectrophotometrically at 480 nm following the reaction of aqueous sulphide with 5 mM $CuSO_4$ in 50 mM HCl solution, to produce a brown CuS precipitate (Cord-Ruwisch, 1985). Sulphate reduction (mM), sulphide production/depletion (mM), and sulphide depletion rates (mM/hrs.) were compared statistically by One-Way Analysis of Variance (ANOVA; Minitab 17, Minitab Ltd., Coventry, UK). Headspace methane was measured following Gas Chromatography using Flame Ionisation Detector (GC-FID) instrument. Gravimetric analysis and corrosion rates were conducted to determine corrosion coupons weight loss in grams (g) and to calculate corrosion rates in millimetres per year (mm/yr), as described elsewhere (Enning, *et al.*, 2012). Surface characterisation of the corrosion coupons was conducted using the Tescan Vega 3LMU scanning electron microscope fitted with a Bruker XFlash® 6 | 30 detector for EDS analysis.

4.2.3 DNA SEQUENCING AND BIOINFORMATICS

Two time-series (hrs.) of 0 hrs. and 5400 hrs. (225 days) were selected for DNA extractions for River Tyne sediment inoculum, 20:80% ISW:PW (NS), and 20:80% ISW:PW (AG) (15°C, 30°C, 45°C, and 60°C) incubations. River Tyne estuarine sediments from the day of sample collection were stored in −20°C and were submitted for sequencing as the negative control for the (20:80% ISW:PW) mix samples. DNA sequencing was carried out at Northumbria university (NU-OMICS) on MiSeq (illumina platform), and *16S rRNA* sequence libraries were analysed using Quantitative Insights into Microbial Ecology 2 (QIIME2) and Statistical Analysis of Metagenomic Profiles (STAMP) software packages. STAMP v2.1.3 (Parks, *et al.*, 2014) was used to determine the overall species level Amplicon Sequence Variants (ASVs) clustering patterns of ISW:PW microcosms across the salinity and temperature gradients (using Principal Component Analysis (PCA)) and generating time-series (hrs.) phylum/ order level heatmaps across the salinity and temperature gradients.

4.3   RESULTS

4.3.1   EFFECTS OF HIGH SALINITY PWRI PRACTICE ON SULFIDOGENESIS AND SULPHATE REDUCTION

Sulphide monitoring revealed a systematic sulphide depletion profile for both 20:80% (ISW:PW) (NS) (127 g/L TDS) and 20:80% ISW:PW (AG) (204 g/L TDS) (with coupon incubations) (15°C–60°C) (Figure 4.1), a minute sulphide production for 20:80% (ISW:PW) (NS) (127 g/L TDS) (no coupon incubation), and negligible sulphide production for 20:80% ISW:PW (AG) (204 g/L TDS) (no coupon incubation) (Figure 4.2), all compared with 100% ISW (30°C) (no coupon incubation) (Figures 4.1 and 4.2).

For the psychrophilic temperature incubations (15°C), in microcosms containing (20:80%) ISW:PW (NS) (127 g/L TDS), a high sulphide depletion of 2.64 mM ± 0.003 was observed. With increasing temperature in the mesophilic temperature incubations (30°C), in microcosms containing (20:80%) ISW:PW (NS) (127 g/L TDS), sulphide depletion decreased to 2.48 mM ± 0.10 (Figure 4.1). With further increase in temperature in the lower thermophilic incubations (45°C), in microcosms containing (20:80%) ISW:PW (NS) (127 g/L TDS), higher sulphide depletion of 2.80 mM ± 0.08 was observed (Figure 4.1). The highest sulphide depletion

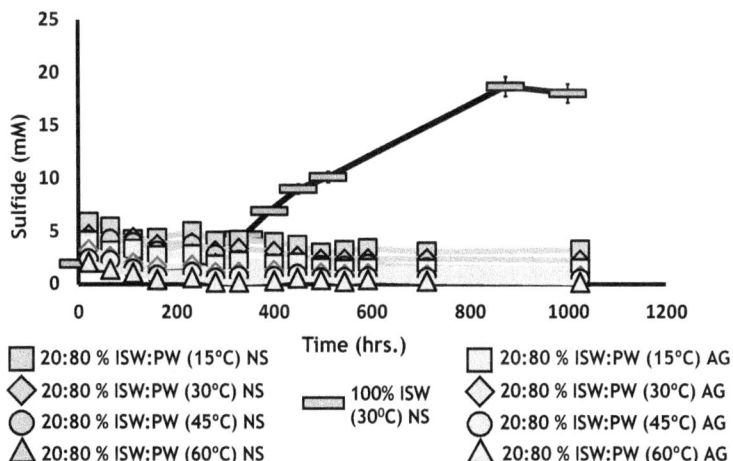

FIGURE 4.1   Sulfidogenesis profiles across the ISW:PW mixes (%) for the temperature gradient of (15°C, 30°C, 45°C, and 60°C) incubations. Symbols with grey lines depict 20:80% ISW:PW (NS) (127 g/L TDS) conditions (with corrosion coupons). Blue symbols and lines depict 20:80% ISW:PW (AG) (204 g/L TDS) (with corrosion coupons). Black line depicts 100% ISW (NS) (42 g/L TDS) (30°C) (without corrosion coupons). Error bars represent 1 × SE.

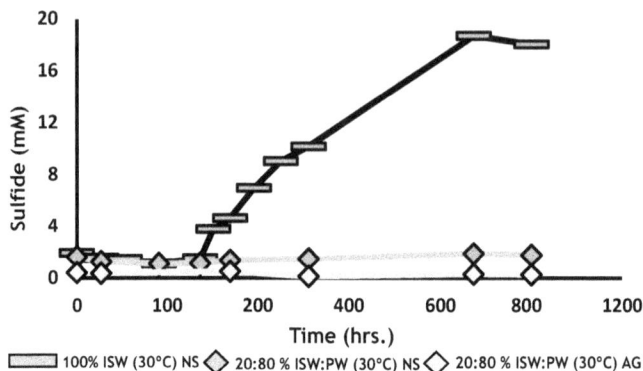

FIGURE 4.2 Sulfidogenesis profiles at 30°C in anoxic microcosms containing 100% ISW (NS), and the ISW:PW mix (%) microcosms from the North Sea and the Arabian Gulf Systems – no coupon incubations. Error bars represent 1 × SE.

of 2.93 mM ± 0.07 was observed under the moderate thermophilic temperature incubations (60°C) in microcosms containing (20:80%) ISW:PW (NS) (127 g/L TDS) (Figure 4.1).

For the psychrophilic temperature incubations (15°C), in microcosms containing (20:80%) ISW:PW (AG) (127 g/L TDS), the highest sulphide depletion of 2.65 mM ± 0.10 was observed. With the increase in incubating temperatures represented in the mesophilic (30°C), lower thermophilic (45°C), and moderate thermophilic (60°C); in microcosms containing (20:80%) ISW:PW (AG) (127 g/L TDS), lower sulphide depletions of 2.49 mM ± 0.06; 1.89 mM ± 0.02; and 1.9 mM ± 0.03, respectively, were observed (Figure 4.1). In contrast, the 100% ISW (30°C) (no coupon incubation) showed a sulphide production rather than depletion of 18.05 mM ± 0.11 (Figure 4.1). One-way ANOVA on Minitab was used as a statistical tool to determine the significance of the sulphide depletion profiles reported for the (NS-PW) system and the (AG-PW) system. This revealed that the differences in sulphide depletion between the (NS-PW) and the (AG-PW) system were non-significant ($P = 0.21$). However, when comparing the sulphide depletion profiles from the (NS-PW) and the (AG-PW) system, with the sulphide production profile reported in 100% ISW (30°C) (no coupon incubation), the statistical tool determined the sulphide production profile was significant ($P = 0.003$).

Next, the sulfidogenesis profiles in no coupon incubations and within the mesophilic incubating temperature (30°C); for both the NS water production system (20:80%) ISW:PW (NS) (127 g/L TDS); and the AG water production system (20:80%) ISW:PW (AG) (204 g/L TDS) were evaluated (Figure 4.2). As illustrated, a minuet sulphide production profile for (20:80% ISW:PW) (30°C – NS) of 1.77 mM ± 0.32 was observed (Figure 4.2), and a negligible sulphide production profile for (20:80% ISW:PW) (30°C- AG) of 0.31 mM ± 0.15 was observed (Figure 4.2). In contrast, the 100% ISW (30°C) (no coupon incubation) showed the highest sulphide production profile of 18.05 mM ± 0.11 (Figure 4.2).

Sulphate reduction profiles were monitored in parallel with sulphide depletion profiles and throughout the microcosm incubation period up to (545 h–18 days).

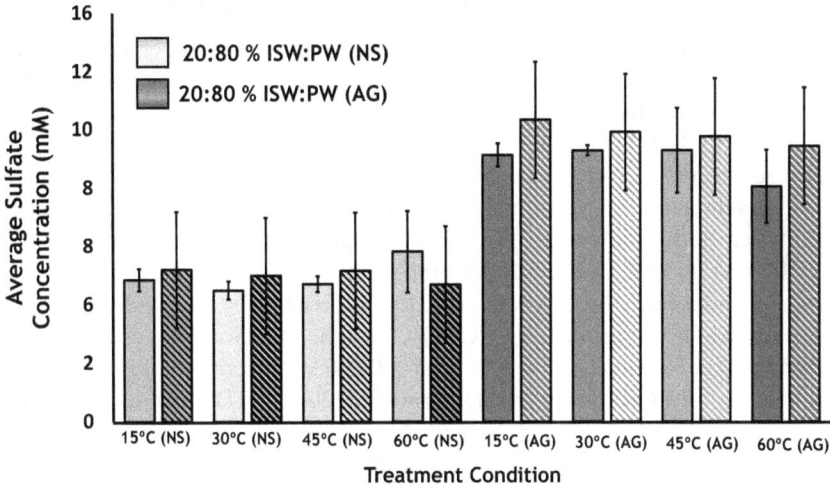

FIGURE 4.3   Effect of High Salinity PWRI Practice and temperature range (15°C–60°C) on sulphate reduction in microcosms containing 20:80% ISW: PW (NS 127 g/L TDS), and 20:80% ISW: PW (AG 204 g/L TDS). Lighter colour gradient represents average sulphate reduction (mM) for 20:80% ISW:PW (NS) treatments. Darker colour gradient represents average sulphate reduction for 20:80% ISW:PW (AG). Average sulphate concentrations (mM) were calculated for (21–161 h) timeframe (Open bars), and (329–545 h) timeframe (Shaded bars); for comparisons. Error bars represent 1 × SE. SD (NS 127 g/L) (15°C: 0.38 mM ± 0.12; 30°C: 0.74 mM ± 0.21; 45°C: 0.52 mM ± 0.15; 60°C: 0.13 mM ± 0.10;), SD (AG 204 g/L) (15°C: 0.86 mM ± 0.25; 30°C: 0.68 mM ± 0.20; 45°C: 0.65 mM ± 0.20; 60°C: 0.53 mM ± 0.15).

Sulphate reduction profiles were determined in high salinity ISW:PW and across a temperature gradient (Figure 4.3). The effect of ISW:PW (%) salinities on sulphate concentration revealed no sulphate reduction, and therefore no changes in the concentration of sulphate initially added to the microcosms (Figure 4.3). Initial concentrations of sulphate in microcosms were 4.6 mM sulphate in 20:80% ISW:PW (NS) (127 g/L TDS) incubations, and 9.8 mM sulphate in 20:80 ISW:PW (AG) (204 g/L TDS) incubations. Sulphate concentrations remained stable and close to initial microcosm concentrations by the end of the incubation period, showing no evidence of sulphate reduction across the temperature range (NS, average 4.6 ± 0.38 mM sulphate; AG, 9.8 ± 0.60 mM sulphate, Figure 4.3). No significant differences in sulphate concentrations were observed in microcosms prepared with NS or AG production waters, respectively, across the temperature range ($P = 0.32$).

4.3.2   Timeframes for Sulphide Depletion, Maximum Rate of Sulphide Depletion, and Maximum Concentration of Sulphide Depleted

To further determine the influence of the selected thermal gradient (15°C–60°C) and the high salinity PWRI mix (NS: 127 g/L TDS; AG: 204 g/L TDS), the following parameters were analysed: A) Sulphide depletion rate (mM/hrs.), B) Time to reach a sulphide depletion of 2 mM, and C) Maximum sulphide depletion concentration (mM).

4.3.2.1 Thermal Gradient Impact on Rates of Sulphide Depletion

Sulphide depletion rates in (mM/hrs.) for the salinity and temperature ISW:PW microcosms were determined (Figure 4.4). Rates of sulphide depletion (mM/hrs.) increased with the increase in the thermal gradient for (NS) the water production system (127 g/L TDS) (Figure 4.4a) but decreased with the increase in the thermal gradient for the (AG) water production system (204 g/L TDS) (Figure 4.4b). At the psychrophilic and mesophilic incubating temperatures (15°C–30°C) for the (NS) water production system, the sulphide depletion rates (mM/hrs.) were lower (0.0044 mM ± 0.00013 and 0.0045 mM ± 0.000073, respectively) compared to higher sulphide depletion observed in the lower thermophilic and moderate thermophilic incubating temperatures (45°C–60°C) for the (NS) water production system (0.0056 mM ± 0.000052 and 0.0054 mM ± 0.0002, respectively) (Figure 4.4a). In the higher salinity (AG) water production system (204 g/L TDS), the sulphide depletion rates (mM/hrs.) for the psychrophilic and mesophilic incubating temperatures (15°C–30°C) were higher (0.0044 mM ± 0.000097 and 0.0037 mM ± 0.00017) compared to the lower thermophilic and moderate thermophilic incubating temperatures (0.0029 mM ± 0.000019 and 0.0025 mM ± 0.000062, respectively) (Figure 4.4b). The sulphide depletion data (mM/hrs.) showed that at the psychrophilic incubating temperature (15°C), the sulphide depletion rates (mM/hrs.) were similar (NS: 0.0044 mM ± 0.00013; AG: 0.0044 mM ± 0.000097). However, with the increase in the thermal gradient at (30°C, 45°C, and 60°C); sulphide depletion for the (NS) water production system increased (**NS 30°C**: 0.0045 mM ± 0.000073; **NS 45°C**: 0.0056 mM ± 0.000052; **NS 60°C**: 0.0054 mM ± 0.0002) but decreased in the (AG)

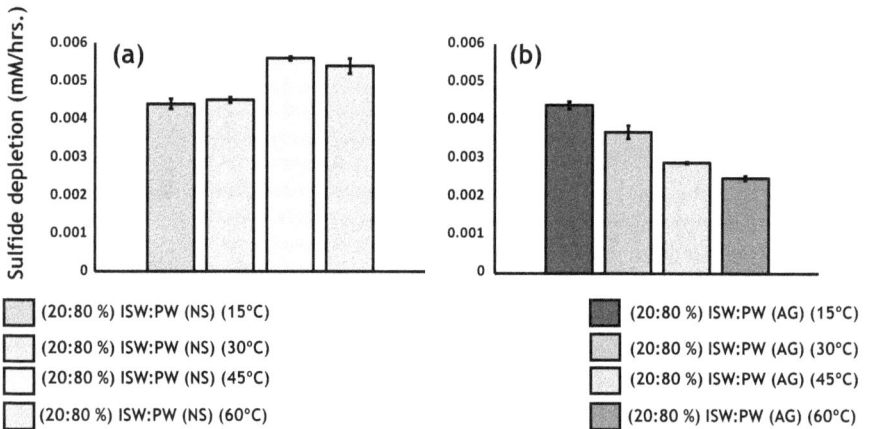

FIGURE 4.4   Sulphide depletion reaction rate (mM/hrs.) in microcosms containing the same ratios of 20% ISW – 80% PW for: (a) The North Sea water production system thermal gradient (127 g/L TDS) (15°C–60°C); (b) The Arabian Gulf (AG) water production system thermal gradient (204 g/L TDS) (15°C–60°C). Lighter colours depict reaction rates for (NS) water production system thermal gradient incubations. Darker colours depict reaction rates for (AG) water production system thermal gradient incubations. Error bars represent 1 × S.E.

water production system (**AG 30°C**: 0.0037 mM ± 0.00017; **AG 45°C: 0.0029 mM** ± 0.000019; **AG 60°C: 0.0025 mM** ± 0.000062) (Figures 4.4a, b). One-way ANOVA on Minitab was used as a statistical tool to determine the significance of the sulphide depletion rates (mM/hrs.) reported for the (NS-PW) system and the (AG-PW) system. This revealed that the differences in sulphide depletion between the (NS-PW) and the (AG-PW) system were non-significant ($P = 0.17$). However, when comparing the sulphide depletion profiles from the (NS-PW) and the (AG-PW) system with the sulphide production profile reported in 100% ISW (30°C) (no coupon incubation), the statistical tool determined the sulphide production profile significant ($P = 0.003$).

4.3.2.2   Time to Reach a Sulphide Depletion of 2mM

Time to reach a sulphide depletion of 2mM for the salinity and temperature ISW:PW microcosms were determined (Figure 4.5). Time to reach a sulphide depletion of 2mM decreased for the NS water production system thermal gradient (127 g/L TDS) (Figure 4.5a) but increased for the Arabian gulf (AG) water production system (204 g/L TDS) (Figure 4.5b). For the NS water production system thermal gradient (127 g/L TDS), the following timeframes (hrs.) to reach a sulphide depletion of 2mM were observed (**15°C**: 381 hrs.–419 hrs.; **30°C**: 427 hrs.–471 hrs.; **45°C**:313 hrs.– 346 hrs.; and **60°C**: 313 hrs.–346 hrs.) (Figure 4.5a), while for the AG water production system thermal gradient (204 g/L TDS) the following timeframes (hrs.) to reach a sulphide depletion of 2mM were observed (**15°C**: 313 hrs.–346 hrs.; **30°C**: 381 hrs.–419 hrs.; **45°C**: 518 hrs.–562 hrs.; **60°C**: 518 hrs.–562 hrs.) (Figure 4.5b).

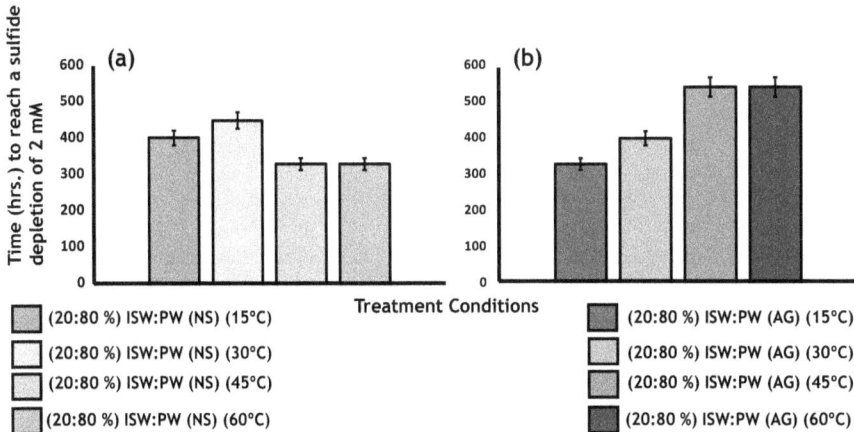

FIGURE 4.5   Time (hrs.) to reach a sulphide depletion of 2 mM across the ISW:PW ratios for: (a) The North Sea water production system thermal gradient (127 g/L TDS) (15°C–60°C); (b) The Arabian Gulf (AG) water production system thermal gradient (204 g/L TDS) (15°C–60°C). Lighter colours depict reaction rates for (NS) thermal gradient incubations. Darker colours depict reaction rates for (AG) thermal gradient incubations. Error bars represent 1 × S.E.

4.3.2.3   Maximum Sulphide Depletion

Maximum sulphide depletion (mM) for the salinity and temperature ISW:PW micro-cosm incubations were determined (Figure 4.6). Maximum sulphide depletion (mM) increased in the NS water production system thermal gradient incubations (Figure 4.6a) but decreased in the AG water production system thermal gradient incubations (Figure 4.6b). With increasing temperature (psychrophilic (15°C)- mesophilic (30°C)) in the (NS-PW) system, an initial slight decrease in maximum sulphide depletion from 2.64 mM ± 0.003 to 2.48 mM ± 0.08 was observed (Figure 4.6a). However, with further increases in temperature (lower thermophilic 45°C- moderate thermophilic 60°C), sulphide depletion increased (2.79 mM ± 0.007; 2.94 mM ± 0.006, respectively) (Figure 4.6a).

For the AG water production system, sulphide depletion decreased with the increase in incubating temperature (Figure 4.6b). The highest sulphide depletion was reported in the psychrophilic (15°C) (AG-PW) incubation of 2.65 mM ± 0.009 (Figure 4.6b). Lower sulphide depletion was reported in the mesophilic (30°C) (AG-PW) incubation of 2.49 mM ± 0.008 (Figure 4.6b). The lowest sulphide depletions were reported in the lower thermophilic (45°C) and moderate thermophilic (60°C) incubations (1.90 mM ± 0.005; 1.90 mM ± 0.002 respectively) (Figure 4.6b). Cross comparisons between the (NS-PW) and the (AG-PW) systems revealed that in the psychrophilic (15°C) and mesophilic (30°C) incubations, similar sulphide depletion concentrations were reported (15°C (NS-PW): 2.64 mM ± 0.003; 15°C (AG-PW): 2.65 mM ± 0.009; 30°C (NS-PW): 2.48 mM ± 0.08; 30°C (AG-PW): 2.49 mM ± 0.008) (Figures 4.6a,b). However, at the higher incubating temperatures of (lower thermophilic (45°C) and the moderate thermophilic (60°C), sulphide depletion was

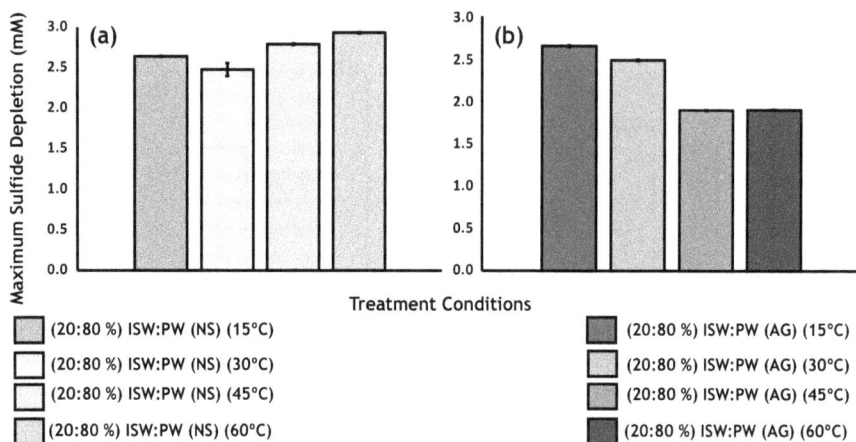

(20:80 %) ISW:PW (NS) (15°C)
(20:80 %) ISW:PW (NS) (30°C)
(20:80 %) ISW:PW (NS) (45°C)
(20:80 %) ISW:PW (NS) (60°C)

(20:80 %) ISW:PW (AG) (15°C)
(20:80 %) ISW:PW (AG) (30°C)
(20:80 %) ISW:PW (AG) (45°C)
(20:80 %) ISW:PW (AG) (60°C)

FIGURE 4.6 Maximum sulphide depletion (mM/hrs.) in microcosms containing the same ratios of 20% ISW – 80% PW for: (a) The North Sea water production system thermal gradient (127 g/L TDS) (15°C–60°C); (b) The Arabian Gulf (AG) water production system thermal gradient (204 g/L TDS) (15°C–60°C). Lighter colours depict reaction rates for (NS) thermal gradient incubations. Darker colours depict reaction rates for (AG) thermal gradient incubations. Error bars represent 1 × S.E.

higher in the (NS-PW) system incubations (45°C: 2.79 mM ± 0.007; 60°C: 2.94 mM ± 0.006) but lower in the (AG-PW) system incubations (45°C: 1.90 mM ± 0.005; 60°C: 1.90 mM ± 0.002) (Figures 4.6a,b). One-way ANOVA on Minitab was used as a statistical tool to determine the significance of the maximum sulphide depletion profiles reported for the (NS-PW) system and the (AG-PW) system. This revealed that the differences in sulphide depletion between the (NS-PW) and the (AG-PW) system were non-significant ($P = 0.21$).

4.3.3 METHANOGENESIS IN HIGH SALINITY PWRI PRACTICE

To determine whether other electron-accepting process than sulphate reduction was occurring under the high salinity 20:80 ISW:PW practice, methanogenesis was deemed a plausible explanation, and therefore microcosms were analysed for headspace methane production. Figure 4.7 illustrates the headspace methane production amounts in the headspace of high salinity 20:80 ISW:PW microcosms. Headspace methane in microcosms incubated at the temperature gradient of (15°C–60°C) for 20:80% ISW:PW (NS) (127 g/L TDS) and 20:80% ISW:PW (AG) (204 g/L TDS) were illustrated (Figure 4.7). Based on the stoichiometric conversions of VFAs to methane, mmoles amounts of methane should be produced from the mmoles amounts of VFAs oxidised (Equations 4.7–4.8). For example, 1mmole acetate oxidation will be coupled to 1 mmole methane (CH_4) reduction by the activity of acetoclastic methanogens (Equation 4.8), and 1mmole reduction of CO_2 by 1mmole H_2; as a result, for the activity of hydrogenotrophic methanogens (Equation 4.7). However, the actual masses of methane detected in the headspace of microcosms, however, were 6 orders of magnitude lower than the envisioned VFA to methane stoichiometry (6.50 nmoles CH_4 ± 0.07; range 6.11–8.02 nmoles CH_4; Figure 4.7), ruling out the possibility for methanogenesis being a potential electron-accepting process under such high salinity ISW:PW systems.

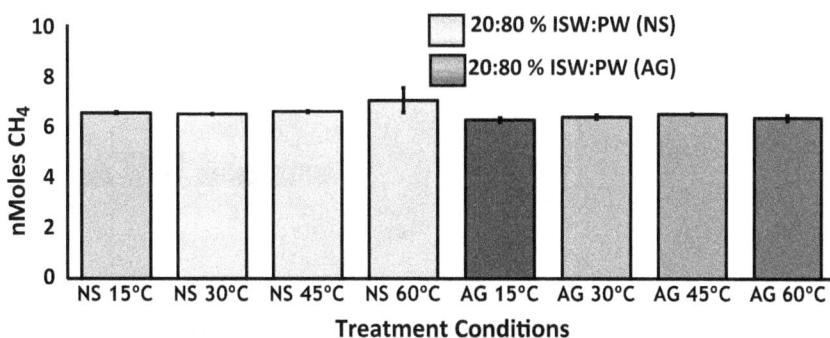

FIGURE 4.7 Mass of methane (nMoles) detected in headspace of anoxic microcosms containing 20:80% ISW: PW (NS) (127 g/L TDS), and 20:80% ISW: PW (AG) (204 g/L TDS). Lighter colour gradient represents headspace methane in 20:80% ISW:PW (NS) (127 g/L TDS) microcosmos. Darker colour gradient represents headspace methane in 20:80% ISW:PW (AG) (204 g/L TDS) microcosms. Error bars represent 1 × SE.

4.3.4 Gravimetric Analyses, Surface Morphology, and Surface Elemental Composition of Corrosion Coupons

Corrosion rates in millimetres per year (mm/yr.) from the high salinity 20:80 ISW:PW practice microcosms were calculated and displayed (Figure 4.8a,b). Overall corrosion rates from both the ISW:PW (NS) and ISW:PW (AG) were low <0.02 mm/ yr. (Figure 4.8a,b). However, corrosion rates were slightly increased at 30°C (0.011 ± 0.0002 mm/yr.; Figure 4.8a) and 60°C (0.018 ± 0.0004; Figure 4.8a) in the NS ISW:PW system, compared to other treatments of NS and AG-PW corrosion coupon samples. (NS: 0.002 ± 0.0003 mm/yr. 15°C; NS: 0.003 ± 0.0006 mm/yr. 45°C) (Figure 4.8a) (AG: 0.002 ± 0.0008 mm/yr. 15°C; 0.002 ± 0.0004 mm/yr. 30°C; 0.003 ± 0.0005 mm/yr. 45°C, and 0.002 ± 0.0002 mm/yr. 60°C (Figure 4.8b).

Selected corrosion coupons (Figure 4.8a stars) were further inspected for surface morphology and elemental composition, using SEM-EDS, to corroborate results of the gravimetric analyses, in comparison to controls. Figures 4.9 and 4.10 illustrate the surface morphology of the corrosion coupons. Corrosion coupon surface morphology for the selected 20:80% (ISW: PW) (NS) (127 g/L TDS) coupon samples highlighted in Figure 4.8 are displayed in Figures 4.9 and 4.10. Surface morphology of coupon surfaces of 20:80% (ISW: PW) (NS) incubated at 30°C and 60°C revealed the development of advanced-stage pitting nucleations (0.17–0.22 pits of 20 μm, and 1400 pits of 50 μm–150 μm, for pits number and diameter per cm^2, respectively for 30°C and 60°C incubations; Figure 4.9). By comparison, the control coupon (chemically treated with HCl, and NaOH, removing the corrosion inhibitor film – no incubation) exhibited a flat and uniform surface morphology and showed no evidence of pitting nucleations (Figure 4.9). Morphological variations in the pits were observed under 20:80% (ISW: PW) at 30°C coupon (wide-shallow pitting nucleation (Figure 4.9); sub-surface pitting nucleation (Figure 4.10). Nano-characterisation of

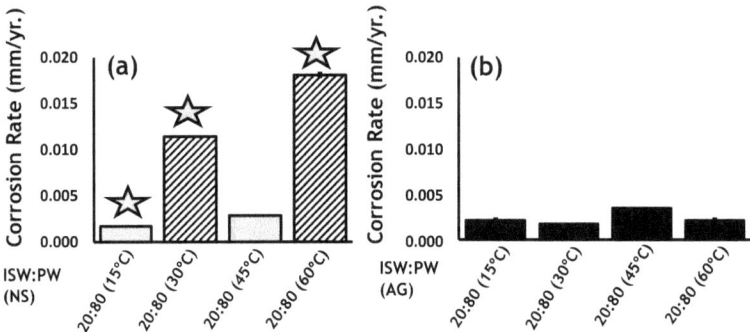

FIGURE 4.8 Corrosion Rates in (mm/yr.) for: (a) The North Sea water production system (127 g/L TDS) temperature gradient incubations (15°C–60°C) (Gray filled bars); and (b) The Arabian Gulf water production system (204 g/L TDS) temperature gradient incubations (15°C–60°C) (Black filled bars). Diagonally striped bars represent ISW:PW incubations with highest corrosion rates (mm/yr.) Selected samples (star) were subjected to SEM/EDS analyses. Error bars represent 1 × S.E.

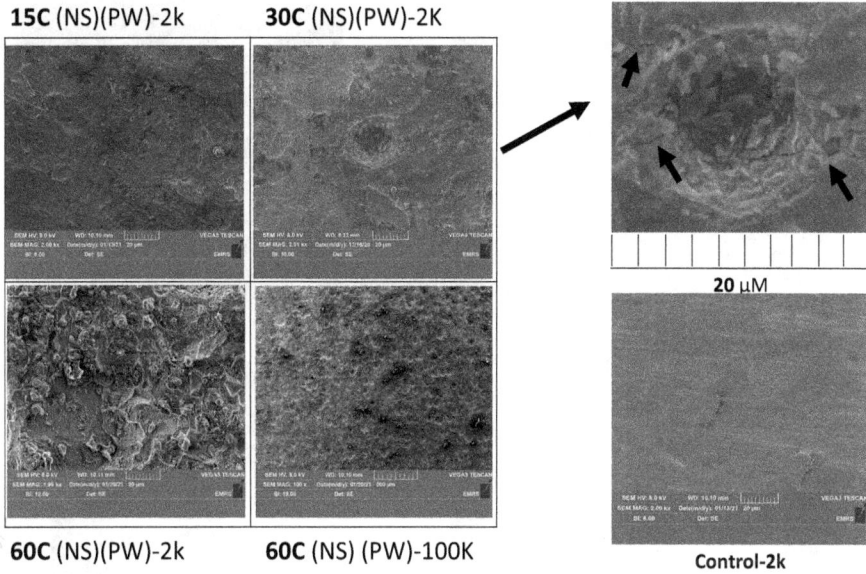

FIGURE 4.9 Scanning Electron Microscopy (SEM) images from steel coupons incubated in anoxic microcosms of 20:80% (ISW: PW) (NS) (127 g/L TDS) at 15°C, and 30°C (2K magnification), 60°C (2K and 100K magnifications), and control coupon (2K and 100K magnifications). Red arrows indicate detected cracks within the wide-shallow pitting nucleation and on coupon surface.

the elemental composition for the pitted surfaces was performed on corrosion coupons of (20:80%) ISW: PW (NS) (30°C) and (20:80%) ISW: PW (NS) (60°C). Table 4.2 illustrate EDS elemental composition of corrosion coupon surface.

Energy Dispersive Spectroscopy of corrosion coupon surface is displayed in Table 4.2. Seven different elements were detected on the coupon surface: carbon, oxygen, sodium, phosphorus, potassium, calcium, and iron. Three elements dominated the atomic composition (%) of pitted surfaces at both 30°C and 60°C incubations – oxygen (30°C Min. 42.41%; Max. 49.61%, 60°C Min. 43.37%; Max. 59.15%), carbon (30°C Min. 9.83%; Max. 41.27% 60°C Min. 11.47%, Max. 39.01%), and finally iron (30°C Min. 7.70%; Max. 32.09%, 60°C Min. 8.48%, Max. 19.28%) (Table 4.2). Iron oxides were envisioned to be one of the main deposits present under both the 30°C and 60°C incubations (20:80%) ISW:PW (NS) coupons. This was evident in the concurrence of the abundant elemental compositions (%) of both iron and oxygen elements, suggesting the corrosion susceptibility of these environments. Furthermore, under both the 30°C and 60°C incubations (20:80%) ISW:PW (NS) coupons, strong evidence for the presence of both phosphate and carbonate deposits was observed. This was reported by looking at the ionic (%) contributions of carbon, oxygen, and phosphorus elements to these respective deposits (i.e., phosphates and carbonates) (1 carbon: 3 oxygen) for carbonates; (1 phosphorus: 4 oxygen) for phosphates (Table 4.2).

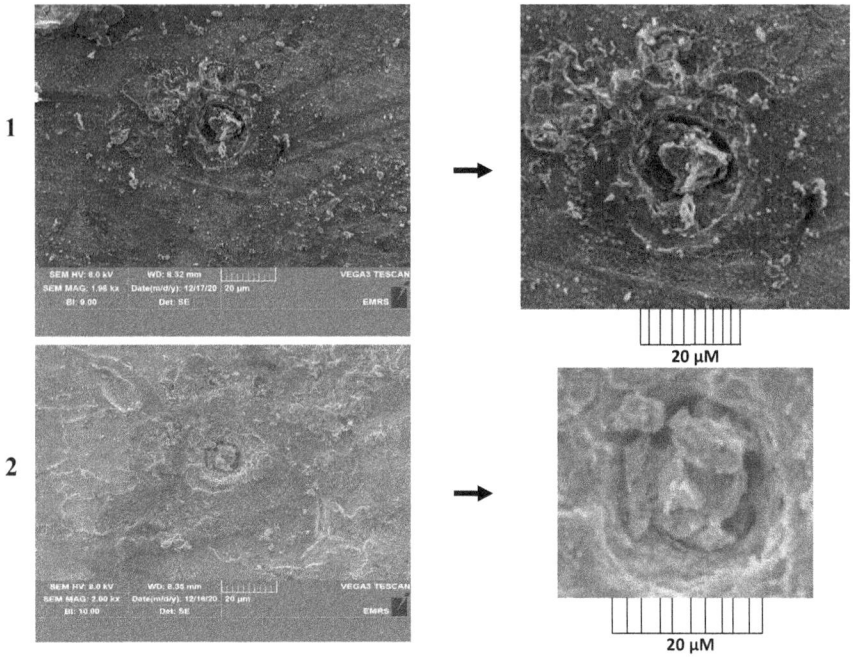

FIGURE 4.10   Other pitting nucleations on steel coupons incubated in anoxic microcosms of 20:80% (ISW: PW) (NS) at 30°C (2K magnification). Red arrows show the zoomed-in view and the morphological characterisation of nucleation (sub-surface nucleation).

TABLE 4.2
Energy Dispersive Spectroscopy (EDS) Elemental Compositions of Corrosion Coupon Surfaces from Anoxic Microcosms 20:80% (ISW: PW) (NS) (30°C) and 20:80% (ISW: PW) (NS) (60°C)

Element	Atomic Composition (%)					
	(20:80%) ISW:PW (NS)-30°C			(20:80%) ISW:PW (NS)-60°C		
	Coupon 1	Coupon 2	Coupon 3	Coupon 1	Coupon 2	Coupon 3
Carbon	13.77	41.27	9.83	11.47	17.74	39.01
Oxygen	45.50	42.41	49.61	59.15	54.02	43.37
Sodium	4.74	1.36	3.11	6.66	3.24	1.48
Phosphorus	11.79	4.91	4.62	7.92	4.27	4.77
Potassium	1.58	N.D.	0.75	0.85	0.44	0.74
Calcium	1.24	2.34	N.D.	0.39	1.01	2.15
Iron	21.35	7.70	32.09	13.56	19.28	8.48
SUM	100	100	100	100	100	100

N.D. Non-detected.

4.3.5 MICROBIAL COMMUNITY ANALYSIS

The microbial communities selected under the different conditions of thermal gradient (15°C–60°C), and the high salinity ISW:PW mix of 20:80 ISW:PW (NS) (127 g/L TDS), and the 20:80 ISW:PW (AG) (204 g/L TDS) were analysed by *16S rRNA* gene-based analysis.

Principal Component Analysis of species level *16S rRNA* gene profiles of microbial communities from anoxic microcosms containing the eight different treatment conditions (with corrosion coupons) in Table 4.1 and representing the (NS) and the (AG) PW system (15°C–60°C) was conducted. The analysis was conducted on the basis of Bray-Curtis similarities of the species level *16S rRNA* gene profiles.

4.3.5.1 Overview of Microbial Community Dynamics for Thermal Gradient (15°C–60°C) Incubations at Different Salinities

Principal Component Analysis of *16S rRNA* gene profiles for microbial community dynamics for the thermal gradient (15°C–60°C), for the 20:80% (ISW:PW) (NS) (127 g/L TDS), and the 20:80% (ISW:PW) (AG) (204 g/L TDS) incubations are illustrated in Figure 4.11. PCA revealed distinct clusters according to temperature (Figure 4.11). Whereby communities from the 20:80% ISW:PW (NS) and the 20:80% ISW:PW (AG) microcosms incubated at the lower temperatures of 15°C and 30°C clustered together, and communities incubated at 20:80% ISW:PW (NS), and 20:80% ISW:PW (AG), higher temperatures of 45°C and 60°C formed a separate cluster (high-temperature cluster), together with River Tyne sediment microbial

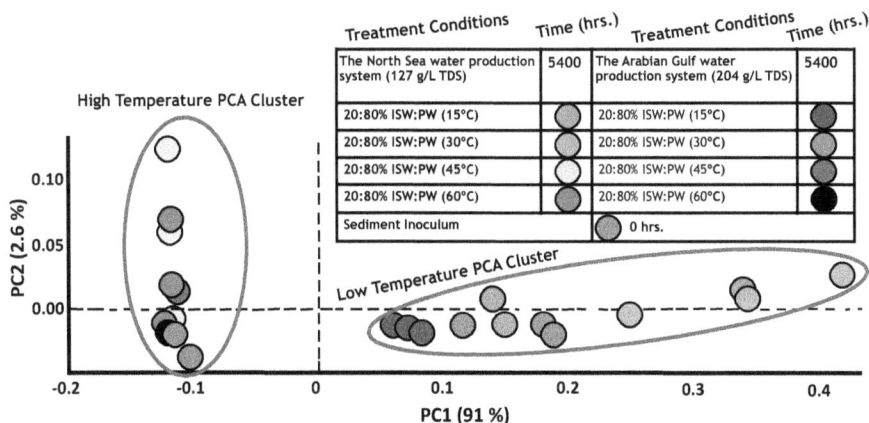

FIGURE 4.11   Principal Component Analysis (PCA) of *16S rRNA* gene profiles from microbial communities in anoxic microcosms at the end of incubations (5400 hrs.–225 days) in ISW: PW incubations of 20:80 ISW:PW from the North Sea water production system (127 g/L TDS), and the Arabian Gulf water production system (204 g/L TDS), incubated at a thermal gradient of (15°C–60°C). End of incubations (5400 hrs.–225 days) (ISW:PW) PCA clusters were compared against the time (0 hrs.–0 days) of sediment inoculum preserved at –20°C prior to utilisation in *16S rRNA* gene profiling analyses. Principal components PC1 and PC2 explained 91.0% and 2.6% of the variance, respectively.

inoculum (Figure 4.11). The first principal component explained 91% of the variation in the dataset (PC1), with 2.6% variation explained on the second principal component dimension (PC2) (Figure 4.11).

Next, heatmaps were generated, to display the relative abundance (%), based on the top 10 most abundant microbial phyla/orders identified across 20:80% ISW:PW (NS), and 20:80% ISW:PW (AG) thermal gradient incubations (15–60°C). Relative abundances (%) for ISW:PW systems were based on end of incubations (5400 hrs.–225 days) samples, compared against the sediment inoculum (0 hrs.–0 days) samples. Figures 4.12 and 4.13 illustrated the generated 20:80% ISW:PW (NS) and 20:80% ISW:PW (AG) heatmaps, respectively.

Low-temperature incubations (15°C and 30°C) appeared to drive the selective enrichment of bacteria from the order *Halanaerobiales* in both the (NS) and (AG) PW systems (Figures 4.12 and 4.13, respectively) – ~35% relative abundance for NS 15°C replicates 1–3, and ~ 50% relative abundance for NS 30°C replicates 1–3 (Figure 4.12). For the Arabian gulf samples however, lower *Halanaerobiales* relative abundance was reported in AG 15°C replicates 1–3 (~ 20% relative abundance), and in AG 30°C replicates 1–3 (~30% relative abundance) (Figure 4.13). By comparison, microcosms at the higher temperatures (45°C, and 60°C) were dominated by the

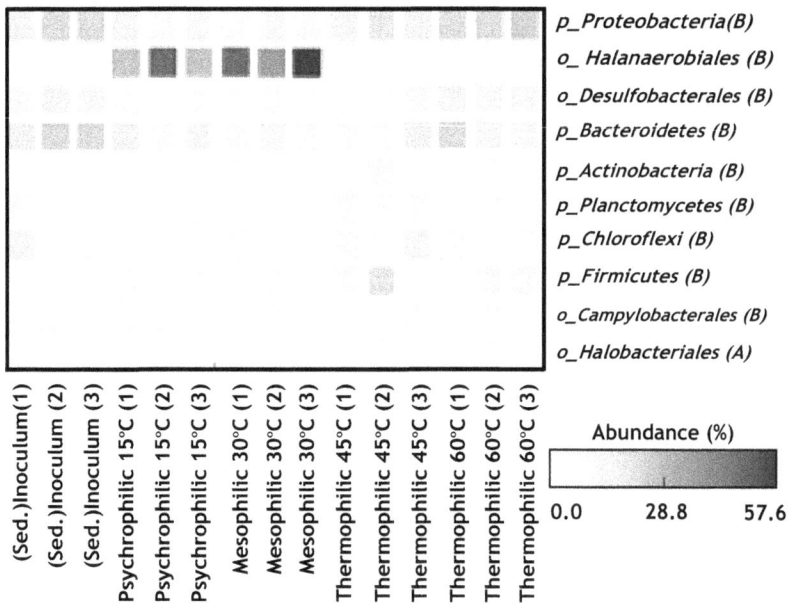

FIGURE 4.12 Heatmap representing the Top 10 bacterial phyla/orders (vertical-axis) detected in (20:80%) ISW:PW (NS) (127 g/L TDS) thermal gradient incubation microcosms (15°C–60°C). Relative abundances are indicated by colour intensity with darker blue shades depicting higher relative abundance and lighter shades depicting lower relative abundance. All end of incubation (5400 hrs.–225 days) ISW:PW microcosms were compared against the Sediment inoculum (0 hrs.) samples. Data from three replicate samples per thermal gradient are shown and the samples were re-ordered based upon increasing order of the thermal gradient (15°C–60°C).

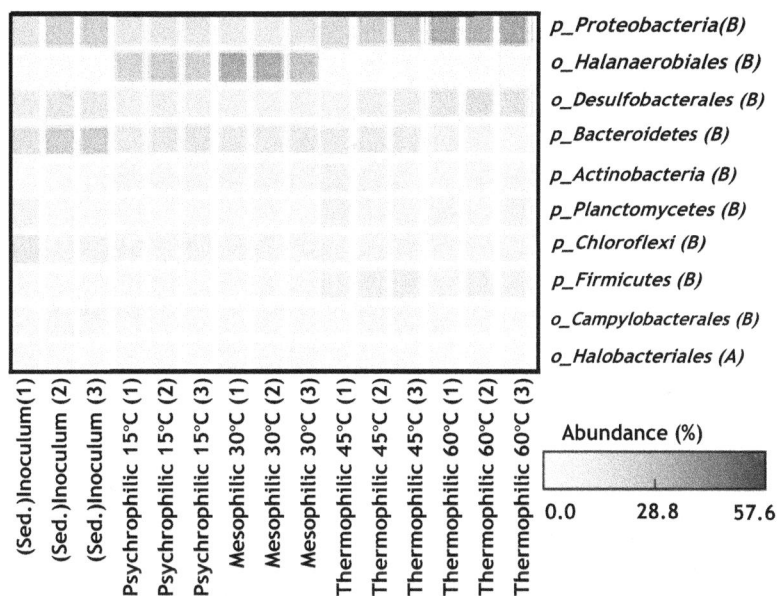

FIGURE 4.13 Heatmap representing the top 10 bacterial phyla/orders (vertical-axis) detected in microcosms containing (20:80%) ISW:PW (AG) (204 g/L TDS) thermal gradient incubation microcosms (15°C–60°C). Relative abundances are indicated by colour intensity with darker blue shades depicting higher relative abundance and lighter shades depicting lower relative abundance. All end-of-incubation (5400 hrs.–225 days) ISW:PW microcosms were compared against the sediment inoculum (0 hrs.) samples. Data from three replicate samples per thermal gradient are shown, and the samples were re-ordered based upon increasing order of the thermal gradient (15°C–60°C).

microbial phyla/orders of *Proteobacteria, Desulfobacterales and Bacteroidetes*, which accounted for ≥10% relative abundance in total in both (20:80%) ISW:PW (NS) and (20:80%) ISW:PW (AG) incubations, and similar to the phylum/order microbial enrichments detected in River Tyne sediment inoculum control (Figures 4.12 and 4.13, respectively).

The species diversity in all thermal gradient (15°C–60°C) incubated microcosms were compared using the Shannon indices based upon the (5400 hrs.) datapoint for ISW:PW incubations, and the (0 hrs.) datapoint for the sediment inoculum (Figures 4.12 and 4.13 respectively). Shannon indices for microbial communities selected at the thermal gradient (15°C–60°C) for (20:80%) ISW:PW (NS) incubations were: psychrophilic incubation (15°C): 7.22 ± 0.55; mesophilic incubation (30°C): 6.77 ± 0.43; lower thermophilic incubation (45°C): 8.52 ± 0.52; and moderate thermophilic incubation (60°C): 8.71 ± 0.69 (Figure 4.12). In comparison to the NS water production system (Figure 4.12), the AG water production system (Figure 4.13) exhibited a similar Shannon indices profile (i.e., lower Shannon indices values for lower temperature incubations (15°C–30°C); and higher Shannon indices values for the higher temperature incubations (45°C–60°C), and were: psychrophilic incubation (15°C): 7.50 ± 0.50; moderate incubation (30°C): 6.71 ± 0.64; lower thermophilic incubation

(45°C): 8.98 ± 0.65; and moderate thermophilic incubation (60°C): 8.96 ± 0.63 (Figure 4.13). For sediment inoculum controls, the Shannon indices were very similar (8.85 ± 0.13), compared to the lower and moderate thermophilic incubations (45°C incubation average: 8.75 ± 0.59; 60°C incubation average: 8.84 ± 0.66) ($P \geq$ 0.05), but quite different (8.85 ± 0.13), compared to the low-temperature incubations (15°C incubation average: 7.36 ± 0.53; 30°C incubation average: 6.74 ± 0.54 selective enrichments ($P \leq 0.05$).

Therefore, based on the Shannon indices, higher microbial diversity, and much lower relative abundance ($\geq 10\%$) for both the high-temperature microbial inoculum (45°C, and 60°C incubations) and River Tyne sediment inoculum control were observed (Figures 4.12 and 4.13 respectively). Additionally, lower microbial diversity and much higher relative abundance (20%–50%) for low-temperature microbial inoculum (15°C–30°C) compared to the sediment inoculum were observed (Figures 4.12 and 4.13, respectively).

4.3.5.2   Detailed Analysis of Microbial Community Dynamics for River Tyne Sediment Inoculum

An illustration of the microbial diversity for River Tyne Sediment inoculum is displayed in (Figure 4.14). The relative abundance (%) of the top 12 species/genera detected ranged from 1 to just over 7% (Figure 4.14). Ten of these taxa were present at relative abundances of (1%–2%) and with a greater relative abundance of two *Woeseia* sp. that had a relative abundance of 7.1 ± 1.16% and 6.6 ± 1.12%, respectively (Figure 4.14).

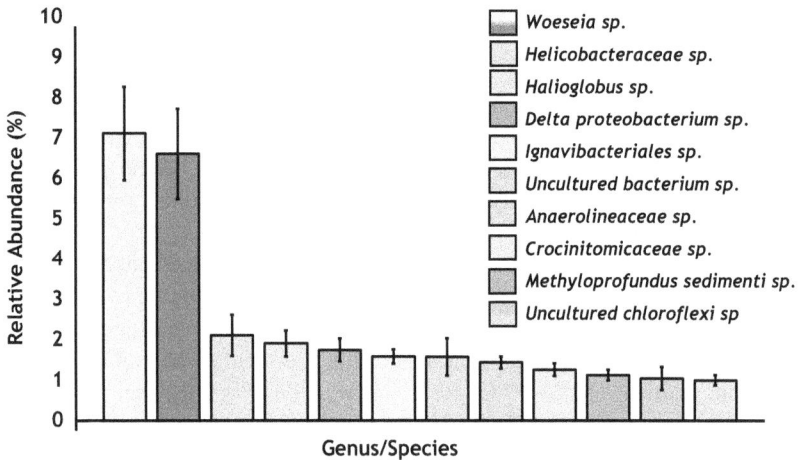

FIGURE 4.14   *16S rRNA* gene copy number adjusted relative abundance (%) at species/genus level for River Tyne Sediments Microbial Inoculum. An illustration for the diversity of microbial genera of *Woeseia* sp. denoted with yellow/blue coloured bars, and the other microbial genera denoted with different coloured bars. Sediment inoculum microbial diversity was based on (0 hrs.) samples preserved at −20°C prior to *16S rRNA* gene profiling analyses. Error bars represent 1 × S.E.

4.3.5.3 Detailed Analysis of Microbial Community Dynamics for the North Sea Water Production System Thermal Gradient (15°C–60°C)

A comparative illustration of the microbial diversity of the enriched *Halanaerobium* sp. across the NS water production, and the AG water production system (Figure 4.15), and other microbial species enrichments for the NS water production system's thermal gradient (15°C–60°C) (Figure 4.16) are displayed. A significant enrichment in relative abundance of the genus *Halanaerobium* was observed under the NS water production system: 20:80 15°C: two different *Halanaerobium* species were enriched at 22.58 ± 1.23% and 11.96 ± 1.74%, respectively, 20:80 30°C: two different *Halanaerobium* species were enriched at 31.02 ± 2.34%, and 13.6 ± 1.87%, respectively (Figure 4.15a), and the AG water production system: 20:80 15°C: only one *Halanaerobium* sp. enriched at 20.66 ± 0.93%, and for 20:80 30°C: three different *Halanaerobium* species were enriched at 24.3± 1.62%, 3.07 ± 0.46%, and 1.13 ± 0.01% respectively (Figure 4.15b). However, *Halanaerobium* sp. sequences were not detected in the lower thermophilic (45°C) and the moderate thermophilic (60°C) incubations from both the NS water production and the AG water production systems (Figure 4.15a,b). No SRP enrichments were observed under the NS water production system (NS) (127 g/L TDS) or the AG water production system (204 g/L TDS) incubations (Figure 4.15a,b, respectively).

In contrast, from the microcosm experiments conducted by Sindi, *et al.* (2021), and under 100% ISW (42 g/L TDS) salinity, the microbial enrichment results

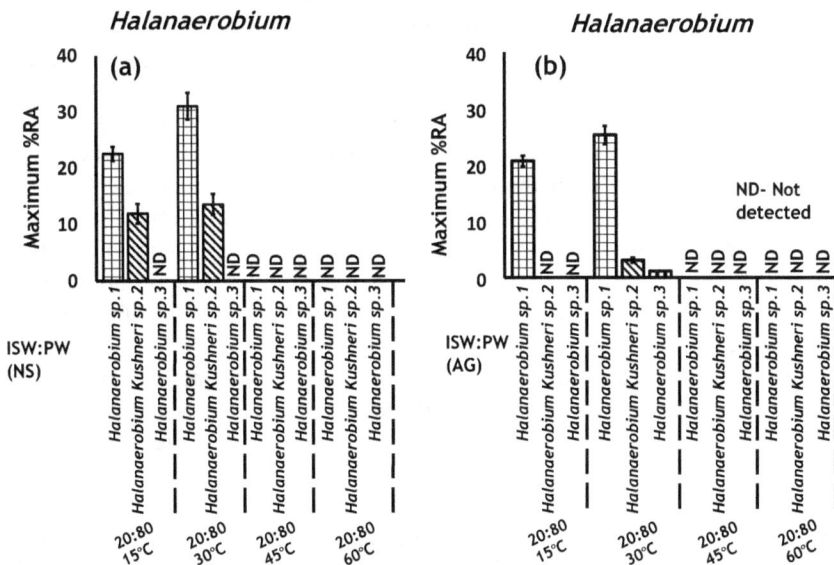

FIGURE 4.15 *16S rRNA* gene copy number adjusted relative abundance (%) at species/genus level for ISW:PW incubations from the North Sea (NS) and the Arabian Gulf (AG) water production systems. A comparative illustration of the diversity of the enriched *Halanaerobium* sp. for: (a) The North Sea water production system (NS) and (b) The Arabian Gulf water production system. Error bars represent 1 × SE.

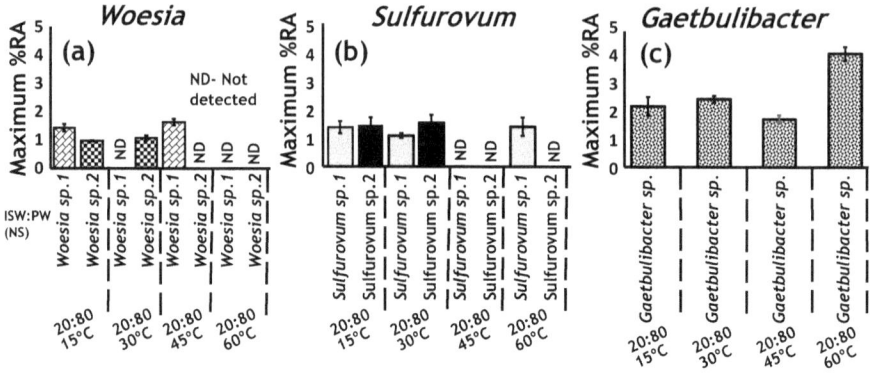

FIGURE 4.16 *16S rRNA* gene copy number adjusted relative abundance (%) at species/genus level for ISW:PW incubations from the North Sea (NS) water production system. An illustration of the diversity of the other enriched or detected species of: (a) *Woesia* sp., (b) *Sulfurovum* sp., and (c) *Gaetbulibacter* sp. Error bars represent 1 × SE.

revealed a significant (Maximum) enrichment of the SRP microbial consortium of *Desulfobulbus* sp., *Desulfobacter* sp., and *Desulfotignum* sp. between (31–75 days) post-incubation (*Desulfobulbus* sp.: 18.62 ± 2.30% at 55 days post-incubation; *Desulfotignum* sp.: 4.33 ± 0.49% at 55 days post-incubation; and *Desulfobacter* sp.: 8.40 ± 1.40% at 55 days post-incubation (Sindi, *et al.*, 2021). Therefore, this experiment revealed that elevated salinities under (30°C) incubations inhibited sulphate reduction, therefore, inhibiting the enrichment of sulphate-reducing microorganisms.

High-temperature incubations (45°C–60°C) coupled with high salinity (i.e., 127 g/L TDS ISW:PW (NS) resulted in no microbial community changes, whereby the microbial communities under the high-temperature incubations (45°C–60°C), and under the NS water production system, were detected at relative abundance (%) of between 1–2% (except for *Gaetbulibacter* sp.) and looked like those of the River Tyne microbial inoculum controls (Figure 4.16a–c). Additionally, no sulphate-reducing microorganisms were enriched under 20:80 45°C and 20:80 60°C incubations (Figure 4.16a–c).

In both the low- and high-temperature microbial enrichments (15°C–60°C), *Woeseia* sp., *Sulfurovum* sp., *and Gaetbulibacter* sp. were consistently detected as the most abundant taxa (Figure 4.16) (*Woesia* sp.: two different species were detected, and maximum %RA of 1.63 ± 0.11 detected at 225 days post-incubation; *Sulfurovum* sp., two different species were detected and maximum %RA 1.52 ± 0.28 detected at 225 days post-incubation; and one *Gaetbulibacter* sp. detected with a maximum %RA of 4.20 ± 0.25 detected at 225 days post-incubation (Figure 4.16a–c)). The prevalence of *Woesia* sp. as more abundant in the sediment control (7.1 ± 1.16% and 6.6 ± 1.12% (Figure 4.14)), compared to the high-temperature incubations point towards the high-temperature systems (i.e., 45°C and 60°C incubations) not showing much microbial activity and therefore the high temperature (45°C–60°C) treatments not eliciting a major population shift (Figure 4.16a–c).

4.3.5.4 Detailed Analysis of Microbial Community Dynamics for the Arabian Gulf Water Production System Thermal Gradient (15°C–60°C)

A comparative illustration of the microbial diversity of the enriched *Halanaerobium* sp. across the NS water production and the AG water production system is displayed in Figure 4.15a,b, and other microbial species enrichments for the AG water production system's thermal gradient (15°C–60°C) are displayed in Figure 4.17.

A comparative illustration for the diversity of the enriched *Halanaerobium* sp. across the NS water production system and the AG water production system is displayed (Figure 4.15a,b). As elicited, the diversity of the enriched *Halanaerobium* sp. was greater for the AG water production system 20:80 30°C incubations but with much lower relative abundance (%) for *Halanaerobium Kushneri sp.2* and *Halanaerobium sp.3* (3.07 ± 0.46%, and 1.13 ± 0.01%, respectively (Figure 4.15b) compared to the 20:80 30°C incubation for the NS water production system (13.6 ± 1.87% and sequences not detected for *Halanaerobium sp.3*, respectively) (Figure 4.15a).

Similar to the NS water production system, high temperature coupled with high salinity (i.e., 204 g/L TDS ISW:PW (AG) resulted in no microbial community changes, whereby the microbial communities under the high-temperature incubations (45°C–60°C) were detected at relative abundance (%) of between 1% and 2% (except for *Sulfurovum* sp. and *Gaetbulibacter* sp.) and looked like those of the River Tyne microbial inoculum controls (Figure 4.17a–c). Additionally, no sulphate-reducing microorganisms were enriched under 20:80 45°C and 20:80 60°C incubations (Figure 4.17a–c).

In both the low- and high-temperature microbial enrichments (15°C–60°C), *Woeseia* sp., *Sulfurovum* sp., *and Gaetbulibacter* sp. were consistently detected as the most abundant taxa (Figure 4.17) (*Woesia* sp.: two different species were detected, and maximum %RA of 2.06 ± 0.23 detected at 225 days post-incubation; *Sulfurovum*

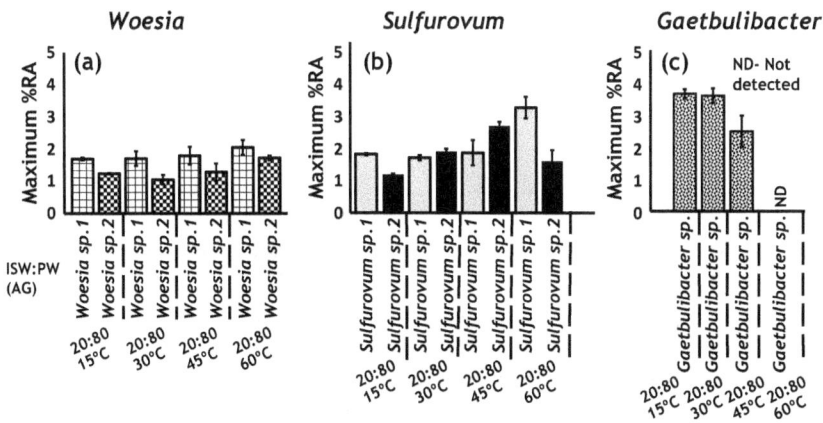

FIGURE 4.17 *16S rRNA* gene copy number adjusted relative abundance (%) at species/genus level for ISW:PW incubations from the Arabian Gulf (AG) water production system. An illustration of the diversity of the other enriched or detected species of: (a) *Woesia* sp., (b) *Sulfurovum* sp., and (c) *Gaetbulibacter* sp. Error bars represent 1 × SE.

sp., two different species were detected and maximum %RA 3.10 ± 0.31 detected at 225 days post-incubation; and one *Gaetbulibacter* sp. detected with a maximum %RA of 3.65 ± 0.15 detected at 225 days post-incubation (Figure 4.17a–c)). The prevalence of *Woesia* sp. as more abundant in the sediment control ($7.1 \pm 1.16\%$ and $6.6 \pm 1.12\%$ (Figure 4.14) compared to the high-temperature incubations point towards the high-temperature systems (i.e., 45°C and 60°C incubations) not showing much microbial activity and the treatment not eliciting a major population shift (Figure 4.16a–c).

4.4   DISCUSSION

4.4.1   TRENDS IN MICROBIAL COMMUNITIES COUPLED WITH ELECTRON ACCEPTOR TO VFA STOICHIOMETRY

The process of sulphate reduction follows the notion of sulphate reduction (electron acceptor), leading to H_2S production (Kaksonen and Puhakka, 2007; (Lens, *et al.*, 2002). Typically, under sulphate-reducing environments, the longer chain VFAs (i.e., butyrate and propionate) are oxidised to acetate coupled to sulphate reduction. 1 mole of butyrate is oxidised to 2 moles of acetate coupled to the reduction of 0.5 moles of sulphate into sulphide (Equation 4.1). Similarly, 1 mole of propionate is converted to 1 mole of acetate at the expense of 0.75 moles of sulphate reduced into sulphide (Equation 4.2). Acetate is oxidised completely to CO_2 with a 1:1 stoichiometry of acetate to sulphate – reduced to sulphide (Equation 4.3). Equations 4.1–4.3 illustrate the VFAs oxidation coupled to sulphate reduction equations utilised by the enriched anaerobic SRP communities:

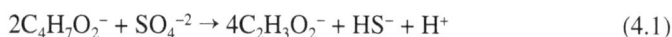

$$2C_4H_7O_2^- + SO_4^{-2} \rightarrow 4C_2H_3O_2^- + HS^- + H^+ \tag{4.1}$$

(Chen, *et al.*, 2017)

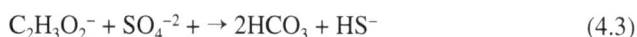

$$4C_3H_5O_2^- + 3SO_4^{-2} + 3H^+ \rightarrow 4C_2H_3O_2^- + 3HS^- + 4CO_2 + 4H_2O \tag{4.2}$$

(Chen, *et al.*, 2017)

$$C_2H_3O_2^- + SO_4^{-2} + \rightarrow 2HCO_3 + HS^- \tag{4.3}$$

(Widdel and Bak, 1992)

Stoichiometric conversions of VFAs coupled to sulphate reduction was displayed (Equations 4.1–4.3). Based on VFAs additions to the NS water production incubations, and the AG water production incubations, it was envisioned that the enriched microbial communities utilised VFAs as the electron donors. Sulphate reduction results revealed no significant sulphate reduction occurring across the NS water production system or the AG water production system from the baseline sulphate (mM) concentrations of (NS $\approx$ 5 mM; AG $\approx$ 10 mM) (Table 4.1) ((NS, average 4.6 ± 0.38 mM sulphate; AG, 9.8 ± 0.60 mM sulphate) measured between day 1 and day 22 post-incubation (Figure 4.3). Therefore, it was envisioned that potentially another electron-accepting process such as hydrogenotrophic or acetoclastic methanogenesis (i.e., utilising H^+ protons or acetate respectively) as potential electron-accepting processes could have been at play under the NS and the AG water production system. Methanogens were reported to utilise VFAs (acetate, butyrate, and propionate) as electron donors to produce headspace methane (Blake, *et al.*, 2020; Tian, *et al.*, 2019). Equations 4.4–4.8 illustrate the chemical reactions in the four steps of anaerobic digestion (AD) of hydrolysis, acidogenesis, acetogenesis, methanogenesis, and highlighting the hydrogenotrophic methanogens activity (utilising H^+ protons as electron acceptors) and the acetoclastic methanogens activity (utilising acetate as electron acceptors):

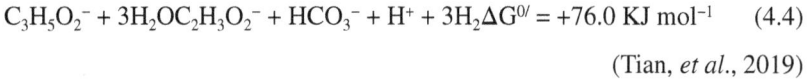

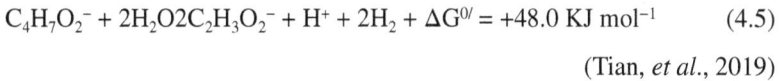

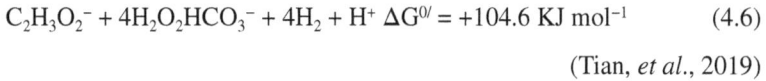

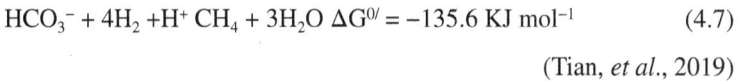

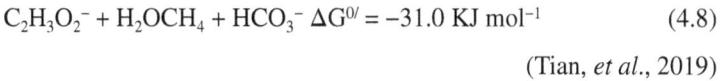

$$C_3H_5O_2^- + 3H_2OC_2H_3O_2^- + HCO_3^- + H^+ + 3H_2 \Delta G^{0/} = +76.0 \text{ KJ mol}^{-1} \quad (4.4)$$

(Tian, *et al.*, 2019)

$$C_4H_7O_2^- + 2H_2O2C_2H_3O_2^- + H^+ + 2H_2 + \Delta G^{0/} = +48.0 \text{ KJ mol}^{-1} \quad (4.5)$$

(Tian, *et al.*, 2019)

$$C_2H_3O_2^- + 4H_2O_2HCO_3^- + 4H_2 + H^+ \Delta G^{0/} = +104.6 \text{ KJ mol}^{-1} \quad (4.6)$$

(Tian, *et al.*, 2019)

$$HCO_3^- + 4H_2 + H^+ CH_4 + 3H_2O \Delta G^{0/} = -135.6 \text{ KJ mol}^{-1} \quad (4.7)$$

(Tian, *et al.*, 2019)

$$C_2H_3O_2^- + H_2OCH_4 + HCO_3^- \Delta G^{0/} = -31.0 \text{ KJ mol}^{-1} \quad (4.8)$$

(Tian, *et al.*, 2019)

Anaerobic digestion (AD) of VFAs is usually mediated by four steps: hydrolysis, acidogenesis, acetogenesis, and methanogenesis (Angelidaki, *et al.*, 2011). The complex VFA polymers (e.g., propionate and butyrate) are initially hydrolysed into the VFA monomer (acetate) during hydrolysis (Equations 4.4–4.5) (Tian, *et al.*, 2019), and consequently converted into acetate during acetogenesis (Equations 4.4–4.6). The released electrons from VFA polymers (propionate and butyrate) are then channelled into H^+ generating hydrogen protons (Equations 4.4–4.5). Although the hydrolysis of propionate and butyrate into acetate generating H^+ is thermodynamically unfavourable, this process would not proceed unless coupled with hydrogenotrophic methanogenesis to consume the generated H^+ protons (Tian, *et al.*, 2019). Finally, for methane to be produced, one of two pathways must be followed: A) oxidation of acetate coupled to methane production (Equation 4.8) or B) reduction of CO_2 by H_2 (Equation 4.7).

From this ISW:PW experimental incubations methanogenesis occurrence as a potential terminal electron accepting process was evaluated under the NS water production system (127 g/L TDS) and the AG water production system (204 g/L TDS) incubations (Table 4.1). Headspace methane measured, however, were in nmoles amounts compared to the envisioned (1:1) ratio for acetate oxidation coupled to methane production by acetoclastic methanogens (Equation 4.7), or the (1:1) ratio of propionate and butyrate oxidation into H^+ protons by hydrolysis for the hydrogenotrophic methanogens then to feed on the generated H^+ (Equations 4.4–4.5). Therefore, for hydrogenotrophic methanogenesis to have been at play consuming the VFAs mixture added to these (NS) and (AG) ISW:PW incubations, 10 mmoles H^+ (from 10 mmoles propionate hydrolysis) (Equation 4.4) and 5 mmoles H^+ (from 5 mmoles butyrate hydrolysis) (Equation 4.5) must have been produced in the headspace of ISW:PW microcosms. However, the previous statement was not supported with the NGS datasets whereby at the end incubations (225 days–5400 hrs.) acetolactic methanogens or hydrogenotrophic methanogens *16S rRNA* gene sequences were not detected (Figures 4.12 and 4.13), and headspace methane in ISW:PW incubations representing the (NS) production water system (127 g/L TDS) and the AG water production system (204 g/L TDS) were seven to nine orders of magnitude lower (in Figure 4.7) than the stoichiometric conversions of VFAs-methane (Equations 4.4–4.8). Compared to the follow-up experiment run with ISW (NS) (42 g/L TDS) incubations, the acetoclastic methanogenic archaeal family of *Methanosaeta 16S rRNA* gene sequences were detected at the end of incubations (225 days–5400 hrs.) (Sindi, *et al.*, 2021).

The microbial community analysis revealed a strong selection of *Halanaerobium* sp. (Figure 4.15) implicated in Guar Gum degradation during drilling and hydraulic fracturing workover procedures (Booker, *et al.*, 2017), and with the reported *Halanaerobium* sp. to utilise thiosulfate/elemental sulphur (i.e., electron acceptors) coupled to VFA metabolism (i.e., mainly acetate) (Booker, *et al.*, 2017). It was hypothesised that the dominant *Halanaerobium* sp. in the fractured shale formation decomposed of Guar gum and produced acetate and sulphide as by-products from the polysaccharide (Guar gum) degradation (Liang, *et al.*, 2016; Lipus, *et al.*, 2017).

However, since thiosulfate ($S_2O_3^-$) or elemental sulphur (S^0) were not added and therefore not measured in the ISW:PW incubations, it would not be possible to quantify the use of these as electron acceptors by the enriched *Halanaerobium* sp., which leads the way to the only other possibility of fermentative metabolism of carbohydrates being the most plausible explanation for the observed enrichments of *Halanaerobium* sp. in this ISW:PW experiment, and consistent with earlier literature (Abdeljabbar, *et al.*, 2013; Booker, *et al.*, 2017, 2019; Kögler, *et al.*, 2021; Lipus, *et al.*, 2017).

Systematic, sulphide depletion profiles observed in both the NS water production system (NS) (127 g/L TDS) and the AG water production system (204 g/L TDS) (Figure 4.1) illustrated that the added oxygen scavenger ($Na_2S.9H_2O$) (5 mM) was rapidly removed from the ISW:PW systems (NS average sulphide depletion: 2.63 ± 0.23 (Figure 4.6a); AG average sulphide depletion: 2.15 ± 0.32) (Figure 4.6b), possibly suggesting the transformation of the depleted sulphide into iron sulphide (FeS) biotically or abiotically. Equations 4.9–4.10 illustrate the SRP cathodic depolarisation procedure by SRP to produce iron sulphide (FeS) black films on coupon surface, and the consequent mineralisation of the produced (FeS) into pyrite (FeS_2):

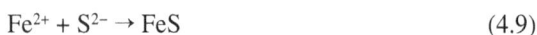

$$Fe^{2+} + S^{2-} \rightarrow FeS \tag{4.9}$$

(El-Hajj, *et al.*, 2013)

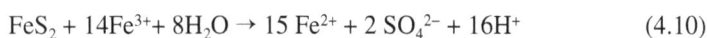

$$FeS_2 + 14Fe^{3+} + 8H_2O \rightarrow 15\ Fe^{2+} + 2\ SO_4^{2-} + 16H^+ \tag{4.10}$$

(Bottrell and Raiswell, 2000)

Consistent with the visual inspections of the added corrosion coupons, black FeS-films were observed to have developed, possibly suggesting the transformation of Fe^{2+} and S^{2-} into FeS (Equations 4.9–4.10). From microbially enhanced oil recovery (MEOR) in reservoir rock sands experiments, a consistent observation to the previous notion of *Halanaerobium* sp. fermentative metabolism pathways, whereby 1% pyrite and calcite enhanced the enriched *Halanaerobium* sp. fermentative metabolism of sucrose yielding sixfold higher acetate production rates compared to quartz sand (Kögler, *et al.*, 2021).

4.4.2   SALINITY EFFECT ON SULPHATE REDUCTION, SULFIDOGENESIS, AND METHANOGENESIS FROM (ISW:PW) – MICROBIAL DYNAMICS PERSPECTIVE

Salinity increases in ISW:PW mixes from the NS water production system (127 g/L TDS) and the AG water production system (204 g/L TDS) across the thermal gradient (15°C–60°C) had a major impact on increasing sulphide depletion rate (mM/hrs.) (Figure 4.4), increasing maximum sulphide depletion (Figure 4.6), and under than the stoichiometric amounts of headspace CH_4 production from VFA metabolism (Figure 4.7). The positive control for sulfidogenesis (100% ISW) (42 g/L TDS), however, exhibited a rapid sulphide production profile (18.05 mM ± 0.11) (Figure 4.1). This was consistent with the follow-up experiment that demonstrated that the salinity threshold for sulfidogenesis induced by SRP under the synthetically prepared water production systems mesophilic incubations (30°C) was 107 g/L TDS and was 64 g/L TDS for moderate thermophilic incubations (60°C) (Sindi, *et al.*, 2021). Additionally, earlier literature also reported that the mesophilic (30°C) acetoclastic methanogens did not grow when NaCl concentrations exceeded 50 g/L (Blake, *et al.*, 2020). The thermodynamics limit for microbial life in the deep biosphere (including sulphate-reducers) under high salinity conditions was proposed by Oren (1999, 2001, 2011). In these literature studies, it was proposed that microbial life at high salt concentrations was energetically expensive. Therefore, for microbial life to thrive, the following factors must be considered: A) The energy amounts generated from the dissimilatory metabolism and B) The mode of osmotic adaptation. These are consistent with some of the findings reported in this experiment; whereby: for instance, under the (100% ISW-30°C) incubations; sulphide production occurred most likely coupled with VFAs metabolism energy channelled into sulphate reduction into sulphide (Figure 4.1) (Equations 4.1–4.3). Additionally, the literature reported mode of osmotic adaptations for the enriched *Halanaerobium* sp., and with which production of osmoprotectant compounds such as (Betanin-glycine) would allow them to pump out the intracellular cations and anions (typically K+ and Cl-), to survive the osmotic pressure (Oren, 2011).

Halanaerobium sp. were enriched in both the NS water production system (127 g/L TDS) and the AG water production system (Max: 31.02 ± 2.34%; Max: 25.3± 1.62%, respectively) (Figure 4.15a,b), with higher diversity in the AG water production system (204 g/L TDS) (Figure 4.15). A literature survey of the higher salinity adaptive capabilities for the enriched *Halanaerobium* sp. is displayed in Table 4.3.

Higher salinity adaptive capability for the selected and literature-reported novel isolated *Halanaerobium* sp. was displayed (Table 4.3). This higher salinity adaptive

TABLE 4.3
Literature Survey on Selected Novel Isolated *Halanaerobium* sp.

Species/Genus	Isolation Habitat	Salinity Range for Growth	Optimum Growth Temperature	VFAs Produced from Carbohydrates Fermentation	References
Halanaerobium sehlinense sp.	surface Sabkha Sediments (Tunisia)	*20% 5–30% 50–300 g/L (NaCl)	**43°C 20–50°C	Lactate, acetate, formate	Abdeljabbar, *et al.* (2013)
Halanaerobium congolense sp.	Oil well-head sample (Republic of Congo)	10% (100g/L) (NaCl)	**42°C	Acetate	Ravot, *et al.* (1997)
Halanaerobium Kushneri sp.	Petroleum reservoir fluid, Oklahoma (U.S.A)	*15–20% (150–200 g/L) (NaCl)	37°C	Acetate and formate	Bhupathiraju, *et al.* (1999)

*   Optimal Growth Salinity
** Optimal Growth Temperature

capabilities for the novel isolated *Halanaerobium* sp. reported in earlier literatures were in range with the reported *Halanaerobium* sp. enrichments from this experiment's (ISW:PW) synthetic mixes mimicking the NS water production system (127 g/L TDS–12.7% TDS), and the AG water production system (204 g/L TDS–20.4% TDS) (Figure 4.15a,b). Additionally, the mesophilic incubation (30°C) parameter was also in range with the literature-reported growth temperature range (20–50°C) for the novel isolated *Halanaerobium* sp. (Table 4.3). At the higher thermophilic incubating temperatures (45°C–60°C) of the (NS-PW 127 g/L TDS) and the (AG-PW 204 g/L TDS) ISW:PW mixes, *Halanaerobium* sp. *16S rRNA* sequences were not detected, with the rest of the detected sequences making up only between (1%–4% RA), and therefore not showing much microbial activity or eliciting a major population shift (Figure 4.16a–c); (Figure 4.17a–c).

4.4.3 Impact of Salinity on *Halanaerobium* sp. Enrichments

Halanaerobium sp. were isolated from field production waters in MEOR evaluation experiments (Kögler, *et al.*, 2021). The isolated *Halanaerobium* sp. grew on nutrient media supplemented with sucrose, dried yeast extract, and molybdate (also SRB growth inhibitor) as carbon source and did not show evidence of sulfidogenesis nor extensive MIC (Kögler, *et al.*, 2021). Indeed, *Halanaerobium* sp. have previously been also detected in a range of medium to high salinity-produced waters (Booker, *et al.*, 2017; Vilcaez, *et al.*, 2018).

In another study, Vilcaez *et al.* (2018) conducted an anoxic microcosm experiment to study methanogenic crude oil biodegrading microorganisms. In their study, the anoxic microcosms were prepared from the Stillwater and Crushing oilfields of Oklahoma, USA, supplemented with a nutrient solution containing a mixture of basal salts, without sulphate. Additionally, a protein-rich matter (i.e., amino acid source) was used as the carbon source for microbial metabolism. Vilcaez *et al.* (2018) reported the detection of *Halanaerobium* sp., in high salinity still water formation (116.7 g/L) and higher salinity formation water (285.76 g/L). Microbial community analysis for the low salinity still water formation revealed 46% relative abundance of *Marinobacter*, and 21.5% relative abundance of *Halanaerobium* and *Acetohalobium* in water sample prior to incubation. However, at the end of the incubation period, *Marinobacter, Halanaerobium*, and *Acetohalobium* could no longer be detected. Instead, *Deferribacter*, (49.9%), *Geotoga, Kosmotoga*, and *Petrotoga* (14.9%) were enriched. These constituted only about 9.3% relative abundance of the microbial communities at the start of incubations. On the other hand, the microbial communities in the high salinity formation water at the start comprised 33.4% of *Actinobacteria* (genus *Propionibacterium*), with 26.9% comprising *Acinetobacter, Marinobacter, and Halomonas*; 14.2% of the genera *Staphylococcus, Streptococcus, Bacillus*, and *Halanaerobium*; and 8.9% *Achromobacter*, and *Pelomonas*. At the end of incubations, however, only, three lineages were detected, which constituted 98.7% of the microbial community. Those lineages belonged to *Deferribacter* (23.4%), *Kosmotoga* (13.5%), and *Candidatus Schekmanbacteria* (61.7%). A point of similarity between the study conducted by Vilcaez, *et al.*, (2018) and our current study was the selective enrichment of *Halanaerobium* sp. under high salinity production water conditions. No sulphate was supplemented to growth media as an electron acceptor (Vilcaez, *et al.*, 2018), and like our experiment, the supplemented sulphate was not reduced under both (20:80%) ISW:PW (NS) and (20:80%) ISW:PW (AG) incubations. The metabolism for *Halanaerobium* sp., which occurred in the study of (Vilcaez, *et al.*, 2018), was possibly due to the consumption of the protein-rich matter and/or the organics in crude oil (without addition of sulphate as a terminal electron acceptor). In contrast, sugar fermentation metabolism pathways for *Halanaerobium* sp. selective enrichment in our PWRI most likely occurred, with VFAs production, and with no significant sulphate reduction (i.e., terminal electron acceptor). However, the contribution of River Tyne organic matter as a carbon source for the enrichment of *Halanaerobium* sp. would require further investigations.

Halanaerobium sp. has been identified in a wide range of oil fields from the Gulf of Mexico (Scheffer, *et al.*, 2021), the Permian Basin-U.S.A (Tinker, *et al.*, 2022) to an offshore oil field in the Republic of the Congo in Africa "*Halanaerobium congolense*" (Ravot, *et al.*, 1997) (Table 4.3). Tinker, *et al.* (2022) have recently demonstrated that *Halanaerobium* sp. was an integral component of the microbiome of the Permian Basin, the highest oil- and gas-producing reservoir in the United States. The *16s rRNA* gene sequencing and metagenomic analysis revealed that microbiome of the Permian basin was dominated by sulphate and thiosulfate-reducing taxa including: *Halanaerobium* sp., *Orenia* sp., *Marinobacter* sp., and *Desulfohalobium* sp., and that there was a high prevalence of sulphate and thiosulfate-reducing genes in metagenome-assembled genomes (MAGs) assembled from the metagenome sequences.

Permian Basin produced water samples had Total Dissolved Solids (TDS) concentrations in the range of (110 g/L–107 g/L), and, therefore, slightly under this experiment's (20:80%) (ISW:PW) (NS) (127 g/L TDS) salinity. From these ISW:PW incubations, the relative abundance of *Halanaerobium* sp. enrichments in the (NS) water production system was: 22.58 ± 1.23% (Max. at 15°C); and 31.02 ± 2.34% (Max at 30°C), and similar to the relative abundance of *Halanaerobium* sp. reported in the Permian basin produced water samples of (RA 33.58%) (Tinker, *et al.*, 2022).

4.4.4 IMPACT OF SALINITY AND TEMPERATURE ON *HALANAEROBIUM* SP. ENRICHMENTS

Booker *et al.* (2017) monitored the enrichment of *Halanaerobium* sp. from the input fluids prior to down-hole injection for hydraulic fracturing operations at the Utica shale well in Ohio, United States. The monitoring revealed strong enrichment of *Halanaerobium* sp. (99% relative abundance in *16S rRNA* gene libraries), at 100 days post-injection. A *Halanaerobium* sp. isolated from the Utica shale well was grown on a defined saltwater liquid medium containing a range of basal salts, making up ~113 g/L salinity. The medium was amended with 10 mM thiosulfate ($Na_2S_2O_3.5H_2O$), as the terminal electron acceptor for *Halanaerobium* sp., growth, and with D-glucose as the carbon source for *Halanaerobium* sp. growth. Further evidence from the literature suggested that *Halanaerobium* sp. is a dominant and colonising microbial taxon across a range of wells and shale plays (Choudhary, *et al.*, 2015; Daly, *et al.*, 2016; Liang, *et al.*, 2016; Lipus, *et al.*, 2017).

Halanaerobium sp. was recently found to be an integral part of the produced water microbiome of the Gulf of Mexico (Scheffer, *et al.*, 2021). Ravot *et al.* (1997) were the first to report the isolation of a *Halanaerobium* sp., from an offshore oil field in the Republic of the Congo in Africa, hence the name "*Halanaerobium congolense.*" *Halanaerobium congolense* grew optimally at 42°C and a pH of 7 (Table 4.3). No growth was observed at temperatures below 20°C, and above 45°C, like the experiment where the selectively enriched *Halanaerobium* sp. grew at 15°C and 30°C but not at 45°C or 60°C (Figure 4.15a,b). The pH range for the growth reported by Ravot *et al.* (1997) was between 6.3 and 8.5, and pH in our experiment was in that range at the pH value of 7.5, and remained constant, throughout the experiment. This strain was able to ferment a range of carbohydrates, including fructose, galactose, D-glucose, maltose, D-mannose, D-ribose, sucrose, trehalose, and bio-tryptase. However, strain SERB 4224, isolated by Ravot *et al.* (1997), did not utilise D-arabinose, lactose, rhamnose, D-xylose, dulcitol, acetate, butyrate, propionate, and lactate. Under a pure culture enrichment experimental setting, the ability of a *Halanaerobium congolense* sp. to produce biofilms and sporulate was reported (Jones, *et al.*, 2021).

4.4.5 ESTUARINE RIVER TYNE SEDIMENTS MICROBIAL PROCESSES

Sulphate reduction and methanogenesis were the electron-accepting processes evaluated under this experiment (Figures 4.3 and 4.7, respectively). This was achieved via aqueous sulphate and the microcosms headspace methane production. Results revealed no sulphate reduction and therefore no sulfidogenesis under both the NS and the AG water production systems (127 g/L) and (204 g/L), respectively, and across the incubation temperature gradients (15°C–60°C). Minimal headspace methane

(CH$_4$) production amounts were observed (nmoles opposed to mmoles) ruling out the contribution of methanogenesis as a potential electron-accepting process.

Methanogenesis is an important microbial process occurring in natural environments, such as River Tyne estuarine sediments (Blake, *et al.*, 2020). Under a range of similar salinity and temperature gradients, and using River Tyne Estuarine sediments for microbial inoculum, the enrichments of methanogens were reported by Blake, *et al.* (2020). The enriched methanogens could grow at both low (5°C–30°C) and high (40°C–70°C) incubating temperatures, and in both (CO$_2$/H$_2$) headspace amended and unamended microcosms. Compared to these ISW:PW experiments, under the NS (PW) and the AG (PW) incubations (127 g/L TDS) and (204 g/L TDS), respectively, minimal sulfidogenesis was occurring under the NS (PW) (2 mM – 40 days), and no sulfidogenesis occurring at the AG (PW) (0 mM – 40 days), no methane was detected in the headspace (Figure 4.7), and no methanogenic enrichments were detected under the tested (NS) and the (AG) ISW:PW mixes incubations (Figures 4.12 and 4.13, respectively).

Recently, the definition of "Palaeopickling" was proposed by Head, *et al. (*2014). This definition is pertinent to reservoir sterilisation because of extreme reservoir temperature and salinity gradients. At such extreme salinity and temperature gradients, the maximum temperature threshold for microbial growth and flourishment under extreme reservoir conditions can be lower at elevated salinities (Head *et al.*, 2014). To survive the extreme reservoir salinity gradients, reservoir microorganisms would deploy one of two mechanisms (Head, *et al.*, 2014). Firstly, the intracellular accumulation of ions (typically K$^+$ and Cl$^-$) countering the extracellular high salinity reservoir conditions. Secondly, the intercellular synthesis of compatible solutes was proposed to counter the salinity effect (Oren, 1999). In both cases, a considerable expenditure of energy will be required, which will be deducted from the energy channelled for primary metabolism (Oren, 2011).

Estuarine sediments are also known to have substantial amounts of elemental sulphur (Viggi, *et al.*, 2017), and River Tyne sediments are no exception. In a more general sense, aquatic marine sediments were reported to have both thiosulfate and elemental sulphur thiosulfate (S$_2$O$_3$$^{2-}$) (0–100 µM) range (Jørgensen, 1990); Sulphur (S^0) (200–1600 µM) (Mitchell, *et al.*, 1988). Consequently, a continuous shuttle of oxidised or reduced sulphur-species forms, including thiosulfate, may occur under such sedimentary systems (Viggi, *et al.*, 2017). Therefore, the S-species reported concentrations in aquatic marine sediments (~2 mM) matched the reported sulphide production (mM) concentrations for (20:80%) ISW:PW (NS) (~127 g/L) reported in this experiment (~2 mM) (Figure 4.2). Additionally, the literature reported S-species in aquatic marine sediments could have provided the required electron acceptor for the low-temperature-driven *Halanaerobium* sp. enrichments in this experiment. Two potential sources of thiosulfate presence in microcosms as a terminal electron acceptor in this study are from: firstly, medium-sediment particulates interactions, which may represent the real-field thiosulfate releasing mechanism from hydraulic shale, as a result of the interactions of hydraulic fracturing fluids with reservoir rocks (Danika, *et al.*, 2021), secondly, the detection of *Sulfurovum* sp. sequences under all low- and high-temperature microcosms (Figure 4.16b); Figure 4.17b is implicated in sulphur oxidation, as reported by Mori *et al.* (2018). This was manifested by the reported

Novel *Sulfurovum* sp. to oxidise sulphide via using thiosulfate and elemental sulphur as electron donors, and with the ability to use oxygen and nitrate as electron acceptors (Mori, *et al.* 2018). An additional thiosulfate source in hydraulic fracturing operations would be the hydraulic fracturing chemical additives added to the mud mixtures in prior to downhole field injections. Sediment-rock interactions represent a long-term reservoir of thiosulfate supporting the microbial metabolism when the thiosulfate from hydraulic fracturing chemical additives is depleted. However, the processes may represent a cryptic cycle and rapid turnover events of sulphur compounds, which consequently would limit the understanding of the true extent of thiosulfate transformation over the monitoring period. The Aarhus Bay marine sediments were a good source of microbial inoculum, harbouring a great microbial diversity and, therefore, used in SRP reaction rates study (Holmkvist, *et al.*, 2011). In that study, a cryptic S-species cycle in both the sulphide and methane zones of marine sediments was proposed (Holmkvist, *et al.*, 2011). This cryptic S-species cycle was manifested in the sedimentary deposition and reaction of Fe (III) minerals with sulphide. Together and in combination with sulphur disproportionation, a quantifiable conversion of sulphide to pyrite was observed (Holmkvist, *et al.*, 2011). Sulphate and thiosulfate are known intermediates of the S-species cryptic cycles and, therefore, marine sediments would provide a plausible source of alternative electron acceptors in marine sediment systems, which might be occurring in parallel with the fermentation of complex organic matter in marine sediments and therefore supporting the low-temperature ($15°C$–$30°C$) microcosms *Halanaerobium* sp. enrichments under in this study (Figure 4.15a,b, respectively).

4.4.6   Implications of *Halanaerobium* sp. in Oil and Gas Industry Processes and Practices

Numerous microbiology studies in hydraulic fracturing fluids have found that microorganisms can grow in the newly fractured hydraulic shales and persist over an extended period, leading to a series of deleterious consequences such as souring (production of H_2S), MIC, production of biogenic gasses, and pore plugging (Murali, *et al.*, 2013a, 2013b). In some cases, *Halanaerobium* sp. were found to be enriched to levels where they contributed 99% of the *16S rRNA* gene sequences in *16S rRNA* gene libraries from flow-back waters and detected as a key member in production fluids (Daly, *et al.*, 2016; Mouser, *et al.*, 2016; Murali, et al., 2013b). It has also been shown that supplementing hydraulic fracturing fluid samples with thiosulfate resulted in enrichment of *Halanaerobium* sp. (Booker, *et al.*, 2017).

Scale formation is another common problem in the oil and gas industry. Initially, it can coat perforations, casing, and tubing. If allowed to further develop, it could lead to limiting the production, and eventually the abandonment of the well. Referring to the enrichment of *Halanaerobium* sp. under high salinity produced water fluids, it has been proposed that this could be happening in the fractured shales in the Appalachian basin, due to the insufficient energy yields (carbon source) to synthesise osmoprotectants and reduce sulphate simultaneously for SRB (high salinity growth requirements), in comparison with the fermentative *Halanaerobium* sp. growth and adaptability under high salinity conditions (Oren, 2011).

To curtail the growth and enrichment of microbes with MIC-potential, biocide chemicals are usually administered in numerous oil and gas industry processes and practices. Current biocidal application practices during the PWRI process, for instance, often involve alternating biocidal treatments using different chemical formulations, and these often need to be reviewed every 2–3 years due to the possible biocidal inefficacies because of microbial resistance. Incorporating field-specific temperature and salinity gradients in biocide efficacy testing before field applications will aid in the better understanding of the possible emergence of biocidal microbial resistance creating multiple lines of benefits for the industry, such as cost avoidance (CAPEX and OPEX), reduced crude processing costs, reduced downtime, and safety benefits.

4.4.7 *Halanaerobium* sp. S-species Respiratory Pathways

Aquatic marine sediments have been reported to have hugely varying concentrations of thiosulfate ($S_2O_3^{-2}$), and elemental sulphur (S^0). Some literatures reported the detection of the high concentrations of: (0–100 μM) range for (Thiosulfate ($(S_2O_3^{2-})$) (Jørgensen, 1990); Sulphur (S^0) (200–1600 μM) (Mitchell, *et al.*, 1988), while other literature reported much lower quantities of: (≥ 0.5 μM) for both thiosulfate and elemental sulphur) (Zopfi, *et al.*, 2004). Potentially, the reported systematic sulphide depletion profiles from this ISW:PW incubations suggested that the depleted sulphide was oxidised; possibly via the manganese oxides (Mn (IV)) and iron oxides Fe (II) oxides), producing preliminary elemental sulphur (S^0) (Pyzik and Sommer, 1981; Yao and Millero, 1993, 1996). Sulphur (S^0) is produced as an intermediate or final product during bacterial oxidation of sulphide and thiosulfate (Schippers and Jørgensen, 2001), and microorganisms produce (S^0) as an intermediate or final product during oxic and anoxic iron sulphide (Fes) oxidation (Schippers and Jørgensen, 2001). The ability of *Halanaerobium* sp. to utilise S-species as a terminal electron acceptor stem from the presence of protein-coding *Rhodanese* enzymes genes in their genome, and their capabilities to upregulate the expression of these *Rhodanese* enzymes coding-genes, including one of the three subunits of the anaerobic sulphite reductase enzyme (*AsrA*). Below equations illustrate the reaction catalysed by *Rhodanese* enzymes (Equation 4.12) and the anaerobic sulphite reductase (Equation 4.13):

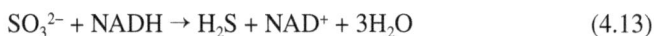

$$S_2O_3^{2-} \rightarrow SO_3^{2-} + S^0 \tag{4.12}$$

$$SO_3^{2-} + NADH \rightarrow H_2S + NAD^+ + 3H_2O \tag{4.13}$$

Halanaerobium sp. capabilities also included their ability to outcompete SRB on energy sources (Booker, *et al.*, 2017). This was due to the need but insufficiency of SRB to produce osmoprotectants from the energetic yields of sulphate reduction to maintain their growth. During *Halanaerobium* sp. growth phase, electron donors are oxidised, and the energy coupled to the desirable for growth electron acceptor (e.g., S^0, $S_2O_3^{2-}$, SO_3^{2-}) Furthermore, Booker, *et al.* (2017) suggested that during the stationary growth phase of the *Halanaerobium* sp., sulphite-reduction would not be coupled to growth but rather as a disposal mechanism for the fermentation reducing equivalents, such as ethanol, acetate, and formate. This was done following a proteomics analysis approach to determine the proteins produced from biomass during the stationary growth phase of *Halanaerobium* sp. The reported *Rhodanese* enzymatic machinery present in *Halanaerobium* sp. genome would explain the enrichment of this species under (20:80%) ISW:PW (NS) (127 g/L TDS) and (20:80%) ISW:PW (AG) (204 g/L TDS) systems, whereby one of three possibilities occurred: A) The literature reported thiosulfate, sulphur and potentially present in River Tyne Estuarine marine sediments were utilised as electron acceptors for *Halanaerobium* sp. Respiration, and B) The observed depleted and potentially turned-over sulphide could have worked as electron acceptor under this specific ISW:PW experimental setup (Figure 4.1), and C) The complex organic matter present in River Tyne Estuarine marine sediments at varying concentrations, including carbohydrates, were fermented into VFAs by the enriched *Halanaerobium* sp. (Figure 4.15a,b).

4.4.8 *HALANAEROBIUM* SP.-INDUCED MICROBIAL CONTAMINATION AND CONTROL MECHANISMS

Halanaerobium sp. can grow and adapt to the rapid shifts in physicochemical conditions of the hydraulic fractured (HF) fluids due to their network of metabolic functions, cantered around the cycling of osmoprotectants and methylamine compounds (Borton, *et al.*, 2018; Liang, *et al.*, 2016). The direct growth advantage of *Halanaerobium* sp. includes their ability to catalyse thiosulfate-dependent sulfidogenesis (Booker, *et al.*, 2017), and growth on HF fluid additives (Borton, *et al.*, 2018; Daly, *et al.*, 2016; Liang, *et al.*, 2016), to potentially form biofilms. Such biofilms could directly lead to bio clogging, and biofouling but could, on the other hand, have a positive impact on sealing the cap-rocks in geologic CO_2-sequestration reservoirs (Mitchell, *et al.*, 2008). Liang *et al.* (2016) demonstrated the ability of *Halanaerobium* sp. strain (DL-01) to consume (degrade) guar gum and produce acetate. A series of biocides efficacies was determined against the isolated and tested *Halanaerobium* sp. (DL-01) and showed that Quaternary Ammonium Compounds (QAC) biocide was the best biocide in controlling the growth of *Halanaerobium* sp. in the presence of thiosulfate, followed by glutaraldehyde, and finally Tetra Hydrokis Phosphonium Sulphate (THPS). These reported observations, and with the salinity and temperature gradients reported from this experiment demonstrate the ability of *Halanaerobium* sp. to utilise the more complex carbon sources, such as guar gum Liang *et al.* (2016), and potentially the more complex organic matter present in River Tyne sediment, either solely or as part of the mixed microbial community's synergist relationships. The latter, however, would require further investigations to confirm.

4.4.9 THE RELEVANCE OF OTHER MICROBIAL SP. ENRICHMENTS UNDER PWRI PHYSICOCHEMICAL PARAMETERS

From the current study, the consistent presence of *Woeseia* sp., under both low-temperature incubations (15°C and 30°C) (Figure 4.16a) and high-temperature incubations (45°C and 60°C) Figure 4.17a suggests that this species may be one of the integral constituents of Tyne sediments microbiome in specific and the global estuarine sediments microbiome in general. Indeed, the literature reported numerous incidences where *Woeseia* sp. sequences were detected in estuarine sediments using *16S rRNA* Next-Generation Sequencing (NGS) in Shuangtaizi River, China (Zhang, *et al.*, 2021), coastal sediments (Malva-rosa beach) (Valencia, Spain) (Vidal-Verdú *et al.*, 2022), and seafloor sediment communities (Hoffmann, *et al.*, 2020), with the utilisation of protein-based microbial cell-components, and other microbial organic remnants as a carbon source for their metabolism. Additionally, the enrichment of *Sulfurovum* sp. under all low- and high-temperature treatments could have potentially led to sulphur oxidation, coupled with the reduction of a suitable electron acceptor (e.g., Oxygen and Nitrate) as reported by (Mori, *et al.*, 2018). Therefore, contributing to a potential S-species cryptic cycle.

4.4.10   THE OBSERVED PITTING NUCLEATIONS

From these ISW:PW incubations, low general corrosion rates of <0.02 (mm/yr.) were reported across the tested thermal gradient of (15°C–60°C) (Figure 4.8a,b). However, the occurrence of advance-staged pitting nucleations was reported for both 20:80% (ISW:PW) (NS) (127 g/L TDS) (30°C) incubations and 20:80% (ISW:PW) (NS) (127 g/L TDS) (60°C) (Figures 4.9 and 4.10). Lower number of pitting nucleations with smaller diameter (per cm²) were reported under the NS water production system (30°C) incubations (0.17–0.22 pits of 20 μm,), compared against the higher numbers of pitting nucleations with larger diameter (per cm²) for under the NS water production system (60°C) incubations (and 1400 pits of 50 μm–150 μm) (Figure 4.9). Additionally, different pitting nucleation morphologies were detected for (30°C) (NS) incubations, possibly suggesting the co-occurrence of multiple pitting corrosion mechanisms (Figures 4.9 and 4.10, respectively) for the reported wide-shallow pitting nucleations and the sub-surface nucleations. Due to the nature of the conducted ISW:PW experiments, however, and whereby a mixture of physical-chemical parameters is tested at once, it would prove difficult and therefore require further investigations to pinpoint the root cause of the reported advance-staged pitting nucleations. In corrosion experiments conducted on stainless steel (more-corrosion-resistant alloy than carbon steel), three key factors were reported to be detrimental for the development of pitting corrosion (chloride, temperature, and pH) (Dastgerdi, *et al.*, 2019). When increasing the concentrations of chloride ions, increasing temperature, and pH; a general decrease in localised corrosion resistance of stainless steel usually occurs (Dastgerdi, *et al.*, 2019; Ramana, *et al.*, 2009). Therefore, in ISW:PW systems whereby a mixture of biotic (i.e., microbes implicated in MIC) and abiotic (i.e., Chloride, temperature, and pH) factors interact, the cumulative and individual contributions of such factors in the potential general corrosion, localised corrosion (e.g., pitting), and the associated corrosion depth (e.g., corrosion pits depth) would require further investigations.

4.5   CONCLUSIONS

A fundamental anoxic microcosm experiment was conducted to understand the effects of higher salinity synthetic ISW and synthetic PW mixing on sulphate reduction, sulphide production, methanogenesis, general corrosion potential, localised corrosion potential, and the associated microbial community dynamics. The fundamentally tested synthetic ISW and synthetic PW mixes represented oil and gas industry mixing scenarios for ISW and PW such as: A) On-shore oil production facility water flooding lines, B) On-shore oil production facility ISW:PW mixing tanks, C) Off-shore platforms ISW:PW mixing tanks, and D) Oil reservoir ISW:PW mixing zone. A summary of the main conclusions from these ISW:PW incubations, followed by the broader implications are shown below:

- Salinity had a major impact on microbial sulphate reduction, whereby under 20:80% ISW:PW (NS) (127 g/L TDS) thermal gradient (15°C–60°C) incubations, and 20:80% ISW:PW (AG) (204 g/L TDS) thermal gradient (15°C–60°C) incubations, sulphate reduction did not occur but rather remained stable throughout incubation timeframes (NS, average 4.6 ± 0.38 mM sulphate; AG, 9.8 ± 0.60 mM sulphate, Figure 4.3).
- A systematic sulphide depletion profile for (with coupon) 20:80% ISW:PW thermal gradient (15°C–60°C) incubations from the NS and from the AG was reported (NS average sulphide depletion: 2.63 ± 0.23 (Figure 4.6a); AG average sulphide depletion: 2.15 ± 0.32) (Figure 4.6b), possibly suggesting the transformation of the depleted sulphide into iron sulphide (FeS) biotically or abiotically.

Under 100% ISW (without coupon) (NS-30°C) (42 g/L TDS) incubations, however, a rapid sulphide production profile was reported (18.05 mM ± 0.11 (Figure 4.1).

- The SRP microbial communities under 100% ISW (without coupon) (NS-30°C) (42 g/L TDS) incubations, most likely coupled their sulphate-reduction respiration into sulphide with the added VFAs metabolism (Figure 4.1).
- The observed shifts in microbial communities from the potential SRP communities under (100% ISW (without coupon) (NS-30°C) (42 g/L TDS) incubations into members of *Halanaerobium* sp. suggest changes in metabolism pathways from (VFAs-sulphate) anaerobic SRP respiration, into thiosulfate reduction or carbohydrates fermentation pathways for *Halanaerobium* sp. (Figure 4.15a,b).
- The possibility for the occurrence of another electron-accepting process (e.g., hydrogenotrophic and acetoclastic methanogenesis) was investigated. Methane headspace (CH_4) mass results ruled out the presence of (mmoles) amounts to drive the (VFAs-methane) stoichiometric conversions (Figure 4.7).
- Gravimetric analysis of corrosion coupons revealed low corrosion rates of <0.02 (mm/yr.) for both the 20:80% ISW:PW (NS) (127 g/L TDS) thermal gradient (15°C–60°C) incubations and the 20:80% ISW:PW (AG) (204 g/L TDS) thermal gradient (15°C–60°C) incubations (Figure 4.8a,b).
- Surface morphology analysis revealed the development of advanced-stage pitting nucleations on corrosion coupon surface for 20:80% ISW:PW (NS) (30°C) (0.17–0.22 pits of 20 µm) and 20:80% ISW:PW (NS) (60°C)

(1400 pits of 50 μm–150 μm), for pits number and diameter per cm^2, respectively) (Figure 4.9).

- Elemental composition of corrosion coupons for 20:80% ISW:PW (NS) (30°C), and 20:80% ISW:PW (NS) (60°C) incubations revealed the potential presence of both phosphate and carbonate deposits, based on the ionic composition (%) of carbon, oxygen, and phosphorus elements (Table 4.2).

The oil and gas industry faces great difficulties in terms of pinpointing field-specific microbiological contamination or microbiological issues. This is caused by reservoir microorganisms' adaptive capabilities to live under harsh reservoir environments (e.g., temperature, salinity, and pressure). Tending to live in microbial communities and clusters (e.g., biofilms), synergistic microbial reaction effects can be evident. To overcome the dilemma of field-specific microbiological contamination, a few pointers were proposed in this chapter. Firstly, multiple layers of geochemical, microbiological, and metallurgical test results must be considered before pinpointing the root cause, and field-specific microbiological issues causative agent. Secondly, the oil and gas industry processes and practices such as PWRI and HF must be considered in these systematic lab experiments, together with the multiple layers of test result evidence. For example, actively growing reservoir microorganisms may be re-injected in an open-loop system during PWRI and HF operations, and whereby continuous provision of suitable nutrients, electron donors, and electron acceptors are achieved. Thirdly, to unravel the mystery of field-specific microbiological contamination/issues, and prolong biocidal efficacies in the field programs, the field-specific salinity and temperature gradients must be implemented within the lab-based biocidal efficacy studies. Therefore, enhancing the effects of the field-applied biocidal chemistries in curtailing the growth and activity of the detrimental microorganisms for the subject field application. Finally, with the advent of NGS and bioinformatics capabilities, field-specific microbiological contamination/issues can be incorporated with the above pointers, aiding towards microbial characterisations and envisioned microbial functions discovery (proteomics). Therefore, a holistic approach aimed at incorporating the pointers would prove vital as proactive approach and may be implemented within microbiological best field practices documents. Therefore, curtailing the growth and activity of microorganisms implicated in oil and gas industry operations.

ACKNOWLEDGEMENTS

We would like to acknowledge the book editors Biwen Annie An Stepec, Torben Lund Skovhus, and Kenneth Wunch for useful discussions on this research during the ISMOS-8 conference and editing of the chapter. The appreciation is extended to colleagues within the Electron Microscopy Research Services, Newcastle University, for conducting the SEM/EDS analyses; Henriette Christensen for conducting ion chromatography for anions; Mark Hedley for microbial community analysis discussions; and Saudi Aramco's Research & Analytical Services Department Management for encouragement and financial support.

REFERENCES

Abdeljabbar, H., Cayol, J.L., Ben Hania, W., Boudabous, A., Sadfi, N. and Fardeau, M.L., (2013). *Halanaerobium sehlinense sp.* nov., an extremely halophilic, fermentative, strictly anaerobic bacterium from sediments of the hypersaline lake Sehline Sebkha. *International Journal of Systematic and Evolutionary Microbiology*, *63*(Pt_6), pp. 2069–2074.

Angelidaki, I., Karakashev, D., Batstone, D.J., Plugge, C.M. and Stams, A.J., (2011). Biomethanation and its potential. In *Methods in enzymology* (Vol. 494, pp. 327–351). Academic Press.

Bader, M.S.H., (2007). Sulfate removal technologies for oil fields seawater injection operations. *Journal of Petroleum Science and Engineering*, *55*(1–2), pp. 93–110.

Bhupathiraju, V., McInerney, M.J., Woese, C.R. and Tanner, R.S. (1999). *Haloanaerobium kushneri sp.* nov., an obligately halophilic, anaerobic bacterium from an oil brine. *International Journal of Systematic Bacteriology*, *49*, pp. 953–960.

Bjørlykke, K., Aagaard, P., Egeberg, P.K. and Simmons, S.P., (1995). Geochemical constraints from formation water analyses from the North Sea and the Gulf Coast Basins on quartz, feldspar and illite precipitation in reservoir rocks. *Geological Society*, London, Special Publications, *86*(1), pp. 33–50.

Blake, L.I., Sherry, A., Mejeha, O.K., Leary, P., Coombs, H., Stone, W., Head, I.M. and Gray, N.D., (2020). An unexpectedly broad thermal and salinity-tolerant estuarine methanogen community. *Microorganisms*, *8*(10), p.1467.

Booker, A.E., Borton, M.A., Daly, R.A., Welch, S.A., Nicora, C.D., Hoyt, D.W., Wilson, T., Purvine, S.O., Wolfe, R.A., Sharma, S. and Mouser, P.J., (2017). Sulfide generation by dominant *Halanaerobium* microorganisms in hydraulically fractured shales. *M Sphere*, *2*(4), pp. e00257–17.

Booker, A.E., Hoyt, D.W., Meulia, T., Eder, E., Nicora, C.D., Purvine, S.O., Daly, R.A., Moore, J.D., Wunch, K., Pfiffner, S.M. and Lipton, M.S., (2019). Deep-subsurface pressure stimulates metabolic plasticity in shale-colonizing *Halanaerobium spp. Applied and Environmental Microbiology*, *85*(12), pp. e00018–19.

Borton, M.A., Hoyt, D.W., Roux, S., Daly, R.A., Welch, S.A., Nicora, C.D., Purvine, S.O., Eder, E.K., Hanson, A.J., Sheets, J.M., Morgan, D.M., Sharma, S., Carr, T.R., Cole, D.R., Mouser, P.J., Lipton, M.S., Wilkins, M.J., Wrighton, K.C., (2018). Coupled laboratory and field investigations resolve microbial interactions that underpin persistence in hydraulically fractured shales. *Proceedings of the National Academy of Sciences of the United States of America*, *115*, pp.E6585–E6594. doi:10.1073/pnas.1800155115

Bottrell, S.H. and Raiswell, R., (2000). Sulphur isotopes and microbial sulphur cycling in sediments. *Microbial Sediments*, pp. 96–104.

Chen, C., Shen, Y., An, D. and Voordouw, G., (2017). Use of acetate, propionate, and butyrate for reduction of nitrate and sulfate and methanogenesis in microcosms and bioreactors simulating an oil reservoir. *Applied and Environmental Microbiology*, *83*(7), pp. e02983–e02916.

Choudhary, L., Macdonald, D.D. and Alfantazi, A., (2015). Role of thiosulfate in the corrosion of steels: a review. *Corrosion*, *71*(9), pp. 1147–1168.

Cord-Ruwisch, R., (1985). A quick method for the determination of dissolved and precipitated sulfides in cultures of sulfate-reducing bacteria. *Journal of Microbiological Methods*, *4*(1), pp. 33–36.

Daly, R.A., Borton, M.A., Wilkins, M.J., Hoyt, D.W., Kountz, D.J., Wolfe, R.A., Welch, S.A., Marcus, D.N., Trexler, R.V., MacRae, J.D. and Krzycki, J.A., (2016). Microbial metabolisms in a 2.5-km-deep ecosystem created by hydraulic fracturing in shales. *Nature Microbiology*, *1*(10), pp. 1–9.

Danika, N., Subasthika, T. and Gieg, L., (2021). Microbial corrosion under thio-sulfate reducing conditions by microbial communities in hydraulically fractured shale flow-back waters. In *8th International Symposium on Applied Microbiology and Molecular Biology in Oil Systems.* p.39. Abstract Book available at: https://ismos-8.org/wp-content/uploads/2021/06/07062021_ISMOS-8_ABSTRACT-BOOKFINALFINAL.pdf

Dastgerdi, A.A., Brenna, A., Ormellese, M., Pedeferri, M. and Bolzoni, F., (2019). Experimental design to study the influence of temperature, pH, and chloride concentration on the pitting and crevice corrosion of UNS S30403 stainless steel. *Corrosion Science, 159*, p.108160.

El Hajj, H., Abdelouas, A., El Mendili, Y., Karakurt, G., Grambow, B. and Martin, C., (2013). Corrosion of carbon steel under sequential aerobic–anaerobic environmental conditions. *Corrosion Science, 76*, pp. 432–440.

Enning, D., Venzlaff, H., Garrelfs, J., Dinh, H.T., Meyer, V., Mayrhofer, K., Hassel, A.W., Stratmann, M. and Widdel, F., (2012). Marine sulfate-reducing bacteria cause serious corrosion of iron under electroconductive biogenic mineral crust. *Environmental Microbiology, 14*(7), pp. 1772–1787.

Head, I.M., Gray, N.D. and Larter, S.R., (2014). Life in the slow lane; biogeochemistry of biodegraded petroleum containing reservoirs and implications for energy recovery and carbon management. *Frontiers in Microbiology, 5*, p.566.

Hoffmann, K., Bienhold, C., Buttigieg, P.L., Knittel, K., Laso-Pérez, R., Rapp, J.Z., Boetius, A. and Offre, P., (2020). Diversity and metabolism of *Woeseiales* bacteria, global members of marine sediment communities. *The ISME Journal, 14*(4), pp. 1042–1056.

Holmkvist, L., Ferdelman, T.G. and Jørgensen, B.B., (2011). A cryptic sulfur cycle driven by iron in the methane zone of marine sediment (Aarhus Bay, Denmark). *Geochimica et Cosmochimica Acta, 75*(12), pp. 3581–3599.

Jones, A.A., Pilloni, G., Claypool, J.T., Paiva, A.R. and Summers, Z.M., (2021). Evidence of sporulation capability of the ubiquitous oil reservoir microbe *Halanaerobium congolense. Geomicrobiology Journal, 38*(4), pp. 283–293.

Jørgensen, B.B., (1990). The sulfur cycle of freshwater sediments: Role of thiosulfate. *Limnology and Oceanography, 35*(6), pp. 1329–1342.

Kaksonen, A.H. and Puhakka, J.A., (2007). Sulfate reduction-based bioprocesses for the treatment of acid mine drainage and the recovery of metals. *Engineering in Life Sciences, 7*(6), pp. 541–564.

Kögler, F., Dopffel, N., Mahler, E., Hartmann, F.S., Schulze-Makuch, D., Visser, F., Frommherz, B., Herold, A. and Alkan, H., (2021). Influence of surface mineralogy on the activity of *Halanaerobium sp.* during microbial enhanced oil recovery (MEOR). *Fuel, 290*, p.119973.

Lahme, S., Enning, D., Callbeck, C.M., Menendez Vega, D., Curtis, T.P., Head, I.M. and Hubert, C.R., (2019). Metabolites of an oil field sulfide-oxidizing, nitrate-reducing *Sulfurimonas sp.* cause severe corrosion. *Applied and Environmental Microbiology, 85*(3), pp. e01891–e01818.

Lens, P., Vallerol, M., Esposito, G. and Zandvoort, M., (2002). Perspectives of sulfate reducing bioreactors in environmental biotechnology. *Reviews in Environmental Science and Biotechnology, 1*, pp. 311–325.

Liang, R., Davidova, I.A., Marks, C.R., Stamps, B.W., Harriman, B.H., Stevenson, B.S., Duncan, K.E. and Suflita, J.M., (2016). Metabolic capability of a predominant *Halanaerobium sp.* in hydraulically fractured gas wells and its implication in pipeline corrosion. *Frontiers in Microbiology, 7*, p.988.

Lipus, D., Vikram, A., Ross, D., Bain, D., Gulliver, D., Hammack, R. and Bibby, K., (2017). Predominance and metabolic potential of Halanaerobium spp. in produced water from hydraulically fractured Marcellus shale wells. *Applied and Environmental Microbiology, 83*(8), pp. e02659-16.

Mitchell, M.J., Schindler, S.C., Owen, J.S. and Norton, S.A., (1988). Comparison of sulfur concentrations within lake sediment profiles. *Hydrobiologia*, *157*(3), pp. 219–229.

Mitchell, A.C., Phillips, A.J., Hamilton, M.A., Gerlach, R., Hollis, W.K., Kaszuba, J.P. and Cunningham, A.B., (2008). Resilience of planktonic and biofilm cultures to supercritical CO2. *The Journal of Supercritical Fluids*, *47*(2), pp. 318–325.

Mori, K., Yamaguchi, K. and Hanada, S., (2018). *Sulfurovum denitrificans sp*. nov., an obligately chemolithoautotrophic sulfur-oxidizing epsilonproteobacterium isolated from a hydrothermal field. *International Journal of Systematic and Evolutionary Microbiology*, *68*(7), pp. 2183–2187.

Mouser, P.J., Borton, M., Darrah, T.H., Hartsock, A. and Wrighton, K.C., (2016). Hydraulic fracturing offers view of microbial life in the deep terrestrial subsurface. *FEMS Microbiology Ecology*, *92*(11).**

Murali Mohan, A., Hartsock, A., Hammack, R.W., Vidic, R.D., Gregory, K.B., (2013a). Microbial communities in flowback water impoundments from hydraulic fracturing for recovery of shale gas. *FEMS Microbiology Ecology*, *86*, pp. 567–580. doi:10.1111/1574-6941.12183

Murali Mohan, A., Hartsock, A., Bibby, K.J., Hammack, R.W., Vidic, R.D. and Gregory, K.B., (2013b). Microbial community changes in hydraulic fracturing fluids and produced water from shale gas extraction. *Environmental Science & Technology*, *47*, pp. 13141–13150. doi:10.1021/es402928b

Oren, A., (1999). Bioenergetic aspects of halophilism. *Microbiology and Molecular Biology Reviews*, *63*(2), pp. 334–348.

Oren, A., 2001. The bioenergetic basis for the decrease in metabolic diversity at increasing salt concentrations: implications for the functioning of Salt Lake ecosystems. In *Saline Lakes: Publications from the 7th International Conference on Salt Lakes, held in Death Valley National Park*, California, USA, September 1999 (pp. 61–72). Springer Netherlands.

Oren, A., (2011). Thermodynamic limits to microbial life at high salt concentrations. *Environmental microbiology*, *13*, pp. 1908–1923.

Parks, D.H., Tyson, G.W., Hugenholtz, P. and Beiko, R.G., (2014). STAMP: statistical analysis of taxonomic and functional profiles. *Bioinformatics*, *30*(21), pp. 3123–3124.

Pyzik, A.J. and Sommer, S.E., (1981). Sedimentary iron monosulfides: Kinetics and mechanism of formation. *Geochimica et Cosmochimica Acta*, 45, pp. 687–698.

Ramana, K.V.S., Anita, T., Mandal, S., Kaliappan, S., Shaikh, H., Sivaprasad, P.V., Dayal, R.K. and Khatak, H.S., (2009). Effect of different environmental parameters on pitting behavior of AISI type 316L stainless steel: Experimental studies and neural network modeling. *Materials & Design*, *30*(9), pp. 3770–3775.

Ravot, G., Magot, M., Ollivier, B., Patel, B.K.C., Ageron, E., Grimont, P.A.D., Thomas, P. and Garcia, J.L., (1997). *Haloanaerobium congolense sp*. nov., an anaerobic, moderately halophilic, thiosulfate-and sulfur-reducing bacterium from an African oil field. *FEMS Microbiology Letters*, *147*(1), pp. 81–88.

Scheffer, G., Hubert, C.R., Enning, D.R., Lahme, S., Mand, J. and de Rezende, J.R., (2021). Metagenomic investigation of a low diversity, high salinity offshore oil reservoir. *Microorganisms*, *9*(11), p.2266.

Schippers, A. and Jørgensen, B.B., (2001). Oxidation of pyrite and iron sulfide by manganese dioxide in marine sediments. *Geochimica et Cosmochi- mica Acta*, 65, pp. 915–922. doi:10.1016/S0016-7037(00)00589-5

Sindi, M. et al., (2021). Effects of extreme physicochemical parameters of injected seawater- produced water (ISW-PW) on sulfidogenesis and Microbiologically-Influenced Corrosion (MIC). In *8th International Symposium on Applied Microbiology and Molecular Biology in Oil Systems*, (p.22). Abstract Book available at: https://ismos-8.org/wp-content/uploads/2021/06/07062021_ISMOS-8_ABSTRACT-BOOKFINALFINAL.pdf

Tian, H., Yan, M., Treu, L., Angelidaki, I. and Fotidis, I.A., (2019). Hydrogenotrophic methanogens are the key for a successful bioaugmentation to alleviate ammonia inhibition in
thermophilic anaerobic digesters. *Bioresource Technology*, *293*, p.122070.

Tinker, K., Lipus, D., Gardiner, J., Stuckman, M. and Gulliver, D., (2022). The microbial community and functional potential in the Midland Basin reveal a community dominated by
both thiosulfate and sulfate-reducing microorganisms. *Microbiology Spectrum*, *10*(4),
pp. e00049–e00022.

Vidal-Verdú, À. Latorre-Pérez, A., Molina-Menor, E., Baixeras, J., Peretó, J. and Porcar, M.,
(2022). Living in a bottle: Bacteria from sediment-associated Mediterranean waste and
potential growth on polyethylene terephthalate. *Microbiology Open*, *11*(1), p.e1259.

Viggi, C.C., Matturro, B., Frascadore, E., Insogna, S., Mezzi, A., Kaciulis, S., Sherry, A.,
Mejeha, O.K., Head, I.M., Vaiopoulou, E. and Rabaey, K., (2017). Bridging spatially
segregated redox zones with a microbial electrochemical snorkel triggers biogeochemical cycles in oil-contaminated River Tyne (UK) sediments. *Water Research*, *127*,
pp. 11–21.

Vilcáez, J., York, J., Youssef, N. and Elshahed, M., (2018). Stimulation of methanogenic crude
oil biodegradation in depleted oil reservoirs. *Fuel*, *232*, pp. 581–590.

Widdel, F. and Bak, F., (1992). Gram-negative mesophilic sulfate-reducing bacteria. In
*The Prokaryotes: A Handbook on the Biology of Bacteria: Ecophysiology, Isolation,
Identification, Applications*; Balows, A., Trüper, H.G., Dworkin, M., Harder, W.,
Schleifer, K.-H., Eds. (pp. 3352–3378) Springer: New York, NY, USA. ISBN:
978-1-4757-2191-1.

Yao, W., and Millero, F.J., (1993). The rate of sulfide oxidation by MnO_2. *Geo- chimica et
Cosmochimica Acta*, 57, p.3359–3365.

Yao, W. and Millero, F.J., (1996). Oxidation of hydrogen sulfide by hydrous Fe (III) oxides in
seawater. *Marine Chemistry*, v 52, p.1–16. doi:10.1016/ 0304-4203(95)00072-0

Zhang, R., Sun, M., Zhang, H. and Zhao, Z., (2021). Spatial separation of microbial communities reflects gradients of salinity and temperature in offshore sediments from Shenzhen,
south China. *Ocean & Coastal Management*, *214*, p.105904.

Zhu, X.Y., Lubeck, J. and Kilbane, J.J., (2003). Characterization of microbial communities in
gas industry pipelines. *Applied and Environmental Microbiology*, *69*(9), pp. 5354–5363.

Zopfi, J., Ferdelman, T. and Fossing, H., (2004). Distribution and fate of sulfur intermediates–sulfite, thiosulfate, and elemental sulfur–in marine sediments. In *Sulfur
Biogeochemistry–Past and Present* (pp. 97–116). Geological Society of America.

5 Metagenomic and Metabolomic Analysis of Microbiologically Influenced Corrosion of Carbon Steel in Produced Water

Susmitha Purnima Kotu*, Fang Yang*,
Cory Klemashevich, M. Sam Mannan,
and Arul Jayaraman
Texas A&M University, College Station, TX, United States

5.1 INTRODUCTION

Microbiologically influenced corrosion (MIC) is a major problem in various chemical processes, as well as in onshore and offshore oil and gas, pipeline, marine, and aviation industries[1]. MIC is an electrochemical process in which microbial metabolism initiates, facilitates, or accelerates the corrosion reactions[2]. The microorganisms associated with MIC exist in complex surface-attached communities (i.e., biofilms) with other microorganisms[3]. Thus, interactions between the corroding metal, the microorganisms that colonize the surface, and the surrounding environment[4] are critical in MIC, and understanding the mechanisms underlying MIC and developing effective mitigation strategies requires investigating all three facets of the problem[5,6]. While the environmental and electrochemical aspects have been relatively well studied[7-9], information on the mechanisms underlying the formation of microbial communities and the molecules potentially associated with MIC is only recently emerging.

Microbial communities associated with MIC are typically analyzed based on different biological macromolecules[10]. Most of the approaches for characterizing microbial communities potentially involved in MIC are based on analyzing DNA from the organisms in the biofilms. These methods include polymerase chain reaction (PCR)/quantitative polymerase chain reaction (qPCR), fluorescent in situ

---

* Both authors contributed equally to this work.

DOI: 10.1201/9781003287056-8

hybridization (FISH), and more recently, 16S ribosomal RNA (rRNA) gene sequencing[11]. While these methods have provided information on the microbial community composition and relative abundances of different microorganisms potentially associated with MIC[12,13], they do not provide much information on the specific biochemical reactions occurring in the community. Furthermore, studies that employed DNA-based methods to study MIC microbial communities have shown that the community composition varies extensively[14], which makes it difficult to identify causative roles for different microorganisms involved in MIC. These observations lead to the hypothesis that factors beyond the community composition, such as the metabolic reactions and their products, are important in MIC.

Small molecules or metabolites are the end products of different biochemical reactions; therefore, analyzing the abundance of these metabolites can provide information on the different biochemical reactions that can occur in the microbial community[15]. Information on the biochemical reactions in a microbial community is important not only for understanding the dynamics of different members of the microbial community but also for the interaction of the community with external substrates, such as the interaction of biofilms with metal surfaces in the case of MIC. Zhang et al.[16] used two metabolomic methods to show that fatty acid metabolism and biosynthesis were upregulated in *Desulfovibrio vulgaris* biofilms relative to planktonic cells and suggested their role in biofilm formation. Similarly, Bonifay et al. [17] used metabolomics to show that crude oil pipelines with higher corrosion had increased levels of succinic acids, and inferred increased anaerobic metabolism of hydrocarbons.

In this study, we characterized the microbial community composition and metabolites present in produced water collected from West Texas oilfield pipeline systems. The microbial community and the metabolome were characterized using 16S rRNA gene sequencing and untargeted liquid chromatography – mass spectrometry (LC-MS/MS), respectively. We also integrated the microbial community composition and metabolome results to identify key microbial contributions to metabolite abundance and MIC in our model system.

5.2 MATERIALS AND METHODS

5.2.1 In vitro MIC Experiment

Produced water (one gallon) collected from an oil well in the Permian basin of west Texas with a known history of MIC was obtained from ChampionX (Sugar Land, TX). The container was transferred immediately upon receipt to a Coy anaerobic chamber and maintained in a 90% nitrogen, 5% hydrogen, and 5% carbon dioxide environment. Eighty milliliters of produced water or autoclaved produced water were dispensed into three 100 mL glass bottles each (Figure 5.1). The autoclaved produced water (i.e., without any live microorganisms) was used as a control to account for abiotic corrosion. The produced and autoclaved produced water stocks were supplemented with 500 ppm sodium sulfate, 500 ppm sodium acetate, 200 ppm sodium propionate, and 10,000 ppm vitamin supplement (ATCC MD-VS) to support microbial growth. To ensure anoxic conditions, 1 ppm of resazurin (a redox indicator) was

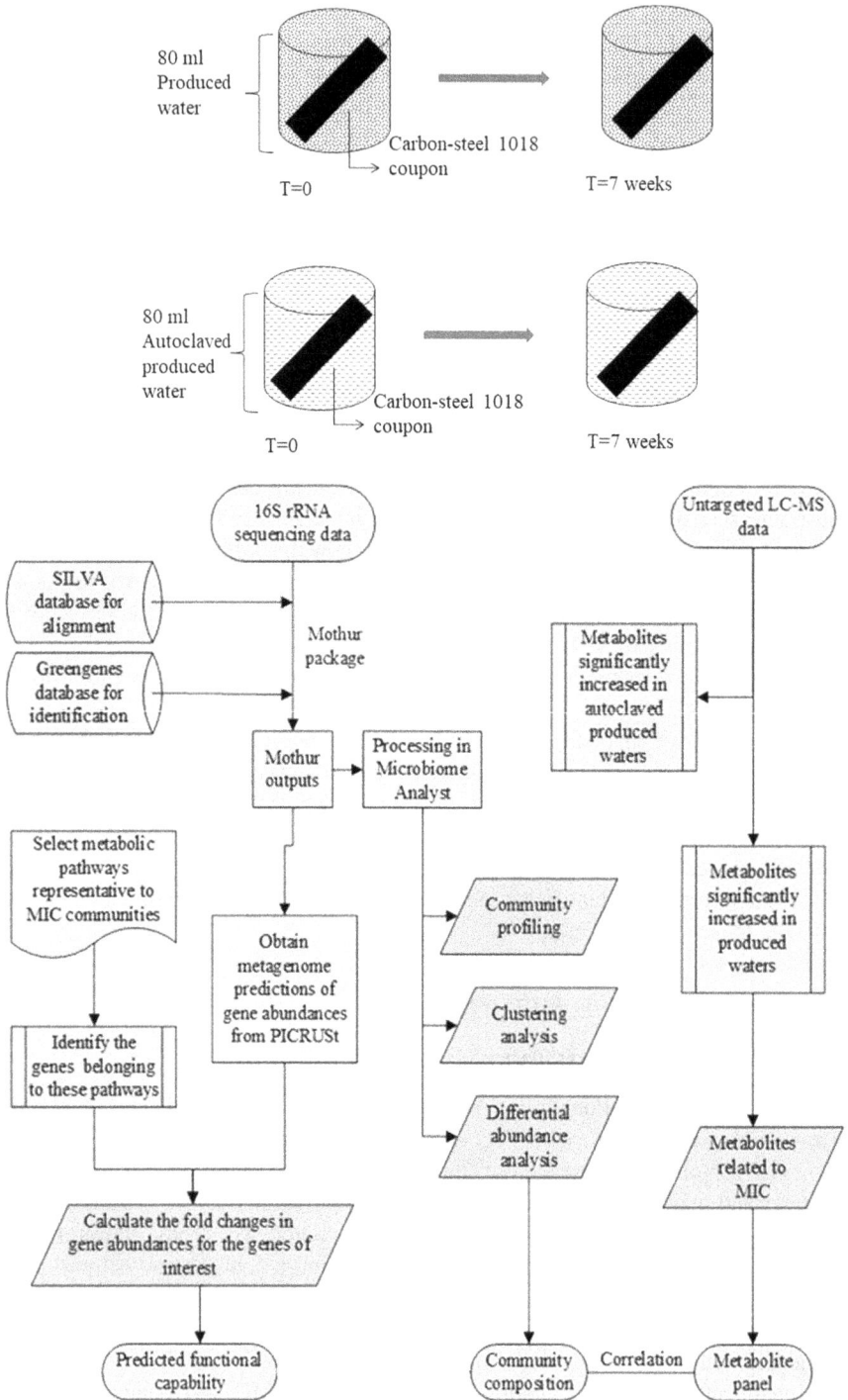

FIGURE 5.1   (a) Schematic of the experiment setup. (b) Workflow for analysis of metage-nomic and metabolomic data from produced water cultures.

added to all bottles. A biotic control with produced water incubated without a carbon steel coupon was not included in the study.

Carbon steel 1018 grade rectangular coupons (Metal Company, Alabama) with dimensions of 3″ length × 0.5″ width × 0.0625″ thickness were used in the study. The coupons were polished using 350 grit sandpaper, cleaned by sequentially sonicating in acetone and methanol for five minutes, and dried with nitrogen prior to the experiment. Coupons were weighed and transferred to either produced water or autoclaved produced water bottles. All bottles were incubated at 37°C inside an anaerobic chamber for seven weeks.

5.2.2 CORROSION RATE MEASUREMENTS

Weight loss measurements were carried out after the coupons were swabbed for metabolite analysis (see below). Carbon steel coupons were then sequentially sonicated in DI water for 15 minutes, followed by sonication for 15 minutes in a solution of 2% antimony trioxide and 5% stannous chloride in hydrochloric acid according to ASTM standard G1-03[18]. The coupons were then mechanically brushed while immersed in this solution to remove any remaining corrosion products. All coupons were air dried with nitrogen gas and their weights were recorded. The difference in the weight loss of the coupons during the experiment was used to calculate the corrosion rate. The corrosion rates of the coupons exposed to autoclaved produced water and produced water were compared to determine if the p-value was <0.05 using Student's t-test.

5.2.3 DNA EXTRACTION

DNA was extracted from produced water before and after exposure to carbon steel coupons using the DNeasy PowerSoil kit (Qiagen). Ten milliliters of the sample were centrifuged at $3000 \times g$ (4°C) and the pellet resuspended in 1 mL of the supernatant. DNA was extracted according to the manufacturer-suggested protocol and eluted in 100 μL of elution buffer (10 mM Tris-Cl, pH 8.5). The extracted DNA was cleaned using the DNeasy PowerClean clean-up kit (Qiagen) and resuspended in 30 μL of elution buffer. Microbiome analysis was performed on the DNA extracted from the produced water before and after exposure. Same coupons were used for both DNA and metabolite analysis. DNA and metabolites were extracted from biofilms on the coupon surfaces by swabbing the coupons with a cotton swab, homogenizing it in a bead beater with 600 μL lysis buffer (provided with DNeasy PowerSoil kit, Qiagen), and using half of the lysate for DNA extraction and the other half metabolite extraction. However, since the DNA yield was very little (<2 ng) from the coupon swabs, microbiome analysis on the biofilms could not be performed.

5.2.4 16S rRNA GENE ANALYSIS

The V4 region of the 16S rRNA gene was sequenced on a 2 × 250 bp cycle using 515F and 806R primers on the MiSeq platform (Illumina) at Microbial Analysis, Resources, and Services, University of Connecticut. The reads were then processed by trimming, screening, and alignment using mothur pipeline[19]. SILVA database v128 was used to align the reads and Greengenes database were used for taxonomical classification. Any operational taxonomic units (OTUs) that were classified to the same taxonomy were merged before they were imported into MicrobiomeAnalyst[20] for community profiling (using Simpson's diversity index for alpha diversity analysis and principal coordinate analysis for beta diversity analysis), clustering analysis (heat map and dendrogram analysis), and differential abundance analysis (linear discriminant analysis effect size, LEfSe method[21]). PICRUSt1 was then used to predict the functional potential (gene abundance) using 16S rRNA gene sequences[22]. Fold-change of the gene abundance was calculated from the abundance of the genes after exposure relative to the abundance before exposure. Differentially significant genes (fold-change > 2.0 or < 0.5 and p-value < 0.05 using Student's t-test) belonging to pathways reported to be involved in MIC were analyzed from the PICRUSt output. The workflow used for data processing and analysis is shown in Figure 5.1b.

5.2.5 Metabolite Extraction

Metabolites were extracted from produced and autoclaved produced water before and after exposure to carbon steel coupons. Metabolite extraction was carried out by solvent extraction[23]. Briefly, 30 mL of methanol and 7.5 mL of chloroform were added to 15 mL of the cultures, vortexed briefly, incubated on ice for 5 minutes, followed by centrifugation at $4000 \times g$ for 10 minutes. The upper phase (~ 48 mL) from each tube was collected and 1.5 mL of water was added. The contents were mixed vigorously and centrifuged again at $4000 \times g$ for 5 minutes. The two phases were separately collected in fresh tubes and stored at −80°C until further processing. The upper phase (~48 mL) was freeze-dried (−120°C and 0.01 mbar) using a lyophilizer (Labconco). The dried material was resuspended by sonication in 1 mL of 1:1 (vol/vol) methanol-water solution. The solution was then transferred to a spin column and centrifuged at $11,000 \times g$ (4°C) for 1 minute and the supernatant was stored at −80°C.

Metabolites on the coupon surfaces from produced and autoclaved produced water tests were collected by swabbing the coupons with a cotton swab and then homogenizing the swab in a bead beater with 300 μL lysis buffer (provided with DNeasy PowerSoil kit, Qiagen) as described earlier. The protocol described above for metabolite extraction was followed except that 600 μL of methanol and 150 μL of chloroform were initially added to the solution, and 600 μL of water was added to the upper phase collected from each tube. Lyophilized metabolites were resuspended in 200 μL of 1:1 (vol/vol) of methanol-water.

5.2.6 Metabolomic Analysis

Untargeted LC-MS/MS metabolite analysis was carried on a Q-Exactive Plus Orbitrap (Thermo Scientific) coupled to a Dionex Ultimate 200 UHPLC at Integrated Metabolomics Analysis Core, Texas A&M University. Metabolite samples from produced water and coupon surfaces were diluted fivefold and twofold respectively, prior to MS analysis. All samples were run using two different LC methods to maximize identification of metabolites. A hydrophilic interaction column (HILIC; SeQuant

ZIC-pHILIC 5um polymeric, EMD Millipore) was used to detect polar metabolites, and a reverse-phase column (Synergi 4um Fusion-RP 80A, Phenomenex) was used to separate a broad range of hydrophobic and polar metabolites. The HILIC column was used in the negative ion mode in MS while the reverse-phase column was used in the positive ion mode in MS. MS analysis was carried out at a resolution of 70,000 for MS1 and 17,500 for MS2 analysis. Compound Discoverer 2.1 (ThermoFisher) was used to process the MS spectra and for metabolite identification using the mzCloud and ChemSpider databases.

The metabolomic data analysis workflow is shown in Figure 5.1b. The positive and negative ion modes metabolite output from Compound Discoverer were combined after removing overlaps. Seven overlaps between positive and negative ion modes were identified, and in such cases, data from the mode in which better ionization was observed was used. The combined concentration data with putatively identified metabolites was then imported into MetaboAnalyst[24] for partial least squares-discriminant analysis (PLS-DA), differential abundance analysis, and pathway analysis. One-way ANOVA with Tukey's multiple comparisons test and fold-change were used to identify differentially abundant metabolites (FDR-adjusted-p-value < 0.05) between different groups. PLS-DA was performed with all samples to visualize the separation between different sample groups. The metabolites identified from produced water after exposure to carbon steel coupons were compared to those present in the produced water before exposure to the coupons. Similarly, supernatants from autoclaved produced water after exposure to the coupons were compared to autoclaved produced water before exposure. Metabolites produced by the microbial communities on coupon surfaces were compared to those in the produced water before exposure. Both these comparisons were also performed for compounds identified in autoclaved produced water and used as controls to identify compounds contributing only to microbial metabolism and not to other abiotic corrosion reactions. Significantly different metabolites in produced water after exposure to carbon steel coupons compared to their abundance prior to exposure were identified as being associated with MIC (i.e., MIC-associated pathways) using the KEGG database.

5.2.7 Integrated Analysis of Microbiome and Metabolome Data

MIMOSA2[25,26] was used to study correlations between microbiome and metabolome data to identify specific taxa contributing to metabolite variation that mediate MIC. MIMOSA uses PICRUSt to predict community gene content from taxonomic composition characterized by 16S rRNA gene sequencing. However, the prediction made by PICRUSt is theoretical without empirical data. The predicted community gene content was combined with reaction information to refer to the metabolic potential of the community. The variation in the predicted metabolic potential was then compared with actual metabolome data to identify specific taxa that contribute to the production and/or consumption of specific metabolites. The MIMOSA2 analysis was performed online (http://elbo-spice.cs.tau.ac.il/shiny/MIMOSA2shiny/). Specifically, microbiome data was provided in the form of a taxa-based table of 16S rRNA microbiome data using Greengenes 13_5 OTUs format. Metabolite names in metabolome data were assigned a KEGG compound identification using Chemical

Translation Service[27] before running it in MIMOSA2. A KEGG metabolic model was used to link microbiome and metabolite data. Rank-based regression was selected to compare metabolic potential and metabolite levels.

5.3 RESULTS

5.3.1 CORROSION RATE

The corrosion rate of the carbon steel coupons exposed to produced water was nearly fivefold higher (Student's t-test p-value <0.05) compared to autoclaved controls (2.6 ± 0.6 mils per year for produced water compared to 0.5 ± 0.1 mils per year for auto-claved produced water, Figure 5.2).

5.3.2 MICROBIAL COMMUNITY ANALYSIS

The DNA from the produced water (planktonic cells) and carbon steel coupon swabs (biofilm cells) was extracted and the OTUs corresponding to the microbial communities were identified and analyzed as described in the Methods. Microbiome analysis on the biofilms could not be performed because of low DNA yield (<2 ng) from the coupons. Simpson's alpha diversity index was computed for all planktonic cell samples to determine the richness (number of distinct OTUs) and evenness (domination of some OTUs over the rest) of the microbial community. Figure 5.3a shows that the alpha diversity of the microbial community in produced water before and after incubation with carbon steel coupon was significantly different (p-value = 0.005). The differences in the microbial community composition in the produced water at the start and end of incubation with carbon steel coupon were determined using principal coordinate analysis (PCoA) ordination with Bray-Curtis distance metric. Hierarchical clustering of samples at the out level showed that produced water before and after incubation with carbon steel clustered together with their respective groups (Figure 5.3b). Moreover, significant compositional variation was observed in the microbial community of produced water after the seven-week incubation with carbon steel coupons (Figure 5.3c).

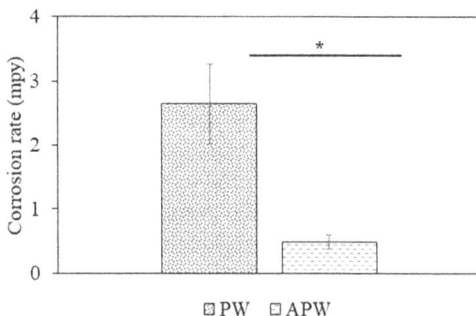

FIGURE 5.2 Corrosion rates of carbon steel coupons immersed in produced water and auto-claved produced water. * Indicates statistical significance at a level of p-value < 0.05 using the Student's t-test.

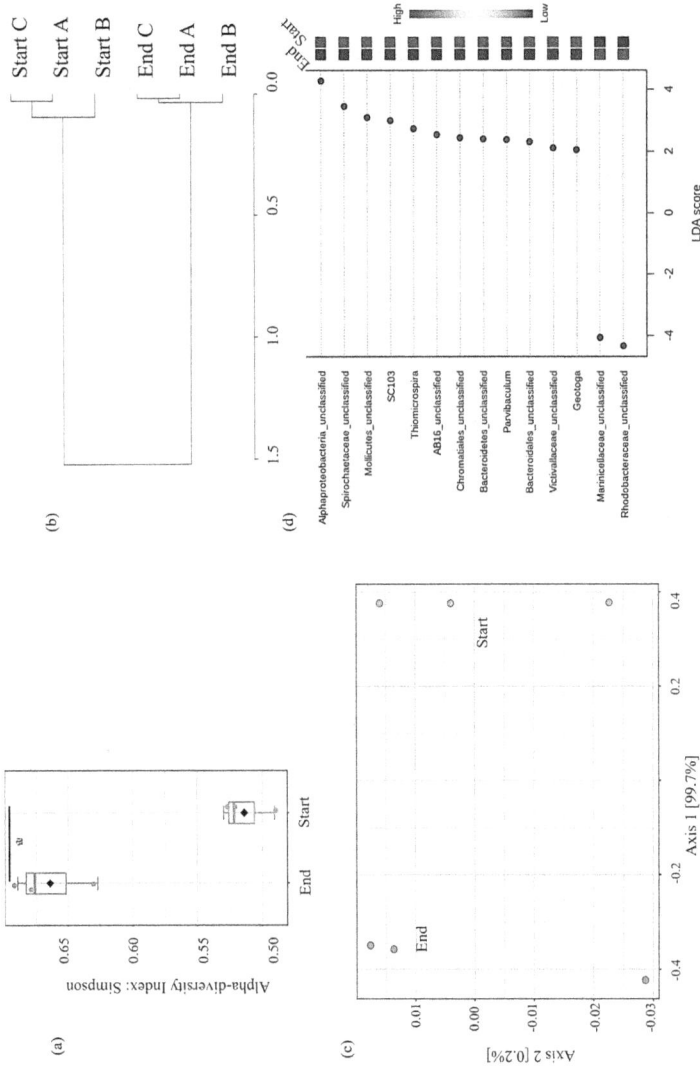

FIGURE 5.3 Microbiome analysis. (a) Simpson's alpha diversity index of microbial communities in produced water at the start and end of incubation with carbon steel coupons; (b) Hierarchical clustering of the microbial communities at OTU level at the start and end of incubation with carbon steel coupons using the Bray-Curtis distance method and Ward clustering algorithm; (c) Principal coordinate analysis (PCoA) of microbial communities at the start and end of incubation with steel coupons computed using the Bray-Curtis distance method; (d) Differential abundance analysis using LEfSe method showing statistically different genera (p-value < 0.05 and −2.0 > LDA score >2.0) at the end of incubation compared to the abundance at the start of incubation.

Differential abundance analysis using LEfSe at the genus level indicated that 14 genera showed a statistically significant abundance at the start and end of exposure to carbon steel coupons. The comparison of the abundances of the 14 genera at the start and end of exposure is shown in Figure 5.3d. The abundances at the OTU level are shown in Table 5.1. Among the differentially abundant genera, two genera (*Marinicellaceae_unclassified* and *Rhodobacteraceae_unclassified*) significantly increased in abundance after exposure to the carbon steel coupons relative to their abundance prior to exposure. The other 12 genera significantly decreased in abundance after exposure to carbon steel coupons relative to their abundance prior to exposure.

Metagenome predictions of the microbial community were carried out using PICRUSt1 to infer alterations in the function of the microbial community after exposure to carbon steel. Fold-changes in abundances of the ~6900 predicted genes in the community between the two time points were determined. Of these, 4.5% of genes were significantly increased in abundance (fold-change > 2.0, p-value < 0.05) while 1.5% of genes were decreased in abundance (fold-change < 0.5, p-value < 0.05). Approximately three times higher number of genes were predicted to have significantly increased abundance in the produced water after exposure to carbon steel coupons, compared to the genes with decreased abundance.

Specific pathways with prior association to MIC[28–30] such as amino acid, energy, carbohydrate, and xenobiotics metabolism were selected for analysis and the predicted abundances of all genes belonging to these pathways were evaluated. These pathways included energy metabolism (nitrogen metabolism, methane metabolism,

---

TABLE 5.1
Differential Abundance Analysis using LEfSe Method Showing Statistically Different OTU Abundance (p-value < 0.05 and −2.0 > LDA score >2.0) at the End of Incubation Compared to the Abundance at the Start of Incubation

Statistically Different OTU	Absolute Abundance (end)	Absolute Abundance (start)	LDA Score
OTU102419_AB16_unclassified	467.67	1165.3	2.54
OTU539878_*Alphaproteobacteria_unclassified*	15054	53596	4.28
OTU100279_*Bacteroidales_unclassified*	42.333	420.33	2.28
OTU105322_*Bacteroidetes_unclassified*	31.667	385.33	2.25
OTU213821_*Chromatiales_unclassified*	141.33	685.33	2.44
OTU252552_*Geotoga_unclassified*	41.333	265.67	2.05
OTU101210_*Marinicellaceae_unclassified*	22883	6.0000	−4.06
OTU1105985_*Mollicutes_unclassified*	559.67	3053.0	3.1
OTU561804_*Parvibaculum_unclassified*	378.67	858.33	2.38
OTU100203_*Rhodobacteraceae_unclassified*	43320	172.33	−4.33
OTU4195023_SC103_unclassified	309.00	2301.7	3
OTU100167_*Spirochaetaceae_unclassified*	721.00	6310.0	3.45
OTU112990_*Thiomicrospira_unclassified*	362.00	1455.3	2.74
OTU100123_*Victivallaceae_unclassified*	68.667	327.00	2.11

and sulfur metabolism), carbohydrate metabolism (propanoate metabolism, pyruvate metabolism, glycoxylate, and dicarboxylate metabolism, butanoate metabolism, and glycolysis degradation), xenobiotics biodegradation and metabolism (toluene degradation, xylene degradation, benzoate degradation, naphthalene degradation, polycyclic aromatic hydrocarbon degradation, and styrene degradation), lipid metabolism (fatty acid biosynthesis and biosynthesis of unsaturated fatty acids), and amino acid metabolism (phenylalanine metabolism, tryptophan metabolism, phenylalanine, tyrosine and tryptophan biosynthesis, glycine, serine and threonine metabolism, and tyrosine metabolism). Note that these pathways are not specific to MIC and are general pathways used by many microorganisms. Additionally, a negative control without the coupons was not designed as part of the study and hence, microbial metabolisms unique to MIC could not be identified. Thirty-four genes belonging to the above pathways of interest (Table 5.2) were predicted with significantly increased

TABLE 5.2
PICRUSt Metagenome Predictions of Genes Belonging to MIC-associated Pathways

Gene	Gene Name	Pathway(s)	Fold-change	p-value
K11263	Acetyl-/propionyl-CoA carboxylase, biotin carboxylase, biotin carboxyl carrier protein; acetyl/propionyl carboxylase subunit alpha	Fatty acid biosynthesis	229.155	0.021
K01031	3-Oxoadipate CoA-transferase, alpha subunit	Benzoate degradation	185.028	0.021
K01032	3-Oxoadipate CoA-transferase, beta subunit	Benzoate degradation	185.028	0.021
K01458	N-Formylglutamate deformylase	Glycoxylate and dicarboxylate metabolism	116.937	0.021
K10218	4-Hydroxy-4-methyl-2-oxoglutarate aldolase	Benzoate degradation	50.005	0.022
K08684	Methane monooxygenase	Methane metabolism	31.719	0.033
K00436	Hydrogen dehydrogenase	Methane metabolism	24.483	0.033
K00370	Nitrate reductase 1, alpha subunit	Nitrogen metabolism	15.952	0.034
K00371	Nitrate reductase 1, beta subunit	Nitrogen metabolism	15.946	0.034
K03315	Na+:H+ antiporter, NhaC family	Methane metabolism	14.094	0.036
K01825	3-Hydroxyacyl-CoA dehydrogenase/enoyl-CoA hydratase/3-hydroxybutyryl-CoA epimerase/enoyl-CoA isomerase	Propanoate metabolism, Butanoate metabolism, Biosynthesis of unsaturated fatty acids, Tryptophan metabolism	13.779	0.036
K03777	D-Lactate dehydrogenase	Pyruvate metabolism	12.760	0.025
K13039	Sulfopyruvate decarboxylase subunit beta	Methane metabolism	10.000	0.029
K01720	2-Methylcitrate dehydratase	Propanoate metabolism	9.179	0.025
K01501	Nitrilase	Nitrogen metabolism, Styrene degradation, Tryptophan metabolism	8.567	0.038

(Continued)

TABLE 5.2   (CONTINUED)

Gene	Gene Name	Pathway(s)	Fold-change	p-value
K13788	Phosphate acetyltransferase	Methane metabolism, Propanoate metabolism, Pyruvate metabolism	8.560	0.038
K12234	Coenzyme F420-0:L-glutamate ligase/ coenzyme F420-1:gamma-L-glutamate ligase; F420-0:gamma-glutamyl ligase	Methane metabolism	8.551	0.021
K03417	Methylisocitrate lyase	Propanoate metabolism	8.005	0.039
K00380	Sulfite reductase (NADPH) flavoprotein alpha-component	Sulfur metabolism	6.902	0.031
K01433	Formyltetrahydrofolate deformylase	Glycoxylate and dicarboxylate metabolism	6.302	0.031
K00114	Alcohol dehydrogenase (acceptor); alcohol dehydrogenase (cytochrome c)	Propanoate metabolism, Glycolysis/ Gluconeogenesis	6.166	0.045
K00248	Butyryl-CoA dehydrogenase; butyryl-CoA dehydrogenase	Butanoate metabolism	5.374	0.042
K06034	Sulfopyruvate decarboxylase subunit alpha	Methane metabolism	5.000	0.039
K00956	Sulfate adenylyltransferase subunit 1	Sulfur metabolism	4.932	0.044
K01640	Hydroxymethylglutaryl-CoA lyase	Butanoate metabolism	4.794	0.046
K12339	Dihydroaeruginoic acid synthetase	Sulfur metabolism	4.729	0.043
K04109	4-Hydroxybenzoyl-CoA reductase subunit beta; 4-Hydroxybenzoyl-CoA reductase subunit 3; 4-Hydroxybenzoyl-CoA reductase subunit beta	Benzoate degradation	4.000	0.049
K00446	Catechol 2,3-dioxygenase	Xylene degradation, Benzoate degradation, Styrene degradation	4.000	0.035
K01658	Anthranilate synthase component II	Phenylalanine, tyrosine, and tryptophan biosynthesis	3.765	0.038
K01571	Oxaloacetate decarboxylase, alpha subunit	Pyruvate metabolism	3.648	0.042
K14155	Cystathione beta-lyase	Nitrogen metabolism, Sulfur metabolism	3.596	0.023
K00651	Homoserine O-succinyltransferase	Sulfur metabolism	3.375	0.022
K04517	Prephenate dehydrogenase	Phenylalanine, tyrosine, and tryptophan biosynthesis	3.333	0.037
K06859	Glucose-6-phosphate isomerase, archaeal	Glycolysis/ Gluconeogenesis	2.400	0.020

Genes with a significantly increased abundance (fold-change > 2.0, p-value < 0.05) after incubation with carbon steel coupons compared to abundance before exposure are shown.

TABLE 5.3
PICRUSt Metagenome Predictions of Genes Belonging to MIC-associated Pathways

Gene	Gene Name	Pathway(s)	Fold-change	p-value
K13745	L-2,4-Diaminobutyrate decarboxylase	Glycine, serine, and threonine metabolism	0.383	0.020
K13810	Transaldolase/glucose-6-phosphate isomerase	Glycolysis/Gluconeogenesis	0.317	0.030
K00131	Glyceraldehyde-3-phosphate dehydrogenase	Glycolysis/Gluconeogenesis	0.275	0.015
K08093	3-Hexulose-6-phosphate synthase	Methane metabolism	0.213	0.020
K04516	Chorismate mutase	Phenylalanine, tyrosine, and tryptophan biosynthesis	0.136	0.048
K03921	Acyl-[acyl-carrier-protein] desaturase	Fatty acid biosynthesis, biosynthesis of unsaturated fatty acids	0.129	0.031
K08097	Phosphosulfolactate synthase	Methane metabolism	0.128	0.037
K13831	3-Hexulose-6-phosphate synthase/6-phospho-3-hexuloisomerase	Methane metabolism	0.123	0.019
K00196	Carbon-monoxide dehydrogenase iron sulfur subunit	Methane metabolism	0.109	0.023

Genes with a significantly decreased abundance (fold-change < 0.5, p-value < 0.05) after incubation with carbon steel coupons compared to abundance before exposure are shown.

abundance (fold-change > 2.0, p-value < 0.05) at the end of incubation with carbon steel while nine genes (Table 5.3) were predicted with significantly decreased abundance (fold-change < 0.5, p-value < 0.05).

5.3.3 METABOLOMIC ANALYSIS

PLS-DA of the extracted metabolites from bulk produced water phase samples indicated a clear separation between the sample groups (Figure 5.4). The separation between metabolites obtained from produced water before and after incubation with carbon steel coupons was significantly larger compared to that observed with autoclaved produced water. This pronounced alteration in the abundance of metabolites is likely due to MIC and abiotic corrosion reactions occurring in the produced water compared to abiotic corrosion reactions occurring in the autoclaved produced water. In addition, the metabolites extracted from the metal coupon surface biofilms were separated from the bulk solution produced water. Similar to the planktonic cells, separation was seen with the compounds extracted from the coupon surface exposed to autoclaved produced water and bulk autoclaved produced water. This suggests

Scores Plot

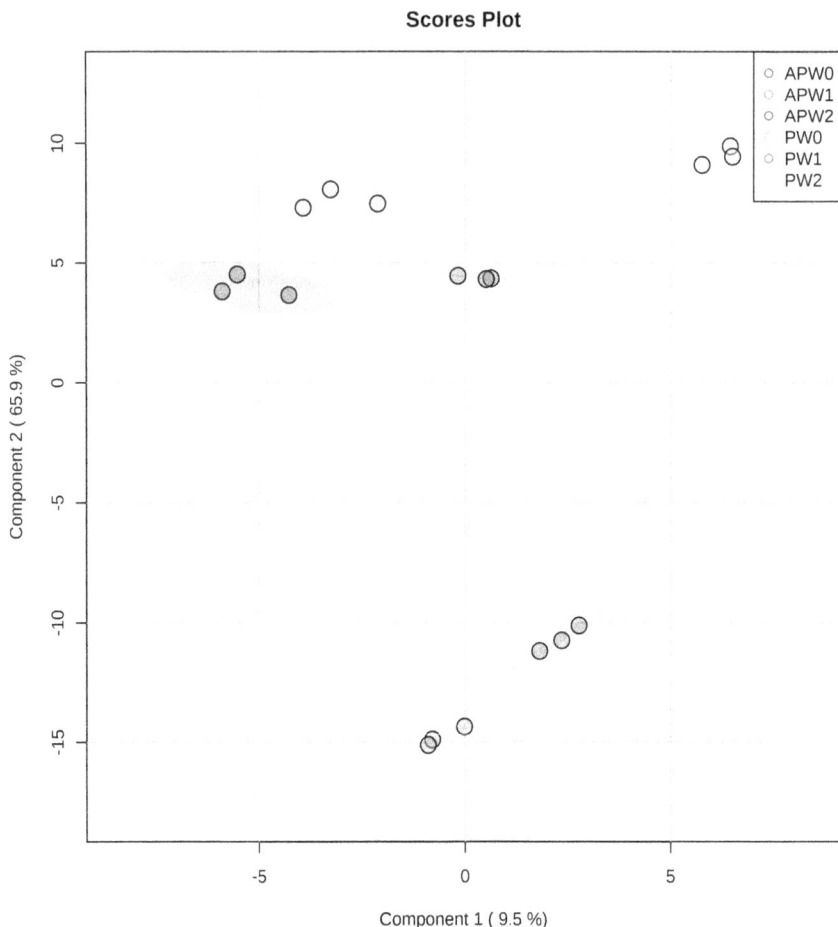

FIGURE 5.4 Partial least squares-discriminant analysis of metabolites extracted from all samples. PW – produced water and APW – autoclaved produced water. APW0 and PW0 represent metabolites extracted from produced water/autoclaved produced water before incubation with carbon steel coupons. PW2 and APW2 represent metabolites extracted from produced water/autoclaved produced water after incubation. PW1 and APW1 represent the metabolites extracted from the metal coupons exposed to produced water/autoclaved produced water.

that the biochemical reactions and the subsequent reaction products in the produced water bulk or suspension phase and on the coupon surface (i.e., biofilm) are different.

Differential abundance analysis (p-value < 0.05 and fold-change > 2.0 or fold-change < 0.5) was used to identify metabolites that significantly correlated to MIC. A total of 50 metabolites were significantly altered in abundance only in produced water, that is, without those arising from abiotic corrosion reactions in autoclaved produced water. Of these, 45 metabolites increased in abundance (Table 5.4), and the other five metabolites decreased in abundance (Table 5.5) after exposure to carbon steel coupons. The significantly increased putatively identified metabolites in the bulk-produced water phase include carboxylic acids and derivatives (cinnamic acid

TABLE 5.4
Significantly Increased Metabolites in Produced Water

Metabolite	Fold-change	Metabolite	Fold-change
Hypoxanthine	274.543	2,4-Xylidine	5.812
Tetrafluorophosphonium	214.035	2,6-Dimethyl-γ-pyrone	4.719
6-Hydroxynicotinic acid	70.473	O-glutaroyl-L-carnitine	4.662
L-Tryptophan	50.218	N-Acetylglycine	4.164
Fluroxene	47.947	Cinnamic acid	3.978
2′-Deoxyadenosine	47.478	5,6,7-Trimethoxy-2H-chromen-2-one	3.715
Encainide	34.322	Tryptamine	3.625
Oxolinic acid	31.966	Bis(2-ethylhexyl) phthalate	3.586
1-Phenylurea	29.975	Dibutyl phenylphosphonate	3.522
Oxadixyl	24.971	N′1,N′3-Bis[(2,2-dimethylpropanoyl) oxy]-5-methylbenzene-1,3-dicarboximidamide	3.472
2-Hydroxyphenylalanine	16.094	Styrene	3.381
L-Phenylalanine	14.881	Dioxohongdenafil	3.023
6-Methylquinoline	14.668	Afegostat	2.899
MDMA Methylene homolog	14.257	Valeric acid	2.856
Lignocaine	12.524	Cinnamyl alcohol	2.594
Ethyl lactate	12.436	(6R,7S)-6,7-Dihydroxy-8-methyl-8-azabicyclo[3.2.1]oct-3-yl (2E)-2-methyl-2-butenoate	2.553
Deschloro-N-ethyl-ketamine	9.755	2,2,6,6-Tetramethyl-4-piperidone	2.497
4-Amino-5-hydroxymethyl-2-methylpyrimidine	9.631	Benzaldehyde dimethyl acetal	2.477
Piracetam	8.598	Vanillyl alcohol	2.433
4-Acetyl-4-phenylpiperidine	8.373	p-Cresol	2.413
2-Methoxyestrone	6.914	2-C-methylerythritol 4-phosphate	2.330
Sinapyl alcohol	6.584	Favan-3-ol	2.000
Amphetamine	6.536		

Metabolites that were increased in abundance (p-value < 0.05 and fold-change > 2.0) in produced water incubated with carbon steel coupons for seven weeks, compared to their abundance prior to incubation with carbon steel coupons are shown.

TABLE 5.5
Significantly Decreased Metabolites in Produced Water

Metabolite	Fold-change
2-Methyl-4-amino-5-(formylaminomethyl)pyrimidine	0.493
Nitrobenzene	0.358
2-sec-Butoxy-4-methyl-1,3,2-dioxaphospholane 2-sulfide	0.299
D-Pantothenic acid	0.238
Cortisol	0.238

Metabolites that were decreased in abundance (p-value < 0.05 and fold-change < 0.5) in produced water with carbon steel coupons for seven weeks, compared to their abundance prior to incubation with carbon steel coupons are shown.

TABLE 5.6

Significantly Increased Metabolites on Carbon Steel Coupons

Metabolite	Fold-change
Tetrafluorophosphonium	158.142
Psoralidin	150.557
Diethyleneglycol dibenzoate	108.059
3-(4-Methoxyphenyl)-5-[(4-nitrophenoxy)methyl]-4,5-dihydroisoxazole	80.776
2,2,4,4,6,6-Hexamethyl-1,3,5-trithiane	36.319
Zopiclone	32.065
Diethylpyrocarbonate	25.066
D-Xylonic acid	16.892
Hexose	14.371
Dibenzoylmethane	13.554
3-Acetyl-2,5-dimethylfuran	10.912
Hypoxanthine	6.661
N~6~-Hydroxy-N~2~,N~2~,N~4~,N~4~-tetramethyl-1,3,5-triazine-2,4,6-triamine	6.629
(5R,6S)-5-Hydroxy-4-methoxy-6-(2-phenylethyl)-5,6-dihydro-2H-pyran-2-one	4.932
4-Ethylbenzaldehyde	4.500
Methyl palmitate	4.384
Stearic acid	3.120
4-tert-Butylcyclohexyl acetate	3.081
Caprylic anhydride	2.254
Triethyl phosphate	2.240
4,6-Dioxoheptanoic acid	2.176
Aminohippuric acid	2.156
2-(4-Carbamimidoylbenzyl)-N'-[(1S)-1-cyclohexyl-2-({4-[(diaminomethylene)amino]butyl}amino)-2-oxoethyl]-N,N-dimethylmalonamide	2.136
Tridecanedioic acid	2.130
2,2,4-Trimethyl-1,3-pentadienol diisobutyrate	2.023

Metabolites that were increased in abundance (p-value < 0.05 and fold-change > 2.0) on carbon steel coupons incubated in produced water for seven weeks, compared to their abundance in the bulk produced water liquid are shown.

and 6-hydroxynicotinic acid), fatty acids and derivatives (valeric acid and (15Z)-9,12,13-trihydroxy-15-octadecenoic acid), and amino acids and derivatives (L-tryptophan, 2-hydroxyphenylalanine, L-phenylalanine, and tryptamine). Thirty-four putatively identified metabolites were significantly altered in abundance on the carbon steel coupon surface compared to the bulk-produced water. Of these, 25 metabolites increased in abundance (Table 5.6) and nine metabolites decreased in abundance (Table 5.7).

5.3.4 CORRELATION BETWEEN MICROBIAL COMMUNITY AND METABOLOMIC DATA

MIMOSA was used to correlate the microbiome and metabolome data and infer the microbiological contribution to the production and utilization of metabolites. Among other community changes, microbiome data revealed that the relative abundance

TABLE 5.7

Significantly Decreased Metabolites on Carbon Steel Coupons

Metabolite	Fold-change
Pivagabine	0.388
Hydroquinone	0.320
Afegostat	0.291
N-Nitrosopiperidine	0.219
Anabasine	0.211
Ethyl acetate	0.193
N-Heptanoylhomoserine lactone	0.144
Phenylisocyanate	0.142
N-[(S)-(+)-1-Ethoxycarbonyl-3-phenylpropyl]-L-alanine	0.036

Metabolites that were decreased in abundance (p-value < 0.05 and fold-change < 0.5) on carbon steel coupons incubated in produced water for seven weeks, compared to their abundance in the water liquid are shown.

of *Alphaproteobacteria_unclassified* (OTU 539878), *Spirochaetaceae_unclassified* (OTU 100167), *Mollicutes_unclassified* (OTU 1105985) and *Parvibaculum* (OTU 561804) decreased while the relative abundance of *Marinicellaceae_unclassified* (OTU 101210) increased in produced water after exposure to carbon steel coupons when compared to before exposure to carbon steel (Table 5.1). Among the metabolite shifts, metabolome data showed that 4-amino-5-hydroxymethyl-2-methylpyrim idine, phenylalanine, tryptophan, and tryptamine were enriched in produced water after exposure relative to prior to exposure to carbon steel coupons (Table 5.4). Interestingly, MIMOSA analysis suggested that the increase of phenylalanine in produced water after exposure to carbon steel coupons was correlated to the decreased degradation potential of three specific taxa – *Alphaproteobacteria_unclassified* (OTU 539878), *Spirochaetaceae_unclassified* (OTU 100167) and *Mollicute_unclassified* (OTU 1105985) – in produced water after exposure to carbon steel coupons (Table 5.8). Similarly, the enrichment of tryptophan in produced water after exposure to carbon steel coupons was correlated to the decreased utilization of tryptophan by the lower abundant *Mollicutes_unclassified* (OTU 1105985) and *Parvibaculum* (OTU 561804) in produced water after exposure to carbon steel coupons (Table 5.8). Increased abundance of tryptamine and 4-amino-5-hydroxymethyl-2-methylpyrimid ine in produced water after exposure to carbon steel coupons were correlated to the decreased consumption of these metabolites resulting from decreased abundance of *Alphaproteobacteria_unclassified* (OTU 539878) in produced water after exposure to carbon steel coupons (Table 5.8). All other metabolite shifts could not be attributed to the changes in the microbial community.

TABLE 5.8

Correlation of Differentially Abundant Microorganisms to the Synthesis or Degradation of Differentially Abundant Metabolites

Metabolite	Species	Synthesis Genes	Degradation Genes
4-Amino-5-hydroxymethyl-2-methylpyrimidine	OTU 539878 "p__Proteobacteria", "c__Alphaproteobacteria"		K00941
Phenylalanine	OTU 539878 "p__Proteobacteria", "c__Alphaproteobacteria"	K04518	K00500; K01593; K01889; K01890; K03782
Phenylalanine	OTU 100167 "p__Spirochaetes", "c__Spirochaetes", "o__Spirochaetales", "f__Spirochaetaceae"		K01889; K0189
Phenylalanine	OTU 1105985 "p__Tenericutes", "c__Mollicutes"		K01889; K01890
Phenylalanine	OTU 101210 "p__Proteobacteria", "c__Gammaproteobacteria", "o__Marinicellales", "f__Marinicellaceae"	K14170	K01889; K01890; K03782
Tryptophan	OTU 1105985 "p__Tenericutes", "c__Mollicutes"		K01867
Tryptophan	OTU 561804 "p__Proteobacteria", "c__Alphaproteobacteria", "o__Rhizobiales", "f__Hyphomicrobiaceae", "g__Parvibaculum"	K01695; K01696	K01867
Tryptamine	OTU539878 "p__Proteobacteria", "c__Alphaproteobacteria"		K00274

5.4 DISCUSSION

In field environments susceptible to corrosion, several abiotic corrosive agents such as carbon dioxide, hydrogen sulfide, and oxygen usually co-exist with microorganisms, corroding metal, water, and electron acceptors/electron donors[31,32]. Since interactions among metal, microorganisms, nutrient sources, and metabolites define MIC, cataloging all these compounds and ascribing them to different sources is important to understand MIC mechanisms. Information only on the dominant community members without observed metabolic activity or cataloging the metabolic reactions without the community composition are both inadequate for understanding MIC. Furthermore, it is critical to examine the activity of microorganisms to specifically understand the contribution of MIC to the observed total corrosion. This has led to integrating metabolome and microbiome analysis to understand the metabolic reactions and eventually MIC mechanisms[33]. Recent studies have employed both metagenomics and metabolomics to identify the metabolites associated with micro-organisms in pipelines transporting drinking water and produced water from a site with MIC[34,35]. In the current study, we used a similar approach to profile and integrate information about the microbial community and its metabolome in produced water from a West Texas oil field.

Carbon steel coupons exposed to produced water showed a fivefold higher corrosion rate with MIC compared to that observed with abiotic corrosion. The higher rate of corrosion with produced water is likely due to the microbial metabolic activity in these samples which was absent in the autoclaved produced water controls. In addition, these observations also suggest that the rate of MIC was significantly higher than the abiotic corrosion rate under the experimental conditions used in this study.

16S rRNA gene analysis of the produced water incubated with carbon steel coupons showed that two genera, *Marinicellaecea_unclassified* and *Rhodobacteraceae_unclassified*, significantly increased in abundance after seven weeks of incubation. Both *Marinicellaceae_unclassified*[36,37] and *Rhodobacteraceae_unclassified*[38–41] have been previously identified as dominant species in diverse MIC environments like seawater, seawater dispersed with crude oil, and injection water from offshore pipelines. *Rhodobacteraceae* species were identified in several studies as dominant early colonizers of steel resulting in increased rates of MIC in marine environments[42,43]. The observed increased abundance of these community members is consistent with the increased corrosion in our study.

Our data also show a decrease in genus SC103, *Thiomicrospira*, *Chromatiales_unclassified*, and *Geotoga* are related to sulfur metabolism after exposure to carbon steel coupons. SC103 is an anaerobic bacterium belonging to the *Thermotoga* phylum[44] and capable of elemental sulfur reduction to hydrogen sulfide[45]. SC103 has been identified as one of the community members in produced water of Netherlands oilfields with MIC issues[46]. The observed decrease in the abundance of SC103 after exposure to carbon steel coupons in this study suggests a decrease in elemental sulfur reduction process. *Thiomicrospira* is a nitrate-reducing, sulfur-oxidizing bacterium previously identified in Canadian oil field consortia that can cause high corrosion rates and produce corrosive by-products like sulfur, thiosulfate, and nitrate[47,48]. The reduced abundance of *Thiomicrospira* after exposure to carbon steel suggests that MIC due to nitrate reduction or sulfur oxidation may not be a dominant corrosion mechanism in this study. *Chromatiales* are sulfur oxidizers previously reported to have decreased in abundance in the tubercles of steel structures compared to the marine sediments collected from coastal Australia[43]. The reduced abundance of *Chromatiales* after exposure to carbon steel coupons in this study suggests a reduced sulfur oxidation process. *Geotoga* is an anaerobic fermentative bacterium capable of elemental sulfur reduction to produce hydrogen sulfide and was previously identified in biofilms of offshore oil production facilities and marine rust tubercles[49,50]. The reduced abundance of *Geotoga* after incubation with carbon steel suggests absence of elemental sulfur reduction as one of the MIC mechanisms in this study.

The decrease in abundance of *Bacteroidetes_unclassified* and *Bacteroidales_unclassified* after exposure to carbon steel coupons is consistent with prior studies. Several

MIC studies have reported the presence of members of *Bacteroidetes* phylum including *Bacteroidales* in samples biofilms from corroded sewer pipelines[51], produced water and pig samples from Nigerian onshore and offshore oil pipelines[52], pigging debris from North Sea oil pipelines[17], and produced water from Canadian oilfields[53]. However, a comparison of the changes in the abundance of these community members was not reported due to the nature of the studies and the experimental design.

16S rRNA gene analysis of the produced water also showed that AB16_*unclassified*, *Alphaproteobacteria_unclassified*, *Mollicutes_unclassified*, *Parvibaculum*, *Spirochaetaceae_unclassified*, and *Victivallaceae_unclassified* decreased in abundance after exposure to carbon steel coupons. While these microbial groups were identified in oilfields and previous MIC studies, their specific role in MIC mechanisms is unknown. *Alphaproteobacteria* has been previously reported as an abundant microbial group in biofilms formed on carbon steel exposed to seawater and crude oil[38]. Several studies investigating anaerobic incubations of Florida seawater[39] and carbon steel incubated with groundwater[54] have observed a decrease in the abundance of *Alphaproteobacteria*, similar to what was seen in the current study. *Sphaerochaeta* belonging to *Spirochaetaceae* family was previously identified as one of the dominant taxa in produced water biofilms on carbon steel[53]. Similarly, *Mollicutes*, which are oil-degrading fermentative bacteria, have been identified in Nigerian onshore produced water samples[52]. However, the roles of *Mollicutes* and *Spirochaetaceae_unclassified* in MIC mechanisms are not reported in the literature.

Differential analysis of the abundant genera between seawaters of coastal India and biofilms of seawaters after exposure to stainless steel revealed the presence of members of AB16 class in the planktonic phase but not in the biofilms[55]. In the current study, microbial community analysis was not performed on biofilms formed on the carbon steel coupons and hence our results cannot be directly compared to this study. However, a decrease in the abundance of AB16_unclassified in the planktonic phase after incubation may result in a decreased abundance in the biofilms on carbon steel relative to their abundance before exposure. *Parvibaculum* is a marine bacterium capable of hydrocarbon and specifically crude oil- and asphaltene-degradation[52] that was decreased in abundance in produced water from oilfields in Nigeria and Saudi Arabia compared to injected seawater, despite the higher corrosion in produced water compared to injected seawater[52,56]. Despite the presence of *Victivallaceae* in petroleum reservoirs and their decreased abundance after stimulation of indigenous communities for enhanced oil recovery[57], *Victivallaceae_unclassified* is not known to be involved in MIC. It should be noted that the observed decrease in the abundance of some of these genera could also be due to temporal variations in the community and not related to reduced involvement in MIC reactions. However, determining the temporal variations in the community in absence of carbon steel was beyond the scope of the study.

Some of the changes in MIC-associated pathways predicted by PICRUSt were also validated by the metabolome data. The predicted increase in abundance of genes belonging to tryptophan metabolism, phenylalanine metabolism, and phenylalanine, tyrosine, and tryptophan biosynthesis pathways from PICRUSt correlated well to the observed increase in abundance of metabolites belonging to these pathways such as L-tryptophan, L-phenylalanine, and tryptamine (Table 5.9).

TABLE 5.9

MIC-associated Metabolic Pathways in Produced Water

Metabolite	Metabolic Pathway
L-Tryptophan	Phenylalanine, tyrosine, and tryptophan biosynthesis; tryptophan metabolism
L-Phenylalanine	Phenylalanine, tyrosine, and tryptophan biosynthesis; phenylalanine metabolism; Glycine, serine, and threonine metabolism
Tryptamine	Tryptophan metabolism
p-Cresol	Toluene degradation

Metabolic pathways involved in the synthesis of the different produced water metabolites identified (see Tables 5.3 and 5.4) are shown.

Several of the putatively identified metabolites that significantly increased in abundance in the bulk produced water after exposure to carbon steel coupons and on the carbon steel coupons compared to the bulk produced water before exposure have been identified as key metabolites in previous MIC studies[17,30,34,58–60]. Metabolites related to amino acid metabolism have been reported on biofilms of carbon steel coupons incubated with seawater and jet fuel[59]. Tryptophan is a common by-product of microbial metabolism and was identified in copper pipelines causing MIC[58]. Phenylalanine and tryptamine derivative (N, N-dimethyltryptamine) were also detected in pigging debris of produced water pipelines of two oilfields from the North Sea[17]. Carboxylic acids and their derivatives are some of the commonly detected metabolites linked to MIC. For instance, metabolomic analysis of pigging debris from produced water pipelines identified tridecanedioic acid and cinnamic acid derivatives (ethyl cinnamate, trans-cinnamate, 3,4,5-trimethoxycinnaic acid, and 4-hydroxycinnamyl aldehyde)[17]. Heptanoic acid was detected as one of the key metabolites in produced water and pig samples collected from Alaskan North Slope oil pipelines[60]. In addition to these metabolites, carboxylic acid derivatives of other carboxylic acids have also been shown to be key metabolites associated with MIC. In particular, succinic acids that indicate anaerobic hydrocarbon metabolism were identified in pigging debris from a North Sea crude oil pipeline with higher corrosion rates compared to one with lower corrosion rates[17].

Most fatty acids identified in this study are reported to be significantly correlated to MIC. Valeric acid or pentanoic acid was previously identified as one of the downstream metabolites in Alaskan North Slope oil field pipelines[60] and in rural water supply networks[34] with MIC concerns. Water distribution pipelines made of copper exhibiting MIC in Australia showed fatty acids such as octadecenoic acid, stearic acid, and palmitic acid[30,34].

MIMOSA analysis correlated the increase in metabolites, tryptophan, phenylalanine, tryptamine, and 4-Amino-5-hydroxymethyl-2-methylpyrimidine, to the decrease in the abundance of taxa that utilize these metabolites. These metabolites potentially contributed to biofilm formation or corrosive reactions and may serve as potential biomarkers for MIC in the West Texas oilfield where the produced waters for this study were obtained. While several earlier MIC studies showed an abundance of these metabolites[17,58,59], the taxa correlated by MIMOSA analysis in our study were not significantly different from the previous studies. Also, the studies[38,39,52,53,56] that observed similar changes to taxa did not use metabolite profiling and, hence, it is unknown whether the metabolites correlated to the taxa changes and the observed MIC in our study may have been present. While more research is needed to determine how these metabolites cause MIC, identifying similar metabolites across multiple MIC studies (samples representative of different service types, operating conditions, and geographical locations) could be promising in the identification of MIC in field samples and the levels of these metabolites in a sample could serve a biomarker for diagnosis of MIC.

Prospectively, correlating the functional output of the microbial community to markers of corrosion at a specific MIC-impacted field location can lead to the identification of integrated microbiota markers (i.e., metabolite with possible bacterial taxa that contribute to their production) that can be used for early detection and monitoring of MIC.

ACKNOWLEDGMENTS

The authors are grateful to Mr. Timothy Tidwell from ChampionX (Sugar Land, TX) for providing the produced water for the experiments. This work is financially supported by Mary Kay O'Connor Process Safety Center at Texas A&M University and the Ray Nesbitt Chair Endowment to Dr. Arul Jayaraman. Texas A&M High Performance Research Computing for providing the necessary resources for data analysis and Integrated Metabolomics Analysis Core (IMAC) at Texas A&M University for metabolite analysis are acknowledged.

REFERENCES

1. Iverson, W. P. Microbial corrosion of metals. *Adv Appl Microbiol* **32**, 1–35 (1987).
2. Videla, H. A. *Manual of biocorrosion*. (CRC Press, Boca Raton, FL, 1996).
3. Little, B., Wagner, P. & Mansfeld, F. An overview of microbiologically influenced corrosion. *Electrochimica Acta* **37**, 2185–2194 (1992), doi:10.1016/0013-4686(92)85110-7
4. Videla, H. A. Microbially induced corrosion: an updated overview. *Int Biodeterior Biodegrad* **48**, 176–201 (2001).
5. Eckert, R. B. & Skovhus, T. L. Advances in the application of molecular microbiological methods in the oil and gas industry and links to microbiologically influenced corrosion. *Int Biodeterior Biodegrad* **126**, 169–176 (2018).
6. Skovhus, T. L., Eckert, R. B. & Rodrigues, E. Management and control of microbiologically influenced corrosion (MIC) in the oil and gas industry—overview and a North Sea case study. *J Biotechnol* **256**, 31–45 (2017).
7. Eckert, R. B. & Amend, B. *Microbiologically influenced corrosion in the upstream oil and gas industry* (eds Torben Lund Skovhus, Dennis Enning, & Jason S. Lee) (CRC Press, Boca Raton, FL, 2017).
8. Enning, D. & Garrelfs, J. Corrosion of iron by sulfate-reducing bacteria: new views of an old problem. *Appl Environ Microbiol* **80**, 1226–1236 (2014).
9. Javaherdashti, R. *Microbiologically influenced corrosion*, pp. 81–97 (Springer, Perth, Australia, 2017).
10. Patti, G. J., Yanes, O. & Siuzdak, G. Innovation: metabolomics: the apogee of the omics trilogy. *Nat Rev Mol Cell Biol* **13**, 263–269 (2012).

11. Muyzer, G. & Marty, F. *Applications of molecular microbiological methods* (eds Sean M. Caffrey, Casey R.J. Hubert & Torben L. Skovhus) (Caister Academic Press, Poole, UK, Adde 2014).

12. Lin, X., Sharma, N. & Lee, C., Development and validation of in-field qPCR methods for water microbial analysis at oil and gas facilities. CORROSION 2015, Dallas, Texas.

13. Okoro, C. C. & Amund, O. O. Microbial community structure of a low sulfate oil producing facility indicate dominance of oil degrading/nitrate reducing bacteria and methanogens. *Petrol Sci Technol* **36**, 293–301 (2018), doi:10.1080/10916466.2017.1421969

14. Vincke, E., Boon, N. & Verstraete, W. Analysis of the microbial communities on corroded concrete sewer pipes–a case study. *Appl Microbiol Biotechnol* **57**, 776–785 (2001).

15. Abram, F. Systems-based approaches to unravel multi-species microbial community functioning. *Comput Struct Biotechnol J* **13**, 24–32 (2015).

16. Zhang, Y., Pei, G., Chen, L. & Zhang, W. Metabolic dynamics of Desulfovibrio vulgaris biofilm grown on a steel surface. *Biofouling* **32**, 725–736 (2016), doi:10.1080/08927014.2016.1193166

17. Bonifay, V. et al. Metabolomic and metagenomic analysis of two crude oil production pipelines experiencing differential rates of corrosion. *Front Microbiol* **8**, 99 (2017), doi:10.3389/fmicb.2017.00099

18. Standard, A. *Standard practice for preparing, cleaning, and evaluating corrosion test specimens, annual book of ASTM standards* Vol. 3, pp. 17–25 (2017).

19. Schloss, P. D. et al. Introducing mothur: open-source, platform-independent, community-supported software for describing and comparing microbial communities. *Appl Environ Microbiol* **75**, 7537–7541 (2009).

20. Dhariwal, A. et al. MicrobiomeAnalyst: a web-based tool for comprehensive statistical, visual and meta-analysis of microbiome data. *Nucleic Acids Res* **45**, W180–W188 (2017).

21. Segata, N. et al. Metagenomic biomarker discovery and explanation. *Genome Biol* **12** (2011), doi:ARTN R6010.1186/gb-2011-12-6-r60

22. Langille, M. G. I. et al. Predictive functional profiling of microbial communities using 16S rRNA marker gene sequences. *Nat Biotechnol* **31**, 814 (2013), doi:10.1038/nbt.2676

23. Sridharan, G. V. et al. Prediction and quantification of bioactive microbiota metabolites in the mouse gut. *Nat Commun* **5**, 5492 (2014).

24. Chong, J., Yamamoto, M. & Xia, J. MetaboAnalystR 2.0: from raw spectra to biological insights. *Metabolites* **9** (2019), doi:10.3390/metabo9030057

25. Noecker, C., Eng, A., Muller, E. & Borenstein, E. MIMOSA2: a metabolic network-based tool for inferring mechanism-supported relationships in microbiome-metabolome data. *Bioinformatics* (2022), doi:10.1093/bioinformatics/btac003

26. Yang, F. et al. Effect of diet and intestinal AhR expression on fecal microbiome and metabolomic profiles. *Microb Cell Fact* **19**, 219 (2020), doi:10.1186/s12934-020-01463-5

27. Wohlgemuth, G., Haldiya, P. K., Willighagen, E., Kind, T. & Fiehn, O. The chemical translation service–a web-based tool to improve standardization of metabolomic reports. *Bioinformatics* **26**, 2647–2648 (2010), doi:10.1093/bioinformatics/btq476

28. Bhatt, P. et al. Major metabolites after degradation of xenobiotics and enzymes involved in these pathways. *Smart Bioremediat Technol: Microbial Enzymes*, 205–215 (2019), doi:10.1016/B978-0-12-818307-6.00012-3

29. Gomez-Alvarez, V., Revetta, R. P. & Santo Domingo, J. W. Metagenome analyses of corroded concrete wastewater pipe biofilms reveal a complex microbial system. *BMC Microbiol* **12**, 122 (2012), doi:10.1186/1471-2180-12-122

30. Beale, D. J., Morrison, P. D., Key, C. & Palombo, E. A. Metabolic profiling of biofilm bacteria known to cause microbial influenced corrosion. *Water Sci Technol* **69**, 1–8 (2014), doi:10.2166/wst.2013.425

31. Beech, I. B. & Sunner, J. Biocorrosion: towards understanding interactions between biofilms and metals. *Curr Opin Biotechnol* **15**, 181–186 (2004), doi:10.1016/j. copbio.2004.05.001

32. Li, K., Whitfield, M. & Van Vliet, K. J. Beating the bugs: roles of microbial biofilms in corrosion. *Corros Rev* **31**, 73–84 (2013).

33. Kotu, S. P., Mannan, M. S. & Jayaraman, A. Emerging molecular techniques for studying microbial community composition and function in microbiologically influenced corrosion. *Int Biodeterior Biodegrad* **144** (2019), doi:10.1016/j.ibiod.2019.104722

34. Beale, D. J. et al. Application of metabolomics to understanding biofilms in water distribution systems: a pilot study. *Biofouling* **29**, 283–294 (2013), doi:10.1080/08927014. 2013.772140

35. Bonifay, V. et al. Metabolomic and metagenomic analysis of two crude oil production pipelines experiencing differential rates of corrosion. *Front Microbiol* **8** (2017), 99, doi:10.3389/fmicb.2017.00099

36. Su, H., Mi, S. F., Peng, X. W. & Han, Y. The mutual influence between corrosion and the surrounding soil microbial communities of buried petroleum pipelines. *Rsc Adv* **9**, 18930–18940 (2019), doi:10.1039/c9ra03386f

37. Celikkol-Aydin, S. et al. 16S rRNA gene profiling of planktonic and biofilm microbial populations in the Gulf of Guinea using Illumina NGS. *Mar Environ Res* **122**, 105–112 (2016), doi:10.1016/j.marenvres.2016.10.001

38. Salerno, J. L., Little, B., Lee, J. & Hamdan, L. J. Exposure to crude oil and chemical dispersant may impact marine microbial biofilm composition and steel corrosion. *Front Mar Sci* **5** (2018), doi:ARTN 196 10.3389/fmars.2018.00196

39. Suflita, J. M., Aktas, D. F., Oldham, A. L., Perez-Ibarra, B. M. & Duncan, K. Molecular tools to track bacteria responsible for fuel deterioration and microbiologically influenced corrosion. *Biofouling* **28**, 1003–1010 (2012), doi:10.1080/08927014.2012.723695

40. Elifantz, H., Horn, G., Ayon, M., Cohen, Y. & Minz, D. Rhodobacteraceae are the key members of the microbial community of the initial biofilm formed in Eastern Mediterranean coastal seawater. *FEMS Microbiol Ecol* **85**, 348–357 (2013), doi:10.1111/1574-6941.12122

41. Thomsen, U. S. & Oehler, M. C., A combination of qPCR, RT-qPCR and NGS provides a new tool for analyzing MIC risk in pipelines. *CORROSION 2018*, Phoenix, Arizona.

42. Moura, V. et al. The influence of surface microbial diversity and succession on microbiologically influenced corrosion of steel in a simulated marine environment. *Arch Microbiol* **200**, 1447–1456 (2018), doi:10.1007/s00203-018-1559-2

43. Phan, H. C., Wade, S. A. & Blackall, L. L. Is marine sediment the source of microbes associated with accelerated low water corrosion? *Appl Microbiol Biot* **103**, 449–459 (2019), doi:10.1007/s00253-018-9455-x

44. Gulhane, M., Pandit, P., Khardenavis, A., Singh, D. & Purohit, H. Study of microbial community plasticity for anaerobic digestion of vegetable waste in Anaerobic Baffled Reactor. *Renew Energ* **101**, 59–66 (2017), doi:10.1016/j.renene.2016.08.021

45. Huber, R. et al. *Thermotoga maritima* sp. nov. represents a new genus of unique extremely thermophilic eubacteria growing up to 90°C. *Arch Microbiol* **144**, 324–333 (1986), doi:10.1007/Bf00409880

46. Bruijnen, P., van Strien, W. & Doddema, S. Integrated approach toward diagnosing microbiologically influenced corrosion in the petroleum industry. *Spe Prod Oper* **35**, 37–48 (2020).

47. Nemati, M., Jenneman, G. E. & Voordouw, G. Impact of nitrate-mediated microbial control of souring in oil reservoirs on the extent of corrosion. *Biotechnol Prog* **17**, 852–859 (2001).
48. Lahme, S. et al. Metabolites of an oil field sulfide-oxidizing, nitrate-reducing Sulfurimonas sp. cause severe corrosion. *Appl Environ Microbiol* **85** (2019), doi:10.1128/AEM.01891-18
49. Usher, K., Kaksonen, A. & MacLeod, I. Marine rust tubercles harbour iron corroding archaea and sulphate reducing bacteria. *Corros Sci* **83**, 189–197 (2014).
50. Vigneron, A. et al. Complementary microorganisms in highly corrosive biofilms from an offshore oil production facility. *Appl Environ Microbiol* **82**, 2545–2554 (2016), doi:10.1128/AEM.03842-15
51. Satoh, H., Odagiri, M., Ito, T. & Okabe, S. Microbial community structures and in situ sulfate-reducing and sulfur-oxidizing activities in biofilms developed on mortar specimens in a corroded sewer system. *Wat Res* **43**, 4729–4739 (2009), doi:10.1016/j.watres.2009.07.035
52. Okoro, C. et al. Comparison of microbial communities involved in souring and corrosion in offshore and onshore oil production facilities in Nigeria. *J Ind Microbiol Biotechnol* **41**, 665–678 (2014), doi:10.1007/s10295-014-1401-z
53. Liu, T., Cheng, Y. F., Sharma, M. & Voordouw, G. Effect of fluid flow on biofilm formation and microbiologically influenced corrosion of pipelines in oilfield produced water. *J Petrol Sci Eng* **156**, 451–459 (2017).
54. Rajala, P. et al. Influence of carbon sources and concrete on microbiologically influenced corrosion of carbon steel in subterranean groundwater environment. *Corrosion-Us* **72**, 1565–1579 (2016), doi:10.5006/2118
55. Rajeev, M., Sushmitha, T. J., Toleti, S. R. & Pandian, S. K. Culture dependent and independent analysis and appraisal of early stage biofilm-forming bacterial community composition in the Southern coastal seawater of India. *Sci Total Environ* **666**, 308–320 (2019), doi:10.1016/j.scitotenv.2019.02.171
56. Nasser, B. et al. Characterization of microbiologically influenced corrosion by comprehensive metagenomic analysis of an inland oil field. *Gene* **774**, 145425 (2021), doi:10.1016/j.gene.2021.145425
57. Gao, P. K. et al. Dynamic processes of indigenous microorganisms from a low-temperature petroleum reservoir during nutrient stimulation. *J Biosci Bioeng* **117**, 215–221 (2014), doi:10.1016/j.jbiosc.2013.07.009
58. Beale, D. J., Dunn, M. S., Morrison, P. D., Porter, N. A. & Marlow, D. R. Characterisation of bulk water samples from copper pipes undergoing microbially influenced corrosion by diagnostic metabolomic profiling. *Corros Sci* **55**, 272–279 (2012), doi:10.1016/j.corsci.2011.10.026
59. Brauer, J. I. et al. Mass spectrometric metabolomic imaging of biofilms on corroding steel surfaces using laser ablation and solvent capture by aspiration. *Biointerphases* **10** (2015), doi:Artn 01900310.1116/1.4906744
60. Duncan, K. E. et al. Biocorrosive thermophilic microbial communities in Alaskan North Slope oil facilities. *Environ Sci Technol* **43**, 7977–7984 (2009), doi:10.1021/es9013932

Section IV

Subsurface Reservoir
Microbiome and Hydrocarbon
Degradation

6 The Ecological Interactions of Microbial Co-occurrence in Oil Degradation

The Intra- and Interspecies Relationships in Hydrocarbon Metabolism

Luciano Procópio
National Institute of Technology, Rio de Janeiro, Brazil

6.1 INTRODUCTION

Soil-inhabiting microorganisms compose a considerable part of the biomass present in this compartment, representing around 1,000 kg of carbon mass per hectare. A rough estimate indicates that more than 10 million different species of bacteria, archaea, fungi, viruses, and eukaryotic microorganisms may inhabit the soil compartment, although only a few hundred thousand have been identified (Peay et al. 2016). This high diversity can be further evidenced in the assessment of the presence of 10e9 of individual bacterial and archaeal taxa described in just one gram of soil. This estimate does not consider other microorganisms, which may be represented in about 200 million fungi, trillions of viruses, and ten thousand species of protists. In fact, the abundance of different taxa is not evenly distributed in different soils, and also in different compartments or layers of a specific soil. Usually, bacteria and fungi are dominant in the analysis of soil microorganisms (Arraes et al. 2005; de Souza and Procópio 2021). These two groups are commonly represented at 10e2 to 10e4 times more than other taxa of archaea, viruses, and protists.

Another fact that draws attention is the fluctuation in the number of microbial individuals and their different taxa that vary in a small portion of the soil. This considerable difference in a reduced space is due to the inherent characteristic of the soil, where different microhabitats can be described. The heterogeneous architecture of the soil allows the existence of microenvironments with a wide range of microspaces that produce microaerobic and anaerobic compartments, allowing the presence or absence of water, influencing pH levels and temperatures (de Souza and Procópio 2022).

DOI: 10.1201/9781003287056-10

123

The characteristics present in these three-dimensional compartments influence the electron transfer, oxidation, and reduction of compounds buried underground. These differences are more pronounced when the microbial composition is considered at a global level, where macro-climatic factors are considered. This indicator seems to be related to the availability of carbon in the evaluated soil. Wetter soils also represent an important predictor of local microbial biomass. Soils with higher water availability typically have higher microbial content than dry environments. Indeed, equatorial biomes are more abundant in numbers and microbial diversity when compared to drier environments. In the same way that the environment directly influences the microbial structures present in the soil, biochemical processes conduced by microorganisms change different aspects of the soil, underground water, and even atmospheric compartments.

Soil-inhabiting microorganisms strongly influence the structures of the environment around them, altering the chemical composition of the soil, its texture, water availability, and the concentration of gases present in microcompartments and even the atmospheric concentrations of CO_2 and nitrogen compounds (Garcia and Procópio 2020). The explanation of these influences in different aspects of the formation and constitution of the soil lies in the fact that microbial agents participate in several terrestrial biogeochemical processes. Microorganisms are responsible for the decomposition of organic matter, which directly affects the concentrations of nutrients, such as carbon, nitrogen, and phosphorus. Furthermore, the input of inorganic/organic matter from anthropogenic sources also has repercussions on soil microbial populations, and these, in turn, influence the fate of allochthonous components. Some of these reactions convert recalcitrant and/or harmful components into biodegradable products or even assimilable to micro- and macroorganisms (Sokol et al. 2022). All these processes driven by microorganisms result in the influence of soil fertility and the sequestration or emission of greenhouse gases, which directly influence the health of this environment, as well as the global climate.

Throughout the development and anthropogenic growth, and consequently, of different economic activities, numerous changes have been added to the soil and its inhabiting microbes (Procópio and Barreto 2021). The input of fertilizer compounds, chemicals, heavy metals, and hydrocarbons exerted changes on the communities of indigenous microorganisms. Among these pollutants, hydrocarbons from the oil industry severely impact countless soil inhabitants. Furthermore, the entry of crude oil into marine environments ends up reaching the ocean floor and/or coastal areas, causing profound environmental questions in these compartments. In general, the lighter portions of the oil components are evaporated into the atmosphere or readily assimilated by microorganisms. However, the heavier portions, owing to their chemical stable nature, are more difficult to degrade by microorganisms, at first. In addition, many components adhere tightly to soil particles, hampering the microbial access to the droplet's oil. As a consequence, in addition to being more persistent in the environment, another concern is that these compounds are more toxic to living beings, many of which have mutagenic and/or carcinogenic properties.

In the past decades, many studies have been conducted to describe the interactions between oil components and microbial communities. Numerous microorganisms were characterized as oil-degrading in distinct environments and over diverse conditions (Procópio 2020). The direct role of microbes in the assimilation of petroleum components has always been from isolated labor conditions. Commonly, from a microorganism isolated in the laboratory, growth tests were carried out with the concomitant degradation of the oil compounds. Since these results, genetic studies have been conducted to describe the genes responsible for converting oil compounds into chemical energy or cellular components (Procópio et al. 2012). Despite the knowledge raised in the past decades have greatly contributed to the understanding of the genetics and physiology of oil degraders, experiments with pure cultures or microbial consortia proved to be limiting in providing a broad and in-depth view. Currently, with the advent of metagenomic sequencing, it is possible to describe entire communities of environmental samples, without the impeding need for cultivation under laboratory conditions. Changes in the microbial structures present in the contaminated environment, in the face of pre-contamination conditions, allowed us to depict a broader scenario of microbial participants in oil degradation.

However, the presence or absence of microorganisms in a contaminated environment does not necessarily confirm their role as an oil-degrading species. Many of these microbes can benefit from metabolites or intermediates in the degradation of hydrocarbons, without their presence or absence having a consequence on the reduction of the levels of these components in the soil (Procópio 2022). Nowadays, the occurrence and active co-participation of microbes in different levels of petroleum degradation in the soil environment has been presented as a new frontier to be crossed. Studies on the interactions between microorganisms in an oil-contaminated site have demonstrated that there is a complex network of work throughout the process of remediation (Figure 6.1). These networks have the participation of different microbial species, which have their population levels varying according to the presence of different fractions of the oil, which also change during the biodegradation process. Here are presented the main concepts about the networks of microbial interactions and the description of how the co-occurrence modifies the processes of biodegradation of petroleum compounds. Also is addressed state-of-the-art studies on the main methodologies used in the understanding of the work network in an environment contaminated by oil.

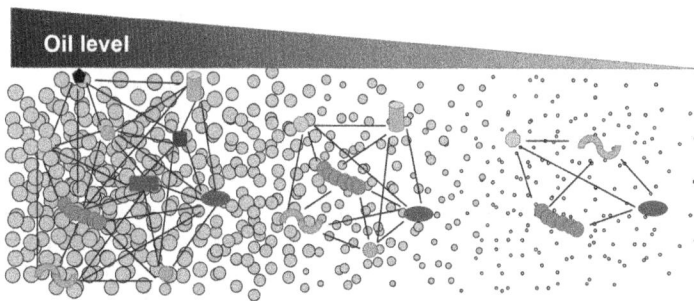

FIGURE 6.1 Schematic illustration suggesting changes in the microbial community during oil degradation over time. Ecological succession is accompanied by the establishment of new networks between microorganisms that participate in the bioremediation process.

6.2 THE MICROBIAL INTERACTIONS INTRA- AND INTERSPECIES RELATIONSHIPS

Microorganisms do not live isolated in the environment, especially in soil environments. In fact, bacteria interact through quorum-sensing processes, where they employ chemical signals such as siderophores, secondary metabolites, and effectors. In these compartments, bacteria, viruses, fungi, archaea, and protists establish positive and negative relationships through complex networks (Deng et al. 2021). These networks can be configured in different ways through mechanisms of negative and positive interactions. The results of these interactions have a profound contribution to changing bacterial communities, as well as aspects of the physical-chemical structures of the soil. Examples of negative interactions can be amensalism, predation, and parasitism. In an amensalism interaction, one partner is harmed to any advantage to the other. This can occur for example when metabolites of a certain bacterial species increase the conditions of the surrounding environment, for example, lactobacillus-induced changes in pH levels. As a result, species sensitive to the decrease in pH in the medium are harmed by the acidity around them. The predation is a disharmonious interspecific ecological relationship where only one species is benefited. In this type of interaction, one species (preda) kills and kills each other (prey). The presence or removal of species at the top of the food web causes important changes in other species.

On the other hand, beneficial interactions are also important in the soil compartment, influencing bacterial species and soil structure. Bacteria from the same species or from different taxonomic groups can cooperate to build biofilms. These structures modify microenvironments, confer a competitive advantage in the acquisition of food or in antibiotic resistance to its members. Another aspect of positive relationships between species is cross-feeding mutualism or syntrophy. In these interspecific relationships, the exchange of metabolite products occurs between the two, both being favored. In contrast, when only one species takes advantage of the metabolic products of another species, commensalism is characterized. In this relationship, one partner benefits without helping or harming the other species. Commensalism is often found in complex networks of complex hydrocarbon degradation, where commensals benefit from compounds produced by other members of the population/community. Ultimately, studying and describing these different intra- and interspecies relationships are a challenge to current methodologies, even the most modern ones such as metagenomic sequencing, flow cytometry, and metabolomic approaches. Establishing models of work network relationships must take into account various aspects of the soil compartment as well as the disturbances exerted on it, which are reflected in the results of the analyses.

The description of microbial association relationships is known as network inference, and essentially takes into account the presence-absence, or co-occurrence, data between microbes. The description of these interactions allows evaluating different hypotheses of relevance in the understanding of microbial ecology. Another interesting aspect is the identification of a specific bacterium with a central position in microbial networks, called the "microbio hub". Depending on the conditions found at the site, the hub bacterium can considerably influence the entire community present at the site, disturbing the abundance and diversity of other microbes, inter- and intra-kingdom. In addition, key species play a role in changes in the metabolism of the microbial community in general, influencing their functions in processes of availability of nutrients, metabolites, and compounds used in competitions, such as bactericides and antibiotics. As mentioned above, many of these bacteria interact with each other to form complex biofilm structures. Bacteria of different species are associated with surface and deep soil compartments. Polymeric structures synthesized by bacteria colonizing biofilms allow numerous ecological advantages, such as resilience to sudden changes in environmental conditions, defense against antibiotics, and cooperation in chain metabolic processes.

The study and description of these inferences are established through computer science and were boosted with the advent of large-scale sequencing. Computational models allow designing networks in omics science areas, for example, genomics, proteomics, and metabolomics. However, the construction of models of complex networks is also widespread in studies of interactions between species in the environment, especially in the different ecological relationships described above. In general, networks are based on the quantification of patterns of co-occurrence/mutual exclusion between two or more species in the studied site. Notwithstanding, at first glance, the soil appears to be a static environment, in fact, this behavior is highly dynamic, with fluctuations in several physical-chemical and biological aspects. In addition, anthropogenic inputs also profoundly influence its dynamics. These disturbances result in nuances and pitfalls that immediately change the relationships of microbial networks. In addition, after evaluating all the possible combinations, the "strengths" of these connections are determined by means of quantification of the similar relationships, and from these data, the work networks are designed.

Another challenge faced in the predictions of network models is the fact that the relationships between two species, or just the presence/occurrence, do not describe more complex forms of interactions in the environment. In fact, most ecological relationships depend on several species in different ecological relationships between them. Thus, methodologies should consider methodologies such as regression and multivariate analysis. For example, in the regression, the abundance of a given species is considered from the measures of the abundances of other species together. This technique is the most used due to its application, although its analysis of results can be difficult to interpret. Another methodology used in the description of complex relationships is association rule mining. This technique considers all relationships supported by a presence-absence or co-occurrence dataset in order to find meaningful rules. Thus, during rule mining, it is possible to conclude that the presence of a certain species is established due to the presence and absence of two other species, respectively. In this way, all rules are determined by enumerating all possible sets of taxa up to a certain size. This allows establishing all possible rules for each set in a certain evaluated condition. In the second moment of the analysis, the use of different filters together is necessary in order to correct several tests and retain only significant rules. This last step constitutes the biggest challenge in this methodology of analysis of ecological interactions.

6.3 THE RELATIONSHIP BETWEEN OIL SPILLS ON SOIL AND ITS MICROBIOME

Crude oil is a complex mixture of hydrocarbons, and other components such as sulfur, nitrogen, and oxygen, which represent less than 3% (v/v) of its components, in addition to elements such as some organic and organometallic constituents, especially the complex ones (1%) (Maia et al. 2019; Thomas et al. 2021). This smaller portion brings together an immense complexity of petroleum compounds, which can be evidenced by the identification of more than 1,700 different chemical components. These data obtained through ultra-high-resolution mass spectrometry make crude oil one of the most complex organic substances present on Earth. The origin of the oil must also be considered, as its composition varies considerably in relation to the site of its formation. Petroleum can be classified into light, medium, and heavy according to the levels of alkanes or paraffin, and compounds of mono- and polyaromatic hydrocarbons (Procópio et al. 2013; SadrAzodi et al. 2019).

Hydrocarbons can be divided into three broad groups: aliphatic, which include open-chain compounds, alicyclic, or cyclans, where their atoms are arranged in a ring structure, and aromatic hydrocarbons, which appear on one or more benzene rings. Environmental disturbances related to the activities of the oil industry arise mainly from the activities of extraction, transportation, refining and use of its derivatives, with only 10% of everything attributed to accidental spills. However, oil spills can severely affect different environments, altering local physical-chemical aspects, destroying biota, incorporating toxic compounds into the environment, and impacting the local economy (McGenity et al. 2012; Procópio 2021). The extent of damage caused by the entry of oil into the environment will depend on its chemical composition, soil characteristics, and the indigenous microbial community. Among organic compounds, polyaromatic hydrocarbons (PAH) are an additional concern as they are more toxic, many of which have mutagenic and/or carcinogenic properties, as well as being more persistent in the environment. Although PAHs are relatively persistent molecules, numerous microorganisms are able to use them as a source of energy and carbon, including bacteria, fungi, and archaea (Álvarez-Barragán et al. 2022).

It is now accepted that hydrocarbon-degrading microorganisms are widely distributed in the environment, regardless of whether their habitat has been previously exposed to human contamination (Ławniczak et al. 2020). Despite this, microorganisms that have genes associated with hydrocarbon degradation represent only 1% of the total population. This low representativeness can be explained because hydrocarbons occur in the environment as a result of natural discharges or biosynthesis by various organisms, which have a low concentration of "natural" hydrocarbons, in addition to the fact that other energy sources are more prevalent in the environment. When the concentration of hydrocarbons increases dramatically, as a result of oil spills, for example, the abundance of hydrocarbon-degrading microorganisms grows exponentially within a few days (Yakimov et al. 2007). However, the exponential growth of the population of hydrocarbon degraders results in the decrease of essential nutrients, nitrogen, and phosphorus, which can result in the decline of microbial proliferation, limiting the processes of biodegradation of hydrocarbons (Táncsics et al. 2015).

The relationship between oil contamination and microorganisms presents in different environments has been known for a long time. The first record of oil-degrading bacteria was more than a century ago. Currently, more than 80 bacterial genera have been described as oil degraders, and the number of fungi is similarly high. In addition, cyanobacteria and algae genera capable of degrading or transforming hydrocarbon compounds (Jiao et al. 2016) were identified. Microorganisms are able to degrade PAH compounds through aerobic or anaerobic metabolism. Under aerobic conditions, the initial step is the chemical activation of PAHs by the enzymatic activity of ring hydroxylating dioxygenases (RDH) (Kaplan and Kitts 2004). In contrast, in the anaerobic catabolism of PAHs, microorganisms use reductive reactions, which are normally coupled with the reduction of nitrate, sulfate, manganese, and ferric iron (Stauffert et al. 2014). However, very little is known about the metabolism used in degradation under anaerobic conditions and, so far, two main mechanisms have been described for the activation of PAHs compounds, both of which are best described for naphthalene (Sakshi and Haritash 2020). The two mechanisms described involve direct carboxylation to naphthoic acid, or methylation to 1-methylnaphthalene followed by the addition of fumarate to naphthyl-2-methylsuccinate, which is catalyzed by naphthylmethylsuccinate synthase (NMS) (Dhar et al. 2020).

Due to the characteristics of hydrocarbons that strongly adhere to solid surfaces, such as soils, the removal and/or degradation of these compounds is an obstacle for the non-biological technologies used. These environmental pollutant remediation technologies that rely on non-biological approaches are often cost-prohibitive and generally do not deliver the expected results, which can lead to the formation of toxic intermediates with the possibility of their perpetuation on site and entry into the food chain or contamination of aquifers (Lovley and Lloyd 2000). Studies of petroleum biodegradation using microorganisms have become the focus of research in several areas of biology, known as bioremediation (Thomas et al. 2020). Bioremediation can be a spontaneous or accelerated process, where, through biological procedures, fundamentally microbiological, the degradation or transformation of contaminants into less harmful or non-toxic forms occurs. Although yeasts and plants have the ability to detoxify, such as mineralizing, transforming, or immobilizing pollutants, bacteria play a crucial role. This ability is mainly attributed to the ability of bacteria to use alternative electron acceptors for oxygen, in addition to their high growth rate, high surface-to-volume ratio, and genomic plasticity (Dias et al., 2004). The successful application of bioremediation technologies in areas contaminated by oil requires knowledge of the parameters that affect the characteristics of the pollutant since the biodegradability of the oil by microorganisms is inherent to its composition.

In the course of studies on the ability to degrade petroleum compounds by microorganisms, numerous bacteria with the capacity to assimilate petroleum compounds have been used as mold organisms. These isolates facilitate the understanding of genetics, genes, and their regulators, and the metabolic pathways involved in hydrocarbon degradation. However, due to the inability of most microorganisms to grow under laboratory conditions, many species are not identified as degrading oil. Thus, the number of possible hydrocarbon degraders is higher than currently known. This barrier has been overcome in recent decades with the advancement of sequencing techniques. The possibility of sequencing DNA directly from environmental samples made it possible to identify microorganisms from taxonomic analyses of the 16S rRNA genes in bacteria and 18S rRNA in fungi.

Thus, it is possible to phylogenetically characterize the microorganisms that make up a complex local microbial community. This discovery was a breakthrough in the field of bioremediation, where through the analysis of these rRNA sequences of microorganisms present in contaminated sites, it was possible to determine the species of microorganisms associated with the bioremediation process. Indeed, monitoring the marker genes of microbial species allows the identification of new species and/or the description of common patterns that are associated with biodegradation. Oxygen levels, the addition of nutrients such as nitrogen and phosphorus, temperature, soil pH, the chemical composition of the oil, and their availability considerably influence the effectiveness of biodegradation processes. The establishment of standards, considering these variables, in turn, will help to develop new bioremediation tools.

Several approaches have been employed to decipher the relationship of a bacterium to the degradation of hydrocarbons in soil and marine sediments. Omics techniques have been widely used in studies of communities present in contaminated sites or micro and mesocosm experiments simulating oil spills. Despite the difficulties imposed by the complexity of these sites and the establishment of the presence of the bacteria as a probable oil degrader, numerous genera have been described repeatedly. *Alcanivorax*, *Bacillus*, *Dietzia*, *Rhodococcus*, *Pseudomonas*, *Mycobacterium*, *Gordonia*, *Acinetobacter*, *Serratia*, *Nocardia*, and *Halomonas* are genera involved in the degradation of different hydrocarbon chains. Although this list of representatives is underestimated, certainly numerous other genera should be described in it, *Mycobacterium*, *Rhodococcus*, and *Pseudomonas* are dominant in different conditions of low or high concentration of hydrocarbons, inclusive in the presence of PAH contaminants.

To understand the relationship between bacteria and oil degradation in soil, one of the most significant approaches is the study of molecular ecological networks. However, a challenge in this approach is that the properties of networks change considerably when samples from different conditions or contaminated sites are analyzed. This concern becomes more evident when changes are noticed in the general structures of the network, indicating a change in the organization of the evaluated microbial communities (Faust and Raes 2012). In addition, the entry of inputs, such as the entry of oil from anthropogenic spills, results in a rise in the levels of complexity of local communities, followed by their structural instability. One of the explanations is the fact that this imbalance is the result of microbial functional instability. These changes were evidenced in a study that showed that the entry of carbon source by oil spill resulted in an imbalance in the C:N ratio, since the carbon was degraded by the microbial community. Indeed, many studies have suggested that oil contamination induces a loss in microbial diversity and changes in the patterns of community structures (Hazen et al. 2010; Lu et al. 2012). These changes severely alter the potential roles of microbial species and the ecological functions of communities. An important and challenging step in the prediction of networks is the identification of key populations in a community. This step becomes even more challenging when considering that most microbes are not cultivable under laboratory conditions. Thus, the task of building a network is dependent on the results of indirect approaches, for example, metagenomic and metatranscriptomic sequencing.

Analysis of ecological collections during the petroleum classification processes previously presented that make the composition of hydrocarbons of microbial groups of specification. For example, the genera *Thalassolituus* and *Roseobacter* were dominant in the presence of alkanes of sizes between 12 and 32 carbons, while *Alcanivorax* prevailed almost exclusively in the presence of branched alkanes (Yu et al. 2011). A survey of microbial contamination in sediments compared communities in a pre-contamination phase and after prolonged exposure to oil. The results showed an increase in the presence of the genera *Alcanivorax, Borrelia, Spirochaeta, and Micavibrio* when compared to the pre-oil spill period. After a longer period, most other genera showed a significant decline, with notable exceptions of *Marinobacter* and *Parvibaculum* remaining in high representation over time (Rodriguez-R et al. 2015; Zhen et al. 2021). In a survey of the microbial community involved in the degradation of PAH in an abandoned oil field with a long history of contamination, the main groups known as degraders were participants in the ecological networks involved in the oil degradation (Geng et al. 2022b). Numerous known degrading representatives have been described from the phyla Proteobacteria, Firmicutes, Bacteroidota, and Actinobacteria, with emphasis on the genera *Marinobacter, Halomonas, Neocosmospora, and Pseudomonas* among bacteria, and *Halogranum* and *Haladaptatus* among archaea. In this study also, when evaluating the participation of fungi in ecological interactions, fungal genera *Cephalotrichum, Chrysosporium*, and *Acremonium* of Ascomycota were described.

Despite studies on the participation of microbes in bioremediation processes involving all the complexity of petroleum hydrocarbons, the co-occurrence relationships are more evident during the degradation of PAHs compounds. This is due to the fact that PAHs are more recalcitrant and more difficult to be degraded. Normally, a single population of bacteria is not able to fully degrade PAHs. The degradation of PAHs can involve different levels, where the metabolite generated by one species will be the substrate of another species present at the same contaminated site. In general, regardless of the methodology used in the construction of networks, the indication of key bacteria is crucial in the construction of relationships between community participants. Several studies have been conducted to identify key populations in soils impacted by the presence of PAHs, and numerous surveys show the participation of representatives of different taxonomic groups in the degradation of PAHs. These representations of microbial populations change over time or changes in environmental conditions during the degradation process. These findings can be evidenced in a study that described the distribution of the indigenous microbial community in deep contaminated soils (Geng et al. 2020). The results indicated the bacteria SAR202 clade, *Thermoanaerobaculum, Nitrospira*, and *Xanthomonadales* as key species in the network. An interesting point was that the edaphic properties (nutrients and pH) had greater significance in the correlations than the concentration of PAHs in the soil. The large and long-lasting oil accident on the Deepwater Horizon (DH) that occurred in 2010 made it possible to evaluate different bioremediation methodologies and to assess microbial responses to hydrocarbon pollution from various aspects (Bacosa et al. 2018; King et al. 2015; Liu et al. 2017). For example, when microbial communities were compared at the time of pre-contamination with communities right after the oil spill, a high predominance of members of the *Alcanivorax*, *Borrelia*, *Spirochaeta*, *Micavibrio*, and *Bacteroides* genera were evidenced. However, after months of contamination, most bacterial populations showed a significant decline, with notable exceptions for the genera *Marinobacter* and *Parvibaculum*, which remained at high levels of representation over time (Rodriguez-R et al. 2015).

The comparison between undisturbed soil and soil contaminated by petroleum compounds also describes different scenarios that allow the identification of key degrading species of hydrocarbons. In a study evaluating patterns of soil microbial co-occurrence driven by oil pollution, bacteria from oil-contaminated soil were shown to be less connected than those in the control group (Huang et al. 2021). The numbers of connections were also lower in contaminated soil when compared to undisturbed soil, suggesting that oil contamination reduced the correlations between bacteria in the bacterial co-occurrence network. Although the key representatives present in the contaminated soil are dominated by microorganisms known as oil degraders, for example, *Pseudarthrobacter, Alcanivorax, Sphingomonas, Chromohalobacter*, and *Nocardioides*, reduced connections may be unfavorable to maintaining the stability of the microbial community. The reshaping of microbial communities by PAH could also be evaluated in a study on the vertical variation in soil, shallow soil, and deep soil of PAH concentrations (Xu et al. 2022). Although the results show differences in bacterial diversity and abundance at the two different depths, some known oil degradation groups were dominant. Members of Proteobacteria, Actinobacteria, and Acidobacteria were described in all soils, but their relative abundances differed significantly. The main genera reported were *Pseudomonas, Ramlibacter, Bacillus, Rhodococcus, Sphingomonas*, and *Acinetobacter*, which were also described in similar studies (Hamamura et al., 2006). These surveys show that although generalist microorganisms are widely distributed in oil-contaminated soils, key species play a fundamental role in the remediation of these contaminated environments. Furthermore, a closer look at the data showed that *Pseudomonas* established relationships with other different genera, such as *Ramlibacter* and *Pseudonocardia*, which were not non-dominant but played an important role in the degradation of PAHs.

When the structures of microbial communities in soils with different levels of pollution by PAHs were evaluated, it was described how the cooperative relationship between community members is more important than the abundance of their representatives (Sazykina et al. 2022). In this survey, in all the studied soils, there was a predominance of the phyla Actinobacteria and Proteobacteria, although with the increase in their pollution representatives of Actinobacteria were negatively affected in their abundance. On the other hand, members of Proteobacteria were increased in their representation with the concomitant increase in PAHs levels. Analysis of the levels of disturbance in soil microbial structures by the addition of total petroleum hydrocarbons (TPH) showed the interactions that occur between bacterial and fungal communities (Geng et al. 2022a). The relationships evaluated between communities showed positive and negative interactions between bacterial and fungal taxa at both levels of contamination by TPHs. The results described that members of Actinobacteria and Firmicutes were dominant in soils lightly contaminated with TPH, with a decrease in fungal species. On the other hand, a profound change took place as the soil was severely impacted by TPHs. In this condition, there was a predominance of representatives of Proteobacteria alongside fungal species of Ascomycota, indicating that its members are more tolerant to oil contamination than the other phyla.

6.4 CONCLUSION

The relationships between microorganisms, whether positive or negative, are shown to have considerable influences on hydrocarbon degradation processes. These interactions could be evidenced in the assimilation of aliphatic compounds over time along with changes in microbial representatives. However, the impacts of networks are more evident during the degradation of more recalcitrant compounds, such as PAHs. These compounds, because they are more difficult to be metabolized by a single species, normally need the participation of different microbial groups for their total metabolism to occur. Studies that describe these metabolic relationships have increasingly been considered in the descriptions of these scenarios, employing methodologies that designed the networks or co-occurrences of microorganisms directly involved. However, challenges still exist in surveying populations participating in oil degradation. The co-occurrence of microbes in the degradation of PAHs is constantly influenced by the chemical nature of the PAH itself, soil conditions, and the composition of the indigenous microbial community. New and constant studies still need to be conducted under different conditions and at different sites disturbed by oil. These surveys will make it possible to discover which are the key microorganisms that play a role in the assimilation of oil and that can be used as tools in more effective bioremediation techniques.

REFERENCES

Álvarez-Barragán J, Cravo-Laureau C, Duran R (2022). Fungal-bacterial network in PAH-contaminated coastal marine sediment. *Environ Sci Pollut Res Int* 29(48):72718–72728. doi: 10.1007/s11356-022-21012-4

Arraes FB, Benoliel B, Burtet RT, Costa PL, Galdino AS, Lima LH, Marinho-Silva C, Oliveira-Pereira L, Pfrimer P, Procópio-Silva L, Reis VC, Felipe MS (2005). General metabolism of the dimorphic and pathogenic fungus Paracoccidioides brasiliensis. *Genet Mol Res* 4(2):290–308.

Bacosa HP, Erdner DL, Rosenheim BE, Shetty P, Seitz KW, Baker BJ, Liu Z (2018). Hydrocarbon degradation and response of seafloor sediment bacterial community in the northern Gulf of Mexico to light Louisiana sweet crude oil. *ISME J* 12(10):2532–2543. doi: 10.1038/s41396-018-0190-1

de Souza LC, Procópio L (2021). The profile of the soil microbiota in the Cerrado is influenced by land use. *Appl Microbiol Biotechnol* 105(11):4791–4803. doi: 10.1007/s00253-021-11377-w

de Souza LC, Procópio L (2022). The adaptations of the microbial communities of the savanna soil over a period of wildfire, after the first rains, and during the rainy season. *Environ Sci Pollut Res Int.* 29(10):14070–14082. doi: 10.1007/s11356-021-16731-z

Deng Y, Jiang YH, Yang Y, He Z, Luo F, Zhou J (2021). Molecular ecological network analyses. *BMC Bioinform* 13:113. doi: 10.1186/1471-2105-13-113

Dhar N, Sarangapani S, Reddy VA, Kumar N, Panicker D, Jin J, Chua NH, Sarojam R (2020). Characterization of a sweet basil acyltransferase involved in eugenol biosynthesis. *J Exp Bot* 71(12):3638–3652. doi: 10.1093/jxb/eraa142

Dias AA, Bezerra RM, Pereira AN (2004). Activity and elution profile of laccase during bio-logical decolorization and dephenolization of olive mill wastewater. *Bioresour Technol* 92(1):7–13. doi: 10.1016/j.biortech.2003.08.006

Faust K, Raes J (2012). Microbial interactions: from networks to models. *Nat Rev Microbiol* 10(8):538–550. doi: 10.1038/nrmicro2832

Garcia M, Procópio L (2020). Distinct profiles in microbial diversity on carbon steel and different welds in simulated marine microcosm. *Curr Microbiol* 77(6):967–978. doi: 10.1007/s00284-020-01898-4

Geng P, Ma A, Wei X, Chen X, Yin J, Hu F, Zhuang X, Song M, Zhuang G (2022a) Interaction and spatio-taxonomic patterns of the soil microbiome around oil production wells impacted by petroleum hydrocarbons. *Environ Pollut* 307:119531. doi: 10.1016/j.envpol.2022.119531

Geng S, Cao W, Yuan J, Wang Y, Guo Y, Ding A, Zhu Y, Dou J (2020). Microbial diversity and co-occurrence patterns in deep soils contaminated by polycyclic aromatic hydrocarbons (PAHs). *Ecotoxicol Environ Saf* 203:110931. doi: 10.1016/j.ecoenv.2020.110931

Geng S, Xu G, You Y, Xia M, Zhu Y, Ding A, Fan F, Dou J (2022b) Occurrence of polycyclic aromatic compounds and interdomain microbial communities in oilfield soils. *Environ Res* 212(Pt A):113191. doi: 10.1016/j.envres.2022.113191

Hamamura N, Olson SH, Ward DM, Inskeep WP (2006). Microbial population dynam-ics associated with crude-oil biodegradation in diverse soils. *Appl Environ Microbiol* 72(9):6316–6324. doi: 10.1128/AEM.01015-06

Hazen TC, Dubinsky EA, DeSantis TZ, Andersen GL, Piceno YM, Singh N, Jansson JK, Probst A, Borglin SE, Fortney JL, Stringfellow WT, Bill M, Conrad ME, Tom LM, Chavarria KL, Alusi TR, Lamendella R, Joyner DC, Spier C, Baelum J, Auer M, Zemla ML, Chakraborty R, Sonnenthal EL, D'Haeseleer P, Holman HY, Osman S, Lu Z, Van Nostrand JD, Deng Y, Zhou J, Mason OU (2010). Deep-sea oil plume enriches indige-nous oil-degrading bacteria. *Science* 330(6001):204–208. doi: 10.1126/science.1195979

Huang L, Ye J, Jiang K, Wang Y, Li Y (2021). Oil contamination drives the transformation of soil microbial communities: co-occurrence pattern, metabolic enzymes and culturable hydrocarbon-degrading bacteria. *Ecotoxicol Environ Saf* 225:112740. doi: 10.1016/j.ecoenv.2021.112740

Jiao S, Liu Z, Lin Y, Yang J, Chen W, Wei G (2016). Bacterial communities in oil contaminated soils: biogeography and co-occurrence patterns. *Soil Biol Biochem* 98:64–73.

Kaplan CW and Kitts CL (2004). Bacterial succession in a petroleum land treatment unit. *Appl Environ Microbiol* 70(3):1777–1786.

King GM, Kostka JE, Hazen TC, Sobecky PA (2015). Microbial responses to the Deepwater Horizon oil spill: from coastal wetlands to the deep sea. *Ann Rev Mar Sci* 7:377–401. doi: 10.1146/annurev-marine-010814-015543

Ławniczak Ł, Woźniak-Karczewska M, Loibner AP, Heipieper HJ, Chrzanowski Ł (2020). Microbial degradation of hydrocarbons-basic principles for bioremediation: A review. *Molecules* 25(4):856. doi: 10.3390/molecules25040856

Liu J, Bacosa HP, Liu Z (2017). Potential Environmental Factors Affecting Oil-Degrading Bacterial Populations in Deep and Surface Waters of the Northern Gulf of Mexico. *Front Microbiol* 7:2131. doi: 10.3389/fmicb.2016.02131

Lovley DR, Lloyd JR (2000). Microbes with a mettle for bioremediation. *Nat Biotechnol* 18(6):600–601. doi: 10.1038/76433

Lu Z, Zeng F, Xue N, Li F (2012). Occurrence and distribution of polycyclic aromatic hydro-carbons in organo-mineral particles of alluvial sandy soil profiles at a petroleum-contaminated site. *Sci Total Environ* 433:50–57. doi: 10.1016/j.scitotenv.2012.06.036

Maia M, Capão A, Procópio L (2019). Biosurfactant produced by oil-degrading *Pseudomonas putida* AM-b1 strain with potential for microbial enhanced oil recovery. *Bioremediat J* 23:302–310. doi:10.1080/10889868.2019

Peay KG, Kennedy PG, Talbot JM (2016). Dimensions of biodiversity in the Earth mycobiome. *Nat Rev Microbiol* 14(7):434–447. doi: 10.1038/nrmicro.2016.59

Procópio, L (2020). Microbial community profiles grown on 1020 carbon steel surfaces in seawater-isolated microcosm. *Ann Microbiol* 70:13. doi: 10.1186/s13213-020-01547-y

Procópio L (2021). The oil spill and the use of chemical surfactant reduce microbial corrosion on API 5L steel buried in saline soil. *Environ Sci Pollut Res Int* 28(21):26975–26989. doi: 10.1007/s11356-021-12544-2

Procópio L (2022). Microbially induced corrosion impacts on the oil industry. *Arch Microbiol* 204(2):138. doi: 10.1007/s00203-022-02755-7

Procópio L, Barreto C (2021). The soil microbiomes of the Brazilian Cerrado. *J Soils Sediments* 21:2327–2342. doi: 10.1007/s11368-021-02936-9

Procópio L, de Cassia Pereira e Silva M, van Elsas JD, Seldin L (2012). Transcriptional profiling of genes involved in n-hexadecane compounds assimilation in the hydrocarbon degrading *Dietzia cinnamea* P4 strain. *Braz J Microbiol* 44(2):633–641. doi: 10.1590/S1517-83822013000200044

Procópio L, Macrae A, van Elsas JD, Seldin L (2013). The putative α/β-hydrolases of *Dietzia cinnamea* P4 strain as potential enzymes for biocatalytic applications. *Antonie Van Leeuwenhoek* 103(3):635–646. doi: 10.1007/s10482-012-9847-3

Rodriguez-R LM, Overholt WA, Hagan C, Huettel M, Kostka JE, Konstantinidis KT (2015). Microbial community successional patterns in beach sands impacted by the Deepwater Horizon oil spill. *ISME J* 9(9):1928–1940. doi: 10.1038/ismej.2015.5

SadrAzodi SM, Shavandi M, Amoozegar MA, Mehrnia MR (2019). Biodegradation of long chain alkanes in halophilic conditions by Alcanivorax sp. strain Est-02 isolated from saline soil. *3 Biotech* 9(4):141. doi: 10.1007/s13205-019-1670-3

Sakshi, Haritash AK (2020). A comprehensive review of metabolic and genomic aspects of PAH-degradation. *Arch Microbiol* 202(8):2033–2058. doi: 10.1007/s00203-020-01929-5

Sazykina MA, Minkina TM, Konstantinova EY, Khmelevtsova LE, Azhogina TN, Antonenko EM, Karchava SK, Klimova MV, Sushkova SN, Polienko EA, Birukova OA, Mandzhieva SS, Kudeevskaya EM, Khammami MI, Rakin AV, Sazykin IS (2022). Pollution impact on microbial communities composition in natural and anthropogenically modified soils of Southern Russia. *Microbiol Res* 254:126913. doi: 10.1016/j.micres.2021.126913

Sokol NW, Slessarev E, Marschmann GL, Nicolas A, Blazewicz SJ, Brodie EL, Firestone MK, Foley MM, Hestrin R, Hungate BA, Koch BJ, Stone BW, Sullivan MB, Zablocki O; LLNL Soil Microbiome Consortium, Pett-Ridge J (2022). Life and death in the soil microbiome: how ecological processes influence biogeochemistry. *Nat Rev Microbiol* 20(7):415–430. doi: 10.1038/s41579-022-00695-z

Stauffert M, Cravo-Laureau C, Duran R (2014). Structure of hydrocarbonoclastic nitrate-reducing bacterial communities in bioturbated coastal marine sediments. *FEMS Microbiol Ecol* 89(3):580–593. doi: 10.1111/1574-6941.12359

Táncsics A, Benedek T, Szoboszlay S, Veres PG, Farkas M, Máthé I, Márialigeti K, Kukolya J, Lányi S, Kriszt B (2015). The detection and phylogenetic analysis of the alkane 1-monooxygenase gene of members of the genus Rhodococcus. *Syst Appl Microbiol* 38(1):1–7. doi: 10.1016/j.syapm.2014.10.010

Thomas GE, Brant JL, Campo P, Clark DR, Coulon F, Gregson BH, McGenity TJ, McKew BA (2021). Effects of dispersants and biosurfactants on crude-oil biodegradation and bacterial community succession. *Microorganisms* 9:1200.

Thomas GE, Cameron TC, Campo P, Clark DR, Coulon F, Gregson BH, Hepburn LJ, McGenity TJ, Miliou A, Whitby C, McKew BA (2020). Bacterial community legacy effects following the Agia Zoni II oil-spill, Greece. *Front Microbiol* 11:1706.

McGenity TJ, Folwell BD, McKew BA, Sanni GO (2012). Marine crude-oil biodegradation: a central role for interspecies interactions. *Aquat Biosyst*, 8:1–19.

Xu G, Geng S, Cao W, Zuo R, Teng Y, Ding A, Fan F, Dou J (2022). Vertical distribution characteristics and interactions of polycyclic aromatic compounds and bacterial communities in contaminated soil in oil storage tank areas. *Chemosphere* 301:134695. doi: 10.1016/j.chemosphere.2022.134695

Yakimov MM, Timmis KN, Golyshin PN (2007). Obligate oil-degrading marine bacteria. *Curr Opin Biotechnol* 18(3):257–266. doi: 10.1016/j.copbio.2007.04.006

Yu SL, Li SG, Tang YQ, Wu XL (2011). Succession of bacterial community along with the removal of heavy crude oil pollutants by multiple biostimulation treatments in the Yellow River Delta, China. *J Environ Sci*. 1;23(9):1533–1543.

Zhen L, Hu T, Lv R, Wu Y, Chang F, Jia F, Gu J (2021). Succession of microbial communities and synergetic effects during bioremediation of petroleum hydrocarbon-contaminated soil enhanced by chemical oxidation. *J Hazard Mater* 410:124869.

Section V

Microbial Based Emerging
Technologies in Energy Systems

7 Improved MIC Management Using Multiple Lines of Evidence Drives Movement toward Sustainability
A Case Study in Heavy Oil Production

Susmitha Purnima Kotu
DNV Energy Systems, Houston, TX, United States

Richard B. Eckert
Microbial Corrosion Consulting, LLC, Commerce Township, MI, United States

Clara Di Iorio
Eni S.p.A., San Donato Milanese, Italy

Mary Eid
Eni US Operating Co., Inc., Anchorage, AK, United States

7.1 INTRODUCTION

The energy demand of the world has been ever increasing because of improved global standards of living. In parallel, the energy industry is currently undergoing a massive transformation to shift from non-renewable fossil fuel sources of energy such as coal, oil, and natural gas to cleaner and greener renewable sources of energy such as wind and solar. Additionally, several large integrated oil and gas companies have pledged net zero carbon emissions by 2050. This energy transition to cleaner and greener energy drives movement toward sustainability across the energy industry.

DOI: 10.1201/9781003287056-12

139

While on the path to developing more sustainable energy sources, it is imperative that the reliability and integrity of existing energy infrastructure continue to be improved. A report (Pörtner et al. 2022) released in March 2022 by the Intergovernmental Panel on Climate Change (IPCC) (the United Nations body for assessing the science related to climate change) outlined the connectivity of infrastructure system impacts resulting from climate change, such as through thawing permafrost, severe storms, and increasing temperatures disrupting energy production and delivery networks. The report identified that "[i]nterdependencies between infrastructure systems have created new pathways for compounding climate risk, which has been accelerated by trends in information and communication technologies, increased reliance on energy, and complex (often global) supply chains." The need for energy industry transformations in the areas of efficient water use, infrastructure resilience, and reliable power systems was identified in the IPCC report for all sources of energy generation. The report emphasizes the fact that managing and extending the lifecycle of existing energy assets is essential to the safe and reliable operations needed to support the movement to sustainable energy. Another benefit of continuing to improve operational integrity in current assets and infrastructure is that the same principles and improvements will largely apply to future renewable energy assets, which are subject to many of the same deterioration mechanisms, including microbiologically influenced corrosion (MIC).

Developing and implementing an effective MIC management program helps extend the life of assets and save resources, while preventing damage to the environment from spills due to corrosion-related loss of containment. An effective MIC management program is based upon a technically rigorous corrosion threat assessment, which optimizes the benefits of any actions taken to prevent and mitigate corrosion. Hence, an effective MIC management program is crucial for efficient, sustainable, and safe operations of oil and gas assets during the energy transition and is also translatable to the management of future renewable energy generation infrastructure.

Historical field sampling and analysis information can be used for MIC management by diagnosing the corrosion threats in different parts of the system, selecting effective corrosion mitigation method, and monitoring the effectiveness of applied mitigation measures. Since MIC is not diagnosed based on just one isolated parameter, multiple lines of evidence are needed to properly characterize the corrosion threats in an asset. A comprehensive analysis of microbiological abundance, microbiological activity, microbiological community composition, corrosion/pitting rates, pit initiation morphology, chemical conditions, design and operating conditions, corrosion products, inspection history, metallurgy, and failure history should be conducted as part of establishing a MIC management program (Kotu and Eckert 2019).

7.2 HISTORICAL OVERVIEW OF MIC ISSUES IN HEAVY OIL PRODUCTION

7.2.1 FIELD OVERVIEW

The case study presented here is of an oilfield located in the United States that began operation in January 2011. This oilfield consists of two drill sites, located onshore and offshore. The offshore production is imported via a 10-3/4″ nominal diameter

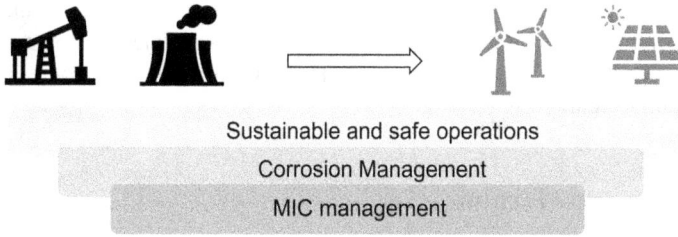

Sustainable and safe operations

Corrosion Management

MIC management

FIGURE 7.1 The role of safe/sustainable operations, corrosion management, and MIC management for energy transition.

pipeline, having a pipe-in-pipe configuration, to onshore, where a multiphase processing facility is also located. Further, a 12″ flowline brings injection water from onshore to offshore. Field production is a viscous crude on the upper end of the heavy oil scale with a gravity of 16–20 °API, together with water, and associated gas. Produced gas contains a moderate amount of CO_2 (<0.5 mole %) and no H_2S. Based on anticipated corrosion conditions, carbon steel has been applied as construction material, for a field design life of 20 years. The reservoir is shallow (4,000 ft True Vertical Depth Subsea, TVDSS) and thin (40 ft.); it consists of good quality sandstone (92–627 mD of reservoir rock permeability), although it is unconsolidated and prone to sand production. Reservoir temperatures are low, about 80°F (26.6°C) to 90°F (32.2°C). Water injection, aimed to sustain oil production, started in March 2011. Injection water is a mixture of separated produced water (PW) and hot source water from a high-temperature aquifer. The temperature of the resulting water mixture varies from 153°F (67°C) to 182°F (83°C). Since heavy oil and the cold reservoir compromise the mobility of produced fluids, the hot water mixture is also commingled with production fluids downstream of production headers to improve oil mobility. By combining produced fluids with circulated injection water, the water cut is brought up to 75% and the temperature of the resulting mixture is increased to a range of 125°F (52°C) to 168°F (75.5°C), thereby lowering fluid viscosity and facilitating flow through flowlines and inlet heat exchangers.

7.2.2 Production Wells

Drilling and completion activities in the field started in 2008 and are scheduled to be completed in 2025 with the drilling of an additional three producer and two injector wells. The total development will consist of 33 producers and 26 injectors. Producers and injectors have been drilled in pairs and located side by side (1,200 ft spacing).

The majority of producing wells consists of multilateral long extended reach horizontal wells, while a monobore lateral configuration has been selected for the injectors. The lateral length of producers varies from 6,000 to 10,000 ft, resulting in total measured depths of 18,000–23,000 ft for the longer wells. The producer wells are artificially lifted with electrical submersible pumps (ESP). The operating temperature and pressure of the production wells vary depending on the well bottomhole temperature and pressure that range from 60°F (15.6°C) to 86°F (30°C) and 132 psi (9.1 bar) to 192 psi (13.2 bar). The construction material is carbon steel, 4½″ diameter API L-80 for producers and injectors.

A polymer Injection trial is ongoing at the onshore drill site for enhanced oil recovery. Polymer is being mixed with injection water and injected down one injection well supporting two producing wells. This trial has been in progress for two, non-consecutive years.

7.2.3 PRODUCTION FACILITIES

Multiphase production from offshore wells is transported by a 10-3/4″ nominal diameter carbon steel pipeline to onshore where it is commingled with produced fluids from onshore wells and sent to the process module. Flow is split between two production trains by flow control valves that maintain equal flow to each train. However, operations usually alternate between trains, flowing one train at a time and preserving the train not in service with treated chemical. Each production train consists of the following in flow sequence: inlet heat exchangers, inlet separator, crude heater, low-pressure separator, and oil treater. After separation, the oil stream from the two trains is combined for oil shipping, the two water streams are combined for further processing, treating, and injection, and the gas streams are combined for compression and fuel gas conditioning. Water that is separated at each separation stage flows to the PW degasser to be treated. Downstream of the PW degasser, water is then sequentially desanded to remove particles ≥25 μm and deoiled to less than 100 ppm oil in water using desanding and deoiling hydrocyclone units. From the deoiling hydrocyclone water flows to the PW accumulator, while the overflow from deoiling hydrocyclone returns to the plant inlet through the oily water accumulator. At the inlet of the PW accumulator, the treated water is commingled with source water (SW), which is produced from three SW wells located onshore, and degassed by the SW degasser. Water from PW accumulator is sent to onshore and offshore injection wells and also used for diluting the produced fluids to increase the temperature. The increase of temperature lowers the overall fluid viscosity and facilitates flow through flowlines and inlet exchangers. Increased water cut improves oil/water separation in the inlet separators. The PW accumulator overflow is recirculated through the plant for different operational purposes. Excluding SW flowlines, where super duplex material has been applied due to the high CO_2 content, the construction material for process facilities (flowlines, piping, and equipment) is carbon steel. Vessels that contact multiphase fluids, PW, SW, or gas downstream of the SW degasser are internally coated over the entire internal surface area.

7.2.4 PHYSICAL-CHEMICAL CHARACTERISTICS

The main corrosive agents in the production facility are chlorides in the PW and source water, as well as residual CO_2 in the source water. CO_2 in the produced gas is lower than 0.5 mole % and H_2S is absent. Produced water composition varies throughout the field; it is primarily influenced by the different chemical composition from well to well. Total dissolved solid (TDS) values are in the range of 21,000–33,000 mg/L, while alkalinity varies from 540 mg/L to 850 mg/L. Sulfates show the highest variability with the concentration ranging from 1 mg/L to 1,200 mg/L. Produced water has a low volatile organic acids (VFA) concentration, with a maximum acetate concentration of a few dozen mg/L. Source water has a similar TDS value to the PW, but VFA content is higher than in the PW. Sulfates are about 30 mg/L, while total alkalinity is higher than in PW, in the range of 1,300–1,600 mg/L.

Significant amounts of sand from the unconsolidated sandstone reservoir are transported with the produced fluids and removed in multiphase separators and the water treatment facility by a desanding hydrocyclone. Fluid velocity throughout surface facilities varies greatly due the inner diameter of the pipelines ranging from 2″ to 14″. Appropriate corrosion inhibitors (CIs), corrosion allowances, cathodic protection, and corrosion-resistant coatings have been selected to prevent corrosion due to chemical-material incompatibilities. No biocide was added until 2017.

7.2.5 PAST CORROSION ISSUES AND FAILURES

Corrosion issues have presented themselves throughout production wells, injection wells, the production facility, and the water injection facility. An increase in failures due to corrosion has been observed since 2017. The costliest failures have been hole-in-tubing failures in the production wells. The temperatures in the production wells are optimal for the growth of mesophilic microbes, ranging from 60°F (15.6°C) to 86°F (30°C). The velocities in the production wells range from 0.3 ft/s to 4.7 ft/s, with 27 out of 31 wells having flow velocities less than 3.28 ft/s (1 m/s). The hole-in-tubing phenomenon is not experienced in all wells or isolated to a specific zone but appears to target specific wells as most failed wells have repeated serious corrosion damage. Most of the corrosion observed has been attributed to MIC either as a dominant mechanism or a supporting mechanism. No biocide treatment is applied to the wells. However, a CI is applied downhole at one of the two drill sites via a capillary string to the intake of ESP. MIC has also been observed on ESP motor bodies.

After seven years in service, the subsea production line that carries production from the offshore drill site to the process facility experienced aggressive internal corrosion with pitting depths up to ~79% of nominal design wall thickness, which was detected via an inline inspection (ILI). A follow-up ILI was conducted six months later indicating that pit depths progressed to 92% of the nominal wall thickness. MIC was identified as the cause of corrosion in the production pipeline. The fluid velocity in this production pipeline was approximately 3.0–3.5 ft/s depending on production rates. Prior to this, a consistent biocide program had not been implemented. This line had been temporarily decommissioned (shut-down) annually for no more than five consecutive days for plant turnarounds. During decommissioning, the oil was flushed from the line and replaced with untreated PW. To mitigate the severe wall loss, this subsea line was re-sleeved by inserting a smaller diameter pipe (10″) inside of the 14″ pipe, which subsequently increased the fluid velocities to 6 ft/s. A targeted biocide program was developed and implemented for this field after the production pipeline was repaired. The program consists of routine pigging and batch biocide treatments. After this implementation, multiple ILI inspections have shown no signs of further internal corrosion. The process experienced multiple failures usually detected on elbows, welds, stagnant valve bodies, etc. Due to the high sand production, historical use of microbially contaminated, lake water during drilling campaigns, and an increase in the levels of microbes during the first few years of production, most of the corrosion seen in the facility is associated with under-deposit corrosion and MIC or erosion and MIC. Heat affect zones of welds have been targeted by microbes in tanks, production well flowlines, and the disposal well flowline.

Oxygen-accelerated corrosion has been observed in the pig launchers and receivers flanking the subsea pipelines. Repeated failures have occurred on the launcher kicker lines and the receiver drains. These failures have all been attributed to MIC.

7.2.6 MONITORING AND MITIGATION STRATEGIES

The facilities are monitored with corrosion coupons, scale coupons, and microbial enumeration methods. There are 76 coupons installed on production, oil process, water process, gas process lines throughout the drill sites and facility. Corrosion coupons are pulled at a frequency ranging from two months to annually, depending on the history of corrosion issues in the line. Corrosion coupons are used to measure general and pitting corrosion rates and to characterize the microbial community via traditional most probable number (MPN) technique, adenosine triphosphate (ATP) testing, and quantitative polymerase chain reaction (qPCR). Planktonic, sessile/biofilm, and solid samples are analyzed throughout the facility most commonly with ATP measurements. Failure locations are swabbed and tested for microbial activity. qPCR surveys were performed in 2018, 2019, and 2021. A decrease in microbial populations was observed in most sample locations post implementation of the biocide program in 2018. Next-generation sequencing has been performed a handful of times of high-risk areas, highlighting the presence of methanogens, sulfate-reducers, iron-reducers, and acid-producers, among other microbes.

Corrosion mitigation programs include cathodic protection, chemical injection, and mechanical operations. External cathodic protection is applied to the subsea bundle, covering the water injection and production lines connecting the offshore and onshore drill sites. Anodes are externally located every 20 feet along the subsea bundle. CI is injected continuously on the top side at the production header at one drill site and down capillaries at the other drill site at 50 ppm. Due to the high amounts of solids in the system, supplemental CI injections are located downstream of the water outlets of the separator(s) and inlets of the produced and source water degassers in the water train at 42 ppm. Pigging operations of the subsea water injection and production pipelines are performed twice per week and followed by a biocide batch treatment (glutaraldehyde-based dual active biocide) at a concentration of 500 ppm based on PW volumes. There is a biocide injection point located at each of the two drill sites. Treatments are applied four hours per location, but the start times are staggered by two hours so that once the biocide reaches the second location, the biocide pump is turned on. The reasoning for this is that an increase in concentration of biocide is seen once injection location 1 fluids reach injection location 2. The water system is one large loop. Dead leg and stagnant line biocide applications are performed quarterly. Scale inhibitor is also injected continuously into production and downstream in the water system to help prevent scale deposition, which could ultimately lead to under deposit corrosion.

Due to the impact of solids on the infrastructure and an increased risk for under deposit corrosion and MIC severity, a cleaning campaign was initiated and completed in 2021. Solids were removed from all vessels and tanks, and the vessels were recoated, or damaged coating repaired. The solids collected were identified as calcite, silica, and iron sulfide. Risk-based inspection (RBI) was performed at the facility in 2020. Critical lines were identified for ultrasonic testing based on the RBI study, historical failures, previous inspections, and line contents.

7.3 DETAILED ASSESSMENT OF CORROSION MECHANISMS AND CONCLUSIONS

7.3.1 SAMPLING STRATEGY

This section discusses a recent comprehensive assessment of the corrosion mechanisms and their severity in this oil production system. Due to the complexity and large differences in operating conditions, microbiological communities, chemical environment, and corrosion rates and severity, the entire production system was grouped into three service type categories to simplify corrosion threat assessment across the entire system. These service type categories are:

a. Production wells: They include the active producing wells and the combined produced fluids until they are separated and treated.
b. PW treatment: It includes equipment/piping used for separation of produced fluids and processing of the PW.
c. Injection water: The separated PW is mixed with source water and used as injection water and dilution of production fluids.

Samples were collected from seven locations of the production system covering the three categories of the system discussed earlier (Figure 7.2 and Table 7.1). To investigate the ongoing corrosion mechanisms and the severity of MIC threat, a comprehensive analysis of the microbiological characteristics (via planktonic and biofilm ATP, qPCR) corrosion/pitting rates, chemical conditions (via cation/anion analysis), design and operating conditions, and corrosion products (via X-ray powder diffraction, XRD) was conducted. The sampling strategy included collecting different types of samples such as coupons (including biofilms on coupons), solids, and water samples for these analyses from seven selected locations in all three service type categories of the system previously described (Table 7.2).

7.3.2 RESULTS OF CORROSION ASSESSMENT IN THE PRODUCTION WELLS

The corrosion and pitting rates of all the coupons are summarized in Table 7.3. These coupons were installed for a time period of 35 days. The corrosion rates at the two production wells (Sample 1 and 2) and the production header (Sample 3) were less than 1 mpy. Pitting rates slightly higher than 5 mpy were seen at one of the production wells (Sample 1) and the production header (Sample 3). Pitting was not observed on the coupon installed at the other production well (Sample 2).

The ATP results from all seven locations are summarized in Table 7.4. The ATP results of the coupons from production well 1 (Sample 1) and production header (Sample 3) were on the order of 10^6–10^7 ME/cm^2. ATP was not detected on the coupon exposed to production well 2 (Sample 2) that also did not show any pitting.

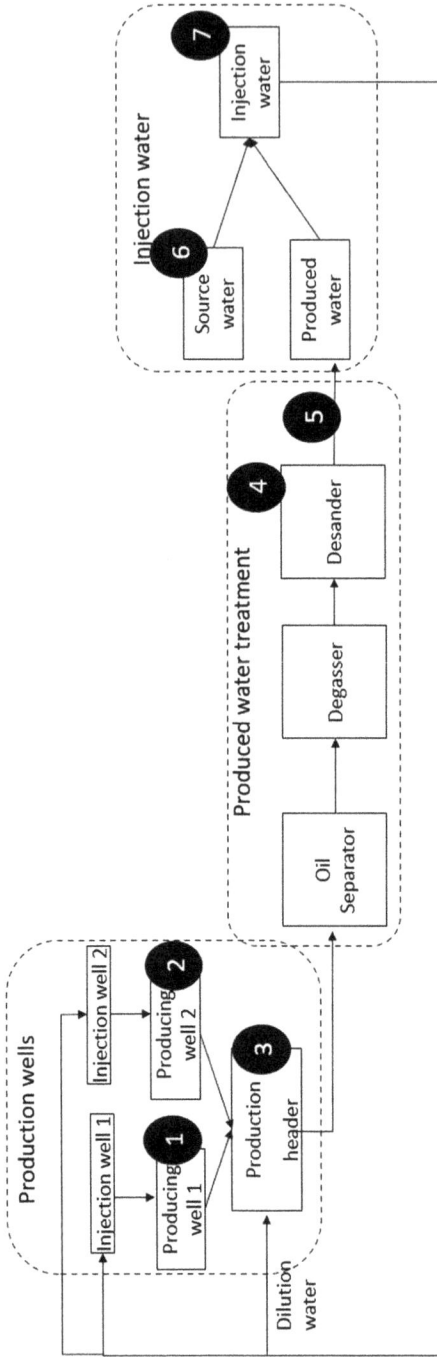

FIGURE 7.2   Schematic of the production system and the sampling locations.

TABLE 7.1
Locations in the Production System Selected for Corrosion Assessment

Category	Description of the Location	Sample ID
Producing wells	Producing well 1 at offshore producing site	1
	Producing well 2 at onshore producing site	2
	Production header	3
Produced water treatment	Desander	4
	Produced water after removing oil, gas, and solids	5
Injection water	Source water	6
	Mixed source water and produced water	7

TABLE 7.2
Sampling Strategy for the Seven Locations in the System Stating the Types of Samples (Coupons, Liquids, or Solids) and Tests Performed

Sample ID	Coupon Analysis				Liquid Analysis			Solids Analysis		
	ATP	qPCR	Corrosion Rate	Pitting Rate	ATP	qPCR	Water Chemistry	ATP	qPCR	XRD
1	x	x	X	x	x	–	x	–	–	–
2	x	x	X	x	x	–	x	–	–	–
3	x	x	X	x	x	–	x	–	–	–
4	–	–	–	–	–	–	–	x	x	x
5	x	x	X	x	x	x	x	–	–	–
6	x	x	X	x	x	x	x	–	–	–
7	x	x	X	x	x	x	–	–	–	–

TABLE 7.3
Corrosion Rates and Maximum Pitting Rates of the Coupons Installed at Various Locations in the Production System

Sample ID	Corrosion Rate (mpy)	Pitting Rate (mpy)
1	0.6	5.6
2	0.3	0.0
3	0.4	6.0
5	0.7	0.0
6	0.7	3.0
7	0.1	5.0

TABLE 7.4

Microbial Activity Results of ATP Tests Conducted on Coupon Biofilms, Liquids, and Solid Samples Collected at Various Locations in the Production System Converted to Cell Numbers Using 1 pg ATP = 1000 Microbial Equivalents (ME)

Sample ID	Coupon (ME/cm²)	Liquid (ME/mL)	Solids (ME/g)
1	5E+06	5E+05	–
2	ND	3E+04	–
3	3E+07	5E+06	–
4	–	–	2E+06
5	4E+05	1E+05	–
6	ND	2E+04	–
7	2E+06	4E+05	–

ND means ATP below the limit of detection of 10^3 ME/units and not detected.

The ATP results of the water sample from the two production wells (Samples 1 and 2) and the production header (Sample 3) were in the range of 10^4–10^6 ME/mL.

The qPCR results from all seven locations are summarized in Table 7.5. The total bacteria and archaea in the biofilms exposed to production well 1 (Sample 1) and production header (Sample 3) were 2–4 orders higher than production well 2. The microbiological community composition of the biofilms at well 1 (Sample 1) and production header (Sample 3) were similar to each other but different from that at well 2 (Sample 2). Biofilms at well 1 (Sample 1) and production header (Sample 3) showed the presence of MIC-causing microorganisms such as methanogens, acetogens, sulfate-reducing bacteria, fermenters, and iron-reducing bacteria. Biofilms at well 2 (Sample 2) only showed the presence of methanogens but none of the other MIC-causing microorganisms.

Liquid chemistry results showed that the water compositions were similar for the wells (Samples 1 and 2) and the production header (Sample 3). Both wells (Samples 1 and 2) showed sulfate concentrations of 200–250 mg/L, but sulfate was not detected at the production header (Sample 3). All three locations showed a similar pH of approximately 7.5, chloride concentrations of 20,000 mg/L, and TDS of 30,000 mg/L. The absence of sulfate in the production header may be attributed to the sulfate consumption by the microorganisms as the samples were not filtered to remove microorganisms before shipment to a laboratory for sulfate analysis.

Because of the higher microbial activity, abundance, presence of MIC-causing microorganisms, and pitting rates, MIC was assessed as the dominant corrosion mechanism at production well 1 (Sample 1) and production header (Sample 3). As pitting was not observed at production well 2 (Sample 2), MIC was not diagnosed as a corrosion mechanism. However, the presence of corrosive methanogens suggests that the threat of MIC cannot be ruled out at production well 2 (Sample 2). One important finding from these analyses was the noted differences in PW from the two wells tested, in terms of microbiology and corrosion characteristics but not chemistry. Historical monitoring data (not discussed here) also showed variations in corrosion severity in these wells over time. These differences in the microbiological

TABLE 7.5

Microbial Abundance Results of qPCR Tests Conducted on Coupon Biofilms (Denoted with "C"), Liquids (Denoted with "L"), and Solids (Denoted with "S") Collected at Various Locations in the Production System

Sample ID	Total Bacteria	Total Archaea	Acetogen	Sulfate-reducing Bacteria	Iron Oxidizer	Fermenter	Iron-reducing Bacteria	Methanogens		Sulfur-oxidizing Bacteria
								micH gene	tatC gene	
1C cells/cm²	2E+8	6E+8	2E+4	4E+3	ND	2E+6	2E+6	5E+6	6E+6	ND
2C cells/cm²	1E+4	3E+4	ND	ND	ND	ND	ND	1E+4	2E+4	ND
3C cells/cm²	5E+7	1E+6	4E+4	5E+3	ND	1E+5	1E+6	7E+3	2E+4	ND
4S cells/g	7E+9	2E+8	1E+6	6E+5	ND	2E+7	2E+6	4E+5	1E+5	ND
5C cells/cm²	8E+4	4E+5	3E+3	ND	ND	7E+3	3E+2	ND	ND	ND
6C cells/cm²	6E+5	9E+5	7E+3	ND	ND	3E+3	1E+3	2E+3	ND	ND
6L cells/mL	2E+5	5E+2	ND	ND	7E+3	ND	ND	ND	ND	2E+4
7C cells/cm²	4E+6	8E+5	8E+3	1E+5	ND	2E+4	9E+3	ND	7E+3	ND
7L cells/mL	4E+8	2E+5	2E+3	9E+4	9E+6	ND	5E+3	8E+2	1E+3	2E+7

ND means not detected.

populations and pitting rates may be attributed to the production well depths and history of operating conditions such as age of wells and use of injection water.

7.3.3 RESULTS OF CORROSION ASSESSMENT IN THE PRODUCED WATER TREATMENT

The coupon installed in the PW stream downstream of oil and gas separation (Sample 5) showed corrosion rate of less than 1 mpy and no pitting. Coupons were not analyzed at other locations of the PW treatment.

The ATP results of the solids at the desander (Sample 4) were in the order of 10^6 ME/g. The ATP results of the biofilms on the coupon exposed to PW and water sample (Sample 5) were 10^5 ME/cm^2 and 10^5 ME/mL, respectively. The total bacteria and archaea in the desander solids (Sample 4) were in the order of 10^8–10^9 cells/g and 4 orders higher than the biofilms on the coupon exposed to PW (Sample 5). The high surface area in the solids is the likely cause for elevated microbial abundance. Microbial communities at both locations showed the presence of MIC-causing microorganisms such as acetogens, fermenters, and iron-reducing bacteria. In addition to these microorganisms, the desander solids (Sample 4) also showed the presence of sulfate-reducing bacteria and methanogens.

The XRD results of the desander solids (Sample 4) did not show any specific corrosion products but instead showed mostly sand and mineral scales. The identified compounds were silica (SiO_2) at 70%, calcite ($CaCO_3$) at 10%, dolomite ($CaMg(CO_3)_2$) at 10%, and albite ($NaAlSi_3O_8$) at 10%.

Liquid chemistry results of the PW (Sample 5) showed a pH of approximately 7.8, TDS of approximately 32,000 mg/L, and chloride concentration of 19,000 mg/L. Sulfates were not identified in this sample.

Due to the absence of corrosion testing results, it is unknown whether the desander (Sample 4) experienced corrosion. However, the very high abundance of MIC-associated microorganisms suggests that MIC is a relevant threat. The PW sample (Sample 5) showed low general corrosion rate and no pitting, but the presence of MIC-associated microorganisms indicates the threat of MIC cannot be dismissed.

7.3.4 RESULTS OF CORROSION ASSESSMENT IN THE INJECTION WATER

The coupon installed downstream of the source water degasser (Sample 6) showed a corrosion rate of less than 1 mpy and a maximum pitting rate of 3 mpy. The coupon exposed to the injection water (Sample 7) showed a corrosion rate of 0.1 mpy and a maximum pitting rate of 5 mpy.

ATP was not detected in coupons exposed to the source water (Sample 6). The ATP results of the source water (Sample 6) showed microbial activity of 10^4 ME/mL. This low microbial activity can be attributed to the high temperatures ($>200°F$) of the source water. The ATP results of the biofilms and the water sample for injection water (Sample 7) were 10^6 ME/cm^2 and 10^5 ME/mL, respectively.

The total bacteria and archaea in the biofilms at both locations in the injection water system (Samples 6 and 7) were similar in the range of 10^5–10^6 cells/cm^2. The total bacteria and archaea in source water (Sample 6) were three orders lower than that in the injection water (Sample 7). The biofilms exposed to source water

(Sample 6) showed the presence of MIC-causing acetogenic, fermentative and iron-reducing bacteria, and methanogens, while the water sample (Sample 6) only showed the presence of iron-oxidizing and sulfur-oxidizing bacteria. The biofilms exposed to the injection water (Sample 7) showed the presence of methanogens, acetogenic, sulfate-reducing, fermentative and iron-reducing bacteria, while the water sample (Sample 7) showed the presence of methanogens, acetogenic, sulfate-reducing, iron-oxidizing, iron-reducing and sulfur-oxidizing bacteria. Both source water and injection water samples (Samples 6 and 7) showed the presence of oxidizing bacteria that were not seen in the respective biofilms and other parts of the system, suggesting the role of environment in promoting the growth of these oxidizing bacteria.

Liquid chemistry results of the source water (Sample 6) showed a pH of approximately 7.4, TDS of approximately 21,000 mg/L, chloride concentration of 12,000 mg/L, and sulfate concentration of 33 mg/L. Historical gas composition of the gas stream exiting the Source Water Degasser showed 14% of carbon dioxide, suggesting a role of carbon dioxide corrosion. Liquid chemistry results of the injection water (Sample 7) showed a pH of approximately 7.5, TDS of 30,000 mg/L, a sulfate concentration of 1.6 mg/L, and chloride concentration of 19,000 mg/L.

Based on the results, carbon dioxide corrosion was identified as the dominant threat. However, the threat of MIC cannot be ruled out in the source water (Sample 6), especially when the operating conditions are more conducive for microbial growth, for example, when operating temperatures are lower than the usual ~200°F. The presence of MIC-associated microorganisms and moderate pitting corrosion rates suggests that MIC was an active corrosion mechanism in the injection water (Sample 7). Hence, both MIC and carbon dioxide corrosion are relevant corrosion threats in the injection water system.

7.3.5 Conclusions of the Corrosion Assessment

Integrating information on the presence and abundance of MIC-associated microorganisms, observed corrosion/pitting rates, chemical conditions, operating conditions, and historical information, MIC was concluded as a potential corrosion mechanism at all three parts of the system, production wells, PW treatment, and injection water. However, the differences in the operating conditions, corrosion/pitting rates, failures, microbiological environment, and water chemistry suggest that the potential for MIC may not be the same at all locations in the production system. The large amounts of solids being produced by the wells support microbiological activity and hence all locations with the potential for solids accumulation are susceptible to MIC. Carbon dioxide corrosion was identified as a potential mechanism for source water and injection water. However, the observed corrosion cannot solely be attributed to carbon dioxide corrosion.

7.4 RECOMMENDATIONS FOR FUTURE MONITORING AND MITIGATION

The results of the microbiological, corrosion, and chemical analyses showed that MIC was a dominant threat at all three parts of the production system. However, this comprehensive analysis only reflects the corrosive conditions at one point in time. Hence, longer duration monitoring was recommended to monitor changes in the corrosion threats over time and to allow necessary changes in treatment strategy to achieve effective corrosion mitigation.

The strategic locations recommended for additional comprehensive monitoring for the production wells category included all of the wells with past failures. The strategic locations recommended for the PW treatment category included PW at the inlet and outlet of oil-water separators and degasser. The strategic locations recommended for the injection water category included the inlet and outlet of source water degasser and the injection water. These locations were recommended for monitoring with the following tests for two monitoring cycles of six months each before further changes are made to the long-term monitoring program and mitigation treatments.

- Corrosion and pitting rate measurements using coupons
- Microbial activity using ATP of coupon biofilms and water samples
- Microbial abundance using qPCR of coupon biofilms and water samples
- Chemical composition of all water samples

7.5 APPLYING THIS STRATEGY TO OTHER ENGINEERED ENVIRONMENTS AND ASSETS

Corrosion, including MIC, is an operational and integrity threat to nearly any engineered material operating in an environment containing some form of water. For example, in offshore wind turbines, MIC (Larsen 2020) has been observed on the submerged surface of closed-compartment foundations and in the buried external-facing portions of the monopile exposed to sediment. Alternating aerobic and anaerobic conditions in these environments could lead to high localized corrosion rates. This may occur when iron sulfide films, produced by sulfide-producing prokaryotes (SPP) under anaerobic conditions, are exposed to dissolved oxygen under aerobic conditions such as contact with aerated water. Sulfide films exposed to intermittent oxygen are known to result in very high corrosion rates, for example, over 1 mm/year (Boivin and Oliphant 2011). Similarly, the potential effects of microbial activity resulting from hydrogen being injected in natural gas storage fields are currently being investigated by several researchers to determine if microbial diversity and the potential for MIC and souring changes (Dopffel et al. 2021). Many of the assets used in renewable energy generation have been in service for less than a decade and since corrosion is a time-dependent threat, they have not yet had corrosion damage as experienced by 50-year-old (or more) oil and gas assets. The lessons learned from MIC management of existing assets will certainly translate well to renewable energy assets of the future, even if operating conditions are somewhat different, and since microorganisms are found in nearly every environment on the planet, the potential for MIC cannot be easily dismissed.

The corrosion control activity cycle is a continuous process of threat assessment, prevention and mitigation, and monitoring, with each step providing essential information to the other steps. MIC management follows this same process, and MIC must be assessed in the context of other possible abiotic (non-biological) corrosion threats as well. Further, corrosion control activities should ideally be supported by a management system (i.e., a corrosion management system or CMS) to ensure they are coordinated, efficient, and effective. Corrosion management systems have been described in NACE and ISO standards (ISO 55001 2014; NACE SP21430 2019). The corrosion management system essentially coordinates the administrative functions needed to execute the corrosion control activities in a reliable manner. The process of MIC management can be performed most effectively using this approach, leading to safer and more sustainable operations (Figure 7.1).

REFERENCES

Boivin J., and Oliphant S., Sulfur corrosion due to oxygen ingress. In *CORROSION 2011 Conference*, Paper 11120. NACE International, Houston, TX, 2011.

Dopffel N., Jansen S., and Gerritse J., Microbial side effects of underground hydrogen storage – knowledge gaps, risks and opportunities for successful implementation. *International Journal of Hydrogen Energy*, 46, 8594–8606, 2021.

ISO 55001, Asset management – Management systems – requirements, ISBN 978 0 580 86467 4, 2014.

Kotu S.P., and Eckert R.B., A framework for conducting analysis of microbiologically influenced corrosion failures. *Inspectioneering* 25(4), 30–35, 2019.

Larsen K.R., Corrosion risks and mitigation strategies for offshore wind turbine foundations. *Materials Performance*, AMPP, May 2020.

NACE SP21430, *Standard Framework for Establishing Corrosion Management Systems*. NACE International, Houston, TX, 2019.

Pörtner H.O., Roberts D.C., Tignor M., Poloczanska E.S., Mintenbeck K., Alegría A., Craig M., Langsdorf S., Löschke S., Möller V., Okem A., Rama B., and IPCC, *Climate Change 2022: Impacts, Adaptation, and Vulnerability*. Contribution of Working Group II to the Sixth Assessment Report of the Intergovernmental Panel on Climate Change, 2022.

8 Halophyte-based Biocides for Mitigation of Microbiologically Influenced Corrosion (MIC) in Industrial Water Systems

Jakob Lykke Stein and Tanmay Chaturvedi
Aalborg University, Aalborg, Denmark

Torben Lund Skovhus
VIA University College, Horsens, Denmark

Mette H. Thomsen
Aalborg University, Aalborg, Denmark

8.1 HALOPHYTES AND THE PROBLEM OF SOIL SALINIZATION

High salt concentrations in the soil will decimate crop yield – this fact has been known for millennia. Since the Roman era, it has even been used as a war strategy [1, 2]. In 1298, Pope Boniface VIII had everyone in Palestrina, Italy, killed, after which the Vatican plowed salt into the town's farmland "so that nothing, neither man nor beast, be called by that name, Palestrina" [2].

Salt-affected soil is, by definition, soil with more than 200 mM NaCl (11.7 g/L). Only a few plants can tolerate or grow in these highly saline environments [3]. Current estimates indicate that 20% of all farmland and 33% of the world's irrigated farmland are salt-affected, and those percentages are expected to increase fast. Salt concentrations in agricultural land are expected to increase by 10% annually, and by 2050, 50% of all arable land on Earth will be salt-affected [4, 5].

Salinities worldwide are increasing due to overutilizing arable land, groundwater irrigation, depleting water tables, and climate change causing increased evapotranspiration (water removed by plants or evaporation) [6, 7]. Increases in salinity cause damage to plants in direct and indirect ways. The direct damage to plants is caused

DOI: 10.1201/9781003287056-13

by increased osmotic stress, causing the roots to dry out. Furthermore, halogen-based free radicals, which are plentiful in saline water, disrupt the plant's functions on a molecular level. Indirect salinity-related damage is caused by halotolerant microorganisms trying to break down weakened plants.

Halophytes are plant species that thrive or tolerate growing in salt-affected soil or being irrigated by seawater. Some halophyte species not only tolerate the high salt concentrations but will accumulate salt in their tissues. A highly saline environment is a hostile environment that poses challenges to the survival of the plants. However, halophytes have evolved to produce various phytochemicals, including antioxidants, phenolics, vitamins, and antimicrobial compounds that help them survive and thrive [8, 9].

As salt concentrations increase, the agricultural yield of conventional crops will decrease. Halophytes are likely to become more and more prevalent among crops for food and fodder or as rotational crops that can draw salt out of the farmland [10]. Thus, the availability of halophyte-based biomass is expected to rise.

Past and current studies have investigated halophytes' utilization for food crops and fodder [11, 12]. In 1992, a study substituted the Rhodes grass (*Chloris gayana*) that goats were usually fed for *Salicornia bigelovii* [13]. Both washed (lowering salt content) and unwashed *Salicornia* were fed to the goats. Washing the *S. bigelovii* did not change the goats' forage patterns. However, the goats that ate the unwashed *Salicornia* did drink slightly more water [13]. Crushed *Salicornia bigelovii* has also been mixed into chicken feed as an alternative protein source [14].

Salicornia has been used in traditional cooking in many coastal areas. Their spring season succulent tips have been boiled similarly to spinach, mixed into salads, and used as a salty garnish for various dishes. Recently, *Salicornia* has made a resurgence in the kitchen because of its nutrient composition. It has been put in a group of "superfoods" alongside other foods like quinoa, chia seeds, and dark leafy greens like kale [15–21].

Various species of *Salicornia* are the most common agricultural halophyte [22, 23]. However, *Salicornia* and other succulent halophytes pose a challenge for the farmers, who are left with salty, lignified plants at the end of each harvest cycle [24]. During lignification, water is drawn out of the succulent tissue; the plant turns into straw, and the salt concentration increases. If halophyte farming is going to be used as a salt remediation strategy, lignified plants must be removed from fields. Anaerobic digestion (AD) of biomass into biogas is a commonly used route. However, the saline biomass poses challenges to the AD process [10, 25, 26]. Most biomass used in AD is low in salt content. However, halophyte biomasses usually have NaCl concentrations of 5%–15% (w/w), while methanogenic microorganisms used in AD are significantly inhibited at only 0.8% (w/v) [27].

Furthermore, because salt is not consumed in the AD process, it will accumulate until the digestion process stops or is overtaken by halotolerant microorganisms that may turn the biomass into undesired fermentation products. However, halophytes in AD may be used when kept separate from glucophytic biomass. A study investigated the biomethane yield of *S. europaea* and *S. ramosissima*. 24 m^3 CH_4/ton of fresh biomass was produced. This could be increased to 74 m^3 CH_4/ton by fractionating the plants into a pulp fraction and further to 149 m^3 CH_4/ton if the biomass was dead for one week at room temperature [28].

FIGURE 8.1 The phytochemicals inside halophilic plants (left) can be extracted. Differences in extraction techniques and conditions will change the composition of the extract. Therefore, the extraction process can be optimized to extract specific phytochemicals. The right picture shows a biocrude extracted from a halophyte. The biocrude, like crude oil, can be further refined into sub-fractions of similar phytochemicals or into pure components.

(Source: Stein, chapter author.)

Nevertheless, one could argue that primarily using halophytes for a bulk product like biogas is a wasteful process and a missed opportunity. As mentioned, halophytes produce many phytochemical compounds, such as antioxidants, vitamins, phenolics, and antimicrobial compounds [16–21, 29]. Due to the complex chemical structure of some of these compounds, many of these are expensive or impractical to produce through traditional synthesis.

The pretreatment to prepare the saline biomass for an AD process requires energy, enzymes, or an inorganic catalyst, which increases the process' cost. Extracting these compounds and the salt before or during the pretreatment could be a way to valorize the process before the biomass is used to produce biogas and biofuel.

The ongoing Horizon project, Aquacombine,[1] investigates how valuable phytochemicals and salt can be extracted from various halophilic biomasses, as shown in Figure 8.1 [12].

8.2 MICROBIOLOGICALLY INFLUENCED CORROSION

Microbiologically influenced corrosion (MIC) is an umbrella term for various mechanisms in which bacteria, such as sulfate-reducing bacteria (SRB), influence corrosion. MIC as a corrosion mechanism has been known for over a century [30, 31], but its diagnosis and mitigation are still ongoing areas of research and development [32–34]. The bacteria use metals such as iron as part of their primary or secondary metabolism, producing H_2S and degrading the structural integrity of the metal over time [33]. MIC happens in many places where non-sterilized water runs in pipelines, including cooling systems, fire suppression, district heating, drinking water, and water treatment facilities [33, 35]. However, the most studied area of MIC research is the oil and gas industry, where leaks caused by corrosion can have substantial environmental impacts [32]. Estimates of MIC prevalence vary greatly. MIC's prevalence is often reported as anywhere between 10% and over 40% of all corrosion without citation of the original source [33]. A specific MIC analysis in Alberta, Canada's oil and gas sector, puts the number at 13.4% of internal corrosion cases [33]. The significant variances in estimates stem from variations in mitigation measures, process conditions, and a lack of standardized MIC failure analysis [33]. Furthermore, MIC is sometimes used as a catch-all for unexplained corrosion, and, in some cases, MIC is diagnosed based purely on surface morphology [33].

Poorly managed MIC can have profound consequences, both economically and environmentally. Two examples include an oil pipeline in Prudhoe Bay, Alaska, in 2006 and a methane leak in Aliso Canyon, California, in 2015. Prudhoe Bay MIC developed an almond-sized hole, causing a 750,000 L oil spill covering about 8,000 m^2 [34, 36]. The subsequent investigation concluded that internal corrosion caused by SRB was the leading cause of the leak, and 27 km of the pipeline had to be rebuilt, temporarily halting production operations. In Aliso Canyon, California, a natural gas storage field leak caused a huge single-source climate impact, releasing 100,000 tons of methane into the atmosphere. Subsequent investigations concluded that MIC caused by methanogens was the leading cause of the incident [37]. Alaska's fines and maintenance costs exceeded $750 million [34, 36]. The costs in Aliso Canyon were a $120 million fine that was settled with the state of California. Additionally, over $1 billion has been spent on repairs and several ongoing (as of 2019) civil lawsuits related to claims of adverse health impacts due to toxic gas emissions [37–40].

The two examples emphasize how mismanaged MIC can have significant financial and environmental consequences. Catastrophic failure of infrastructure is often much more expensive than the precautions needed to prevent the failures, and therefore, measures are taken to mitigate the causes of MIC. Currently, there are no predictive tools for detecting MIC in the field, so the best practice for MIC prevention is a proactive approach [34].

Therefore, MIC mitigation is done in two ways that complement each other: chemically with biocides and physically with mechanical cleaning [32]. Biocides generally function by suspending specific metabolic activities of organisms, ultimately leading to the death of the cells [41]. Tetrakis Hydroxymethyl Phosphonium Sulfate (THPS) and glutaraldehyde are the most common biocides in the literature to prevent MIC in the oil and gas industry. However, these biocides have drawbacks. For example, some microorganisms, such as SRB from the *Desulfovibrio* genus, can work synergistically with other bacteria and archaea to create biofilms, which are notoriously difficult to eradicate using biocides [42]. The biofilm is an amorphous gel-like substance made from excreted biopolymers such as polysaccharides, proteins, and DNA. It acts as a protective barrier while allowing essential nutrients to the microorganisms [43]. Over time, some biofilms may absorb inorganics such as calcium, hardening the outer layers [43]. Thus, biofilms provide a protective environment for microorganisms to settle on the surface of infrastructure undisturbed.

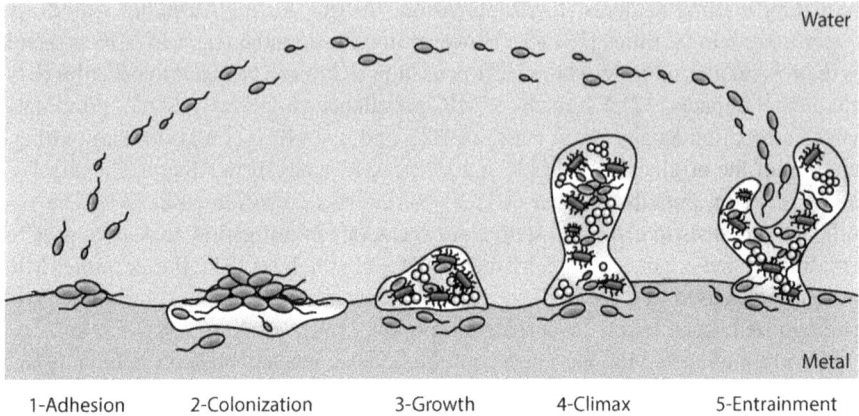

FIGURE 8.2   Biofilm formation schematic. (1-Adhesion) Planktonic microorganisms adhere to the surface of the pipeline. (2-Colonization) Microorganisms start reproducing. At some point, the microorganisms signal to each other to start excreting a polymeric substance made from sugars, proteins, and DNA – the biofilm. (3-Growth) The biofilm continues to grow. It shelters the microorganisms from the surrounding flow. SRB uses iron in their anaerobic metabolism at the interface between biofilm and wall, causing pitting corrosion. (4-Climax) The biofilm continues to grow until the shear stress of the flow overcomes the biofilm. (5-Entrainment) The liquid flow will tear away biofilm, carrying it downstream, where the process will repeat.

(Source: Skovhus, chapter author.)

The environment inside a biofilm may differ significantly from the environment outside. Differences in pH, temperature, pressure, dissolved oxygen, and chemical and microbial composition between biofilms and the bulk solution have been observed. Even significant differences within the same biofilm can occur [34]. The localized nature of the biofilm formation and the micro-environment it creates is why MIC is often characterized by localized pitting corrosion [32]. Figure 8.2 depicts how biofilm forms and causes localized pitting corrosion.

THPS, the biocide most commonly used in the oil and gas industry, effectively inhibits the planktonic SRB but struggles to impact the bacteria embedded in biofilm due to poor penetration of the biofilm and its inability to destroy it. Once biocides are ineffective, scraping or "pigging" is needed. Pigging is done by inserting and pushing a device that fills the entire pipeline cross-section. The pig physically scrapes off the biofilm as it is moved through the pipeline. Figure 8.3 shows a pig for a 71-cm pipeline.

8.3   THE USE OF MULTIPLE BIOCIDES AND THE "HURDLE EFFECT"

The best way to prevent MIC would be to prevent biofilms from forming. However, despite current measures to slow this process, biofilms are still forming, suggesting that some MIC-causing species survive the biocides and reproduce between biocide injections. One solution is to increase the dosage of biocides. However, a single anti-microbial treatment in a high dose is only effective until the microorganisms develop

FIGURE 8.3   A "Pig" used for scraping or "pigging" inside a pipeline. The disks seal tight against the pipeline walls. The top is equipped with magnets to attract loose iron. Different pig sizes exist for different pipeline diameters.

(Source: Harvey Barrison. [66], Creative Commons: CC BY-SA 2.0.)

tolerance to the active compound [44], after which an alternative antimicrobial treatment must be applied when the maximum legal or economically viable dosage is reached.

On the contrary, if a combination of antimicrobials is used, the same microorganism must simultaneously develop tolerance to a mix of antimicrobials. This relatively new approach to antimicrobials has been named "hurdle technology" because the organisms have to overcome many different hurdles instead of a single big one [45]. The approach is used in the food industry, where increasing the dosage of an antimicrobial treatment makes the food product unsafe for consumption [45, 46]. The "hurdle effect" approach appears in medical research on methicillin-resistant *Staphylococcus aureus* (MRSA) and other emergent multi-resistant pathogens [44, 47–49]. These multi-resistant pathogens have adapted to most currently used antimicrobial drugs through mutations. Using a range of antimicrobials on pathogens is thought to reduce the risk of antimicrobial resistance significantly.

Halophytes and other plants have evolved to protect themselves against many natural threats. They have exploited the "hurdle effect" by producing more than one antimicrobial compound [50]. The fact that these plants exist suggests that many microorganism species that attempt to consume the plants have not developed a tolerance to one or more antimicrobials produced by plants. The following section explores how natural biocides could be an alternative or a supplement to the current biocides used in water treatment today.

8.4 NATURAL BIOCIDES

Halophilic biomass is a source of inexpensive biomass, which is expected to increase in availability in the coming years due to soil salination and because halophytes, such as *Salicornia europaea*, have recently gained public attention. Early-season sprouts from *S. europaea* have several nutritional and pharmaceutical health benefits [51]. However, later in the season, during late summer, the plants partly lignify and become inedible to humans. The late harvest can still be used in the diet for ruminating animals, like cattle and goats, but at the end of the season, the farmers are left with a lot of wholly lignified and highly saline straw [13, 52]. Despite this, the lignified straw still contains high amounts of bioactive phytochemicals, such as potent antimicrobials [19, 29].

An ongoing applied research project, Clean Biocide in Denmark, investigates whether phytochemicals extracted from halophyte farms' leftover lignified biomass can mitigate MIC in upstream oil and gas systems [29, 53, 54]. The project investigates how salt-free biomass can be further utilized for its biocidal effect before the remainder is used in biofuel and biogas production.

Test results on lab-grown biofilms show that a halophyte extract effectively inhibited most SRB, such as *Desulfovibrio sp.* Furthermore, the halophyte biocide could dissolve an already-established biofilm, as shown in Figure 8.4. The ability to dissolve or penetrate biofilm is not a capability that the currently used glutaraldehyde or THPS biocides possess, even at 4,000 ppm – eight times the dosage used in the European oil and gas industry [42, 55, 56]. Sequencing data from 16S rRNA Amplicon Sequencing of the samples treated with halophyte-based biocide showed that the relative abundance of SRB species had diminished from 8% of the microbiome to a single surviving *Shewanella* genus species comprising 0.6% of the total microbial culture. Weight loss measurement on 1018 mild steel showed an 84% reduction in MIC and much shallower and spread-out pitting corrosion [57, 58]. Current research investigates the eco-toxicity effects of halophyte-based biocides, optimizes the extraction process, and applies halophyte-based biocides in different engineering systems [29].

Despite halophytes being the focus of this chapter, these are not the only promising studies of plant-derived natural biocides. Essential oils from many plants have inherent antimicrobial properties. Within the field of oilfield MIC, a recent study from India investigated the use of lemongrass essential oil as a biocide [59]. Hydrosolized[2] lemongrass oil was field-tested for three months. After the first three-phase separator, it was added at 50 ppm to a produced water line. The hydrosol form of lemongrass oil prevented the propagation of planktonic SRB when added at 50 ppm [59]. However, the study did not explore how the lemongrass oil interacted with biofilms and the microorganisms embedded within it.

Another study investigated using Aloe vera (*Aloe barbadensis*) extracts to prevent MIC [60, 61]. Like *Salicornia*, Aloe vera is a succulent plant traditionally used for medicinal purposes, though primarily for skin care. It contains several biologically active compounds, such as anthraquinones, lectins, and polysaccharides, known to possess antimicrobial, anti-inflammatory, and antioxidant properties. However, unlike *Salicornia*, Aloe vera is not halotolerant.

In the context of microbiologically influenced corrosion (MIC), Aloe vera effectively inhibits the growth of certain microorganisms, such as SRBs and iron-oxidizing

FIGURE 8.4 Pictures of the biofilm reactors. (a) Emptied biofilm reactor after 14 days of biofilm formation just before adding the biocide to the growth medium. The bioreactor is wholly covered in biofilm. (b) Emptied biofilm reactor 14 days after adding the biocide to the growth medium. The biocide managed to remove most of the biofilm.

(Source: Stein, chapter author.)

bacteria, that contribute to MIC [60, 61]. To use Aloe vera as a natural biocide for MIC mitigation, the plant extract was prepared by crushing the leaves and extracting the juice. [60, 61] The addition of Aloe vera reduced MIC on stainless steel coupons by 25.2% [60].

These plant-based MIC mitigation strategies could potentially become viable substitutes for chemicals used in the oil and gas sector. Nonetheless, technology is only as good as its scalability, and this can be an issue for uses of less common biomasses. Most biomasses must compete for agricultural land with traditional food and fodder crops or grassing areas, which will drive up the price of production. However, halophytes do not have to compete with conventional agriculture because they grow in intertidal zones, salt marshes, salt flats, and even saline depressions in the Saharan desert [62, 63]. All of these are areas that cannot be used for conventional crop production or animal husbandry. Furthermore, some halophytes, such as *Salicornia* sp. can also be grown in vertical aquaponics or at sea, which completely negates the requirement for land [64, 65].

8.5   CONCLUSIONS

Corrosion is a problem in most applications where water interacts with iron and steel. Over the past decades, the effect microbiologically influenced corrosion (MIC) plays on overall corrosion has become a focal point. Contrary to other types of corrosion, MIC causes localized accelerated corrosion due to biofilm formation. The localized nature of MIC and the poorly understood mechanisms of the symbiotic relationships between biofilm formation and microorganisms make MIC challenging to predict and detect. Nevertheless, MIC can drastically shorten the expected lifetime of the infrastructure.

Current biocides are effective against planktonic and sessile bacteria but struggle to deal with biofilms and the bacteria embedded within them. However, the inherent antimicrobial effect of certain halophyte extracts has shown promising results. Extracts from halophytic plants, Aloe vera, and lemongrass oil have selectively inhibited SRB, which are thought to be one of the primary causes of MIC in some systems. Furthermore, the halophyte extracts seem able to break down established biofilms, an ability the currently used biocides, THPS, and glutaraldehyde lack.

Despite the apparent success of these initial biocide tests, the technology is still in its infancy, and attempts to optimize the production process are underway. Additionally, strict legislation in the oil and gas industry currently prevents the use of chemically complex products such as halophyte extracts. Lists of approved chemicals exist to protect the marine environment into which additives and produced water from the oil and gas operations are discharged. Owing to stringent regulations in the oil and gas sector, other industries with significant water demands and where biofilm outbreaks are persistent might become entry points and initial markets for plant-based biocides.

ACKNOWLEDGMENT

This chapter was written by researchers from the Clean Biocide project, which has received funding from the Danish Offshore Technology Centre (DTU Offshore) under the CTR.2 Green Chemicals program.

NOTES

1   https://www.aquacombine.eu/.
2   A stable colloidal suspension of microscopic oil droplets in water.

BIBLIOGRAPHY

1. Ridley, R.T. To be taken with a pinch of salt: The destruction of carthage. *Class. Philol.* 1986, *81*, 140–146.
2. Martin, M. *The Decline and Fall of the Roman Church*; Putnam, 1981; ISBN 9780399126659.
3. Arora, S., Rao, G.G. Bio-amelioration of salt-affected soils through halophyte plant species. *Bioremediat. Salt Affect. Soils Ind. Perspect.* 2017, 71–85, doi:10.1007/978-3-319-48257-6_4
4. Jamil, A., Riaz, S., Ashraf, M., Foolad, M.R. Gene expression profiling of plants under salt stress 2011, *30*, 435–458, doi:10.1080/07352689.2011.605739

5. Shrivastava, P., Kumar, R. Soil salinity: A serious environmental issue and plant growth promoting bacteria as one of the tools for its alleviation. *Saudi J. Biol. Sci.* 2015, *22*, 123, doi:10.1016/J.SJBS.2014.12.001

6. Beresford, Q., Bekle, H., Phillips, H., Mulcock, J. *The Salinity Crisis: Landscapes, Communities and Politics*; Univ. of Western Australia Press: Crawley, 2001; ISBN 9781876268602.

7. Rengasamy, P. Soil salinization. Available online: https://www.oxfordbibliographies.com/view/document/obo-9780199363445/obo-9780199363445-0008.xml

8. Alassali, A., Cybulska, I., Galvan, A.R., Thomsen, M.H. Wet fractionation of the succulent halophyte Salicornia sinus-persica, with the aim of low input (water saving) biorefining into bioethanol. *Appl. Microbiol. Biotechnol.* 2017, *101*, 1769–1779, doi:10.1007/S00253-016-8049-8

9. Cybulska, I., Brudecki, G., Alassali, A., Thomsen, M., Jed Brown, J. Phytochemical composition of some common coastal halophytes of the United Arab Emirates. *Emirates J. Food Agric.* 2014, *26*, 1046–1056, doi:10.9755/ejfa.v26i12.19104

10. Vicente, O., Gulzar, S., Grigore, M.-N., Mishra, A., Tanna, B. Halophytes: Potential resources for salt stress tolerance genes and promoters. *Front. Plant Sci.* 2017, *8*, 829, www.frontiersin.org doi:10.3389/fpls.2017.00829

11. Chaturvedi, T., Christiansen, A.H.C., Gołębiewska, I., Thomsen, M.H. Salicornia species current status and future potential. In *Future of Sustainable Agriculture in Saline Environments*; 2021; pp. 461–482; ISBN 9781003112327.

12. The AquaCombine Consortium. The AquaCombine Project. Available online: https://www.aquacombine.eu/ (accessed on May 11, 2022).

13. Glenn, E.P., Coates, W.E., Riley, J.J., Kuehl, R.O., Swingle, R.S. Salicornia bigelovii Torr.: A seawater-irrigated forage for goats. *Anim. Feed Sci. Technol.* 1992, *40*, 21–30, doi:10.1016/0377-8401(92)90109-J

14. Attia, F.M., Alsobayel, A.A., Kriadees, M.S., Al-Saiady, M.Y., Bayoumi, M.S. Nutrient composition and feeding value of Salicornia bigelovii torr meal in broiler diets. *Anim. Feed Sci. Technol.* 1997, *65*, 257–263, doi:10.1016/S0377-8401(96)01074-7

15. Rowney, J. Samphire: The next superfood? *Weekend Notes*, 2013.

16. Hulkko, L.S.S., Chaturvedi, T., Custódio, L., Thomsen, M.H. Bioactive extracts from Salicornia ramosissima, Tripolium pannonicum and Crithmum maritimum as value-added products in halophyte-based biorefinery. In *18th Int. Conf. Renew. Resour. Biorefineries*, 2022.

17. Hulkko, L.S.S., Rocha, R.M., Trentin, R., Fredsgaard, M., Chaturvedi, T., Custódio, L., Thomsen, M.H. Bioactive extracts from *Salicornia ramosissima* J. Woods Biorefinery as a source of ingredients for high-value industries. *Plants* 2023, *12*, 1251, doi:10.3390/plants12061251

18. Hulkko, L.S.S., Chaturvedi, T., Thomsen, M.H. Valorisation of Salicornia waste biomass through green biorefinery. In *9th Int. Symp. Energy from Biomass Waste*, 2022.

19. Hulkko, L.S.S., Chaturvedi, T., Thomsen, M.H. Extraction and quantification of chlorophylls, carotenoids, phenolic compounds, and vitamins from halophyte biomasses. *Appl. Sci.* 2022, *12*, 840, doi:10.3390/app12020840

20. Chaturvedi, T., Hulkko, L.S.S., Fredsgaard, M., Thomsen, M.H. Extraction, isolation, and purification of value-added chemicals from lignocellulosic biomass. *Processes* 2022, *10*, 1752, doi:10.3390/pr10091752

21. Hulkko, L.S.S., Turcios, A.E., Kohnen, S., Chaturvedi, T., Papenbrock, J., Hedegaard Thomsen, M. Cultivation and characterisation of Salicornia europaea, Tripolium pannonicum and Crithmum maritimum biomass for green biorefinery applications. *Sci. Rep.* 2022, *12*, 20507, doi:10.1038/s41598-022-24865-4

22. Panta, S., Flowers, T., Lane, P., Doyle, R., Haros, G., Shabala, S. Halophyte agriculture: Success stories. *Environ. Exp. Bot.* 2014, *107*, 71–83, doi:10.1016/J.ENVEXPBOT.2014.05.006

23. Sharma, R., Wungrampha, S., Singh, V., Pareek, A., Sharma, M.K. Halophytes as bioenergy crops. *Front. Plant Sci.* 2016, *7*, 1372, doi:10.3389/FPLS.2016.01372/BIBTEX

24. Halophytes: Classification and characters of halophytes (with diagram) Available online: https://www.biologydiscussion.com/plants/halophytes-classification-and-characters-of-halophytes-with-diagram/6932 (accessed on May 18, 2022).

25. McCarty, P.L. Anaerobic waste treatment fundamentals. IV. Process design. *Publ. Wks. N. Y.* 1964, *95*, 95–99.

26. Kugelman, I.J., McCarty, P.L. Cation toxicity and stimulation in anaerobic waste treatment. *J. (Water Pollut. Control Fed.)* 1965, 97–116.

27. McCarty, P.L., McKinney, R.E. Volatile acid toxicity in anaerobic digestion. *J. (Water Pollut. Control Fed.)* 1961, 223–232.

28. Cayenne, A., Turcios, A.E., Thomsen, M.H., Rocha, R.M., Papenbrock, J., Uellendahl, H. Halophytes as feedstock for biogas production: Composition analysis and biomethane potential of Salicornia spp. plant material from hydroponic and seawater irrigation systems. *Ferment* 2022, *8*, 189, doi:10.3390/FERMENTATION8040189

29. Stein, J.L., Chaturvedi, T., Skovhus, T.L., Thomsen, M.H. Effect of antimicrobial halophilic plant extracts on microbiologically influenced corrosion (MIC). In *Proceedings of the AMPP Annual Conference + Expo 2022; Association for Materials Protection and Performance (AMPP)*; San Antonio, Texas, USA, 2022; pp. 1–15.

30. von Wolzogen Kühr, C.A.H. Sulfate reduction as cause of corrosion of cast iron pipeline. *Water Gas* 1923, *7*, 277.

31. von Wolzogen Kühr, C.A.H., van Der Vlugt, L.S. The graphitization of cast iron as an electrochemical process in anaerobic soil. *Water* 1934, *18*, 147–165.

32. Skovhus, T.L., Enning, D., Lee, J.S. *Microbiologically influenced corrosion in the upstream oil and gas industry*; CRC Press, 2017; ISBN 9781498726566.

33. Eckert, R.B., Skovhus, T.L. *Failure analysis of microbiologically influenced corrosion*; Eckert, R.B., Ed.; First ed.; CRC Press: Boca Raton, 2021; ISBN 9780429355479.

34. Little, B.J., Blackwood, D.J., Hinks, J., Lauro, F.M., Marsili, E., Okamoto, A., Rice, S.A., Wade, S.A., Flemming, H.C. Microbially influenced corrosion—Any progress? *Corros. Sci.* 2020, *170*, 108641, doi:10.1016/j.corsci.2020.108641

35. Corrosionpedia. What is Microbiologically Influenced Corrosion (MIC)? - Definition from Corrosionpedia. Available online: https://www.corrosionpedia.com/definition/773/microbiologically-influenced-corrosion-mic (accessed on Sep 15, 2021).

36. Jacobson, G. Corrosion at Prudhoe Bay - A lesson on the line. *Mater. Perform.* 2007, *46*, 26–34.

37. Conley, S., Franco, G., Faloona, I., Blake, D.R., Peischl, J., Ryerson, T.B. Methane emissions from the 2015 Aliso Canyon blowout in Los Angeles, CA. *Science* 2016, *351*, 1317–1320, doi:10.1126/SCIENCE.AAF2348

38. Garcia-Gonzales, D.A., Popoola, O., Bright, V.B., Paulson, S.E., Wang, Y., Jones, R.L., Jerrett, M. Associations among particulate matter, hazardous air pollutants and methane emissions from the Aliso Canyon natural gas storage facility during the 2015 blowout. *Environ. Int.* 2019, *132*, 104855, doi:10.1016/J.ENVINT.2019.05.049

39. McNary, S. Why the Aliso Canyon gas blowout is still in the news: Firefighters sue alleging health problems. *LAist*, 2018.

40. Barboza, T. SoCal Gas agrees to $119.5-million settlement for Aliso Canyon methane leak — Biggest in U.S. history. *Los Angeles Times*, 2018.

41. Van Hamme, J.D., Singh, A., Ward, O.P. Recent advances in petroleum microbiology. *Microbiol. Mol. Biol. Rev.* 2003, *67*, 503–549, doi:10.1128/MMBR.67.4.503-549.2003

42. Xu, D., Li, X., Gu, T. A synergistic D-tyrosine and tetrakis hydroxymethyl phosphonium sulfate biocide combination for the mitigation of an SRB biofilm. *World J. Microbiol. Biotechnol.* 2012, *28*, 3067–3074, doi:10.1007/S11274-012-1116-0

43. Heggendorn, F.L., Fraga, A.G.M., De Carvalho Ferreira, D., Gonçalves, L.S., De Oliveira Freitas Lione, V., Lutterbach, M.T.S. Sulfate-reducing bacteria: Biofilm formation and corrosive activity in endodontic files. *Int. J. Dent.* 2018, *2018*, doi:10.1155/2018/8303450

44. Dever, L.A., Dermody, T.S. Mechanisms of bacterial resistance to antibiotics. *Arch. Intern. Med.* 1991, *151*, 886–895.

45. Leistner, L. Basic aspects of food preservation by hurdle technology. *Int. J. Food Microbiol.* 2000, *55*, 181–186, doi:10.1016/S0168-1605(00)00161-6

46. Gould, G.W. Biodeterioration of foods and an overview of preservation in the food and dairy industries. *Int. Biodeterior. Biodegrad.* 1995, *36*, 267–277, doi:10.1016/0964-8305(95)00101-8

47. Ryan, R.W., Kwasnik, I., Tilton, R.C. Methodological variation in antibiotic synergy tests against enterococci. *J. Clin. Microbiol.* 1981, *13*, 73–75, doi:10.1128/JCM.13.1.73-75.1981

48. Paull, A., Marks, J. A new method for the determination of bactericidal antibiotic synergy. *J. Antimicrob. Chemother.* 1987, *20*, 831–838, doi:10.1093/JAC/20.6.831

49. Leistner, L. Principles and applications of hurdle technology. *New Methods Food Preserv.* 1995, 1–21, doi:10.1007/978-1-4615-2105-1_1

50. Liu, G., Nie, R., Liu, Y., Mehmood, A. Combined antimicrobial effect of bacteriocins with other hurdles of physicochemic and microbiome to prolong shelf life of food: A review. *Sci. Total Environ.* 2022, *825*, doi:10.1016/J.SCITOTENV.2022.154058

51. Patel, S. Salicornia: Evaluating the halophytic extremophile as a food and a pharmaceutical candidate. *3 Biotech* 2016, *6*, doi:10.1007/S13205-016-0418-6

52. Saadeddin, R., Doddema, H. Anatomy of the "Extreme" halophyte *Arthrocnemum fruticosum* (L.) Moq. in relation to its physiology. *Ann. Bot.* 1986, *57*, 531–544.

53. Stein, J.L., Chaturvedi, T., Skovhus, T.L., Thomsen, M.H. Poster: Clean Biocide Project: Natural corrosion inhibitors halophilic plant extracts for biofilm mitigation. In *Proceedings of the DHRTC Technology Conference 2021 Book of Abstracts*; 2021; p. 6.

54. Chaturvedi, T., Thomsen, M.H., Skovhus, T.L. Investigation of natural antimicrobial compounds for prevention of Microbiologically Influenced Corrosion (MIC). In *Proceedings of the ISMOS-7 Abstract Book*; Caffrey, S., Stone, A., Whitby, C., Skovhus, T.L., Eds.; International Symposium on Applied Microbiology and Molecular Biology in Oil Systems: Halifax, Nova Scotia, Canada, 2019; pp. 53–54.

55. Okoro, C.C. The biocidal efficacy of tetrakis-hydroxymethyl phosphonium sulfate (THPS) based biocides on oil pipeline pigruns liquid biofilms. *Pet. Sci. Technol.* 2015, *33*, 1366–1372, doi:10.1080/10916466.2015.1062781

56. Simões, L.C., Lemos, M., Araújo, P., Pereira, A.M., Simões, M. The effects of glutaraldehyde on the control of single and dual biofilms of Bacillus cereus and Pseudomonas fluorescens. *Biofouling* 2011, *27*, 337–346, doi:10.1080/08927014.2011.575935

57. Stein, J.L., Chaturvedi, T., Skovhus, T.L., Thomsen, M.H. Optimization of extraction conditions for production of halophyte-based biocides for Microbiologically Influenced Corrosion (MIC) mitigation. In *Proceedings of the EUROCORR 2022*; 2022; p. Poster.

58. Stein, J.L., Chaturvedi, T., Skovhus, T.L., Thomsen, M.H. Importance of the Multiple Lines of Evidence (MLOE) approach in diagnosing Microbiologically Influenced Corrosion (MIC). In *Proceedings of the DTU Offshore Technology Conference*; Kolding, Denmark, 2022; p. Poster.

59. Bhagobaty, R.K., Borkataky, M. Application of Lemongrass essential oil as a bacte-ricide for produced water treatment in an Oil Collecting Station of North-East, India. *Upstream Oil Gas Technol.* 2021, *7*, 100050, doi:10.1016/J.UPSTRE.2021.100050

60. Barbadensis, V., Leaf, S. Biocorrosion inhibition efficiency of locally sourced plant extracts obtained from Aloe Vera (Barbadensis miller) and Scent Leaf (Ocimum gratis-simum). *J. Biotechnol. Biomater.* 2019, *9*, 1–8, doi:10.4172/2155-952X.1000289

61. Agwa, O.K., Iyalla, D., Abu, G.O. Inhibition of bio corrosion of steel coupon by sul-phate reducing bacteria and iron oxidizing bacteria using Aloe Vera (*Aloe barbadensis*) extracts. *J. Appl. Sci. Environ. Manag.* 2017, *21*, 833, doi:10.4314/jasem.v21i5.7

62. Sahara - Plant life|Britannica. Available online: https://www.britannica.com/place/Sahara-desert-Africa/Plant-life (accessed on Mar 29, 2023).

63. Flowers, T.J., Hajibagheri, M.A., Clipson, N.J.W. Halophytes. *Q. Rev. Biol.* 1986, *61*, 313–337, doi:10.1086/415032

64. Radulovich, R., Rodríguez, M.J., Mata, R. Growing halophytes floating at sea. *Aquac. Rep.* 2017, *8*, 1–7, doi:10.1016/J.AQREP.2017.07.002

65. Maciel, E., Domingues, P., Domingues, M.R.M., Calado, R., Lillebø, A. Halophyte plants from sustainable marine aquaponics are a valuable source of omega-3 polar lip-ids. *Food Chem.* 2020, *320*, 126560, doi:10.1016/J.FOODCHEM.2020.126560

66. Barrison, H. PipelinePIG.jpg - Wikimedia Commons; Permission: CC BY-SA 2.0. Available online: https://commons.wikimedia.org/wiki/File:PipelinePIG.jpg (accessed on Jun 22, 2023).

9 Response of a Model Microbiologically Influenced Corrosion Community to Biocide Challenge

Damon Brown
Group 10 Engineering Ltd, Calgary, Canada

Raymond J. Turner
University of Calgary, Calgary, Canada

9.1 INTRODUCTION

Biocides are a category of chemicals commonly used as broad-spectrum antiseptics in industrial settings to sterilize surfaces and fluids. Unlike antibiotics, biocides can have a broad range of killing mechanisms and even broader range of chemical structures but generally function either as reactive or static killing mechanisms. Biocides and their mechanisms are reviewed elsewhere [1–3]. Reactive biocides such as chlorine have rapid killing rates but are depleted through their mechanism, while static biocides have slower kill rates but remain active in the system longer. Biocides with static killing mechanisms typically disrupt the structural integrity of membranes, destroy cell bioenergetics, and denature proteins.

Although biocide resistance can be a result of intrinsic cellular mechanisms [4, 5], tolerance toward biocides has been found to be in part a result of multidrug resistance efflux pumps (MDREPs) [6–12]. Though initially discovered for their role in antibiotic resistance, MDREPs have been found to have a broad range of substrates they can transport out of the cell, many of which don't share any physicochemical properties [13–16]. MDREPs can be found on plasmids and mobile genetic elements allowing for horizontal gene transfer (HGT) between unrelated species. This creates the potential to spread resistance genes between community members, improving tolerance toward biocides and other stressors. Table 9.1 provides an overview of well-defined MDREP genes and known substrates of interest. It is relevant to note that few biocides have been explored to see if they are substrates of MDREP, but most can respond to quaternary ammonium compounds and biocides of other chemical structure motifs.

DOI: 10.1201/9781003287056-14

167

TABLE 9.1
Multidrug Resistance Efflux Pump (MDREP) Genes and Their Known Biocide Substrates

MDREP Gene	Substrates	References
acrB	Acriflavine, chloramphenicol, cloxacillin, crystal violet, deoxycholate, erythromycin, ethidium bromide, fluoroquinolone, fusidic acid, glycocholate, nafcillin, novobiocin, rifampin, sodium dodecyl sulfate, taurocholate, tetracycline	[46–48]
emrE	Acriflavine, benzalkonium, betaine, cetylpyridinium chloride, cetyltrimethylammonium bromide, choline, crystal violet, ethidium bromide, methyl viologen, pyronine Y, safranin O, tetraphenylphosphonium	[49, 50]
emrB	Carbonyl cyanide m-chlorophenylhydrazone (CCCP), nalidixic acid, tetrarchlorosalicyl anilide, thioloactomycin	[51, 52]
norM	Acriflavine, doxorubicin, norfloxacin	[53]
qacA	Benzalkonium chloride, chlorhexidine, cetrimide, ethidium bromide	[54]
qacE	Benzalkonium chloride, bromide, cetyltrimethylammonium tetraphenylarsonium chloride, diamidinodiphenylamine dichloride, ethidium bromide, pentamidine isethionate, proflavine, propamidine isethionate, rhodamine	[54, 55]
mexD	Chloramphenicol, ethidium bromide, lincomycin, macrolides, novobiocin, oxacillin, quinolones, sodium dodecyl sulfate, tetracyclines, tetraphenylphosphonium	[56, 57]

In the petroleum industry, biocides are used as a means of preventing or controlling reservoir souring [17, 18] and microbiologically influenced corrosion (MIC) in associated pipelines [19–21]. Biocide treatments are performed as either continuous flow, where biocides are constantly present, or in batch dosing, where high concentrations of biocides are used for short periods of time (measured on the scale of hours). With regard to MIC, these treatments aim to eliminate the sessile microbes, which, while growing within biofilms, cause corrosion directly and indirectly as reviewed elsewhere [22–24]. As with all biofilms, the true issue lies in clearing the microbes embedded within, which become significantly more resilient toward external pressures such as extreme pH [25], starvation [26], predation [27], antibiotics [28], and biocides [29, 30]. Furthermore, the proximity and abundance, as well as phenotypic changes of sessile microbes, provide an ideal environment to allow and promote genetic exchange [31–33]. As proven in the medical field, biofilms are the primary location for the spread of antibiotic resistance genes [34, 35], but this has also been observed in wastewater and agricultural industries [36–38]. Sublethal concentrations of biocides create conditions where the presence and expression of beneficial MDREP genes provide greater fitness, increasing the chance of spreading mobile MDREP genes through HGT events.

Anecdotally, repeated application of a biocide in batches to oil and gas pipelines results in a decreased killing efficiency. While the cause and accuracy of these observations are uncertain, similar observations have been made in other industries where biocides are used including treatment of bacterial infections in humans [3, 39–41]. Therefore, it is reasonable to assume the same or similar mechanisms are present in the oil and gas pipelines. The cost, both in terms of the volume of the biocide and the loss of production (i.e., downtime flowing petroleum products) for the duration of biocide treatments means biocide treatments can be very expensive. Any means to improve biocide treatments could thus have significant impact for the cost of management and improve the lifetime of the asset.

Here, we track a defined microbial community composed of four species chosen to represent different metabolic clades found in an MIC-associated community. We challenge this model community with exposure to two different biocides at sublethal concentrations to provide selective pressure that would select more tolerant bacteria to proliferate. By following the relative abundance of targeted MDREP genes in the community using quantitative PCR (qPCR), we aim to identify MDREP genes responsible for increased biocide tolerance. We hypothesize that the MDREP genes which are responsible for tolerance toward a given biocide would increase in relative abundance within the community. Such genes could then act as genetic markers for increased tolerance toward the tested biocide in the field. A secondary contribution of our experiments is a side-by-side comparison of growth monitoring approaches, which highlight challenges inherent in such monitoring. The result of this work lays the foundation for a method for pre-screening a microbial community to determine the existing MDREP ratios and thus inform the biocide treatment to ensure effective treatment, and through continued monitoring ensure that the employed biocides remain effective and protect the integrity of the pipeline asset longer.

9.2 MATERIALS AND METHODS

9.2.1 BIOREACTOR SETUP AND COMPONENTS

A mixed community was created from four pure cultures: *Desulfovibrio vulgaris, Geoalkalibacter subterraneus, Pseudomonas putida*, and *Thauera aromatica* and grown together in closed-loop reactor systems consisting of a media reservoir, peristaltic pump, CDC Biofilm Reactor (bioreactors), and connecting tubing. Bioreactors were connected to modified 2 L reservoir bottles containing an injection port and effluent port with a stopper as low to the base as possible. Between the reservoirs and bioreactors, a peristaltic pump was employed to flow at 3.5 mL/min, feeding into the top of the bioreactors. The effluent port of the bioreactors was connected back to the reservoir through Tygon® tubing with a three-way stopcock sampling port present to allow sample collection for measuring planktonic growth. 10% CO_2/90% N_2 gas was supplied to both bioreactors to maintain microaerophilic conditions (with the understanding that running this system outside of an anaerobic glove box while ensuring complete anaerobic conditions was unrealistic). Bioreactors were operated on electronic stir plates running constantly at 130 rpm. Six sleeves of coupon holders, each holding three coupons, were prepared, and preweighed 1,018 carbon steel coupons were inserted for use in bioreactors. A representative picture is shown in Figure 9.1 to illustrate the bioreactor setup.

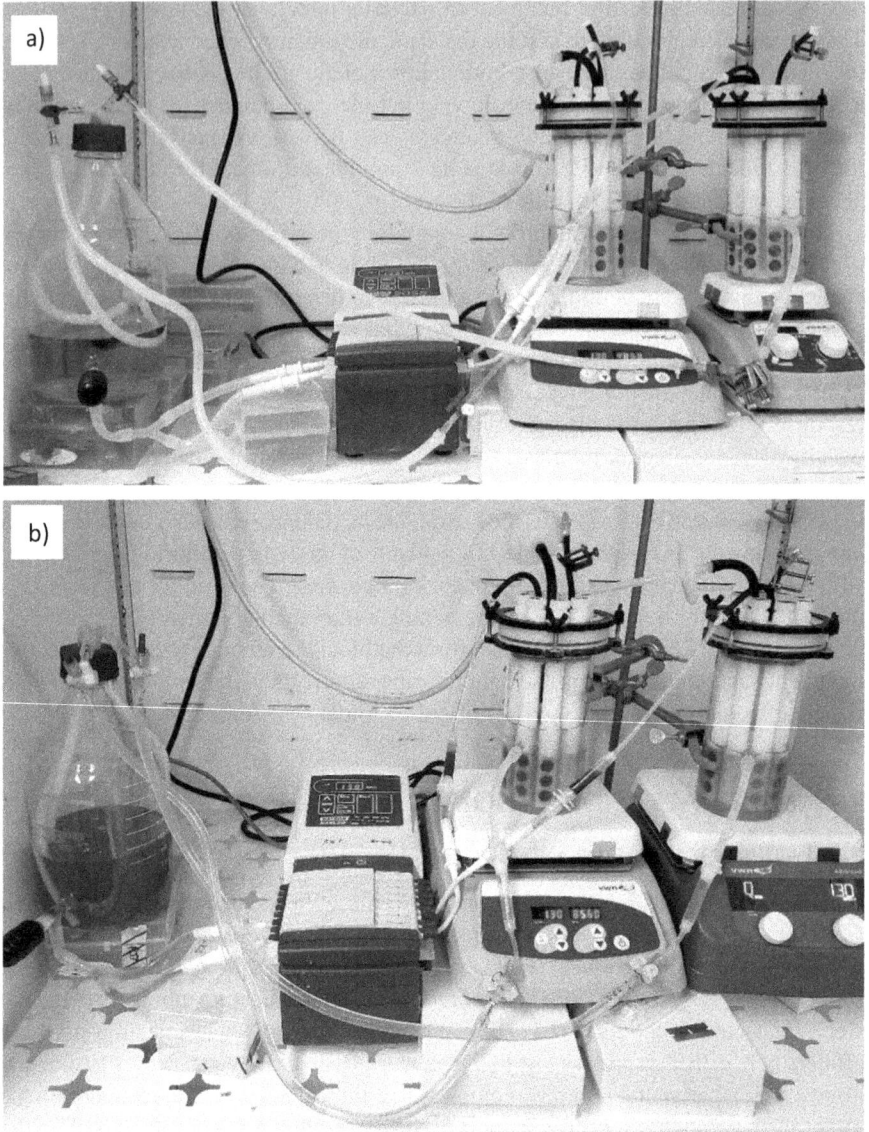

FIGURE 9.1 Example photographs of bioreactors illustrating typical conditions seen during the course of the 21-day biocide trials. Photographs are a) initial setup prior to inoculation b) time zero inoculation.

9.2.2 BIOREACTOR INOCULATION, OPERATION, AND SAMPLING

Two bioreactors were run in parallel for each experiment. Setup and operation of each experiment occurred in three phases, each over a span of seven days. In the first seven days, 1.5 L of artificial seawater (ASW) media (see Table 9.2 for recipe) was autoclaved and allowed to flow through both CDC Biofilm Reactors

TABLE 9.2
Recipe for Artificial Sea Water (ASW) Medium

Chemical	Concentration (g/L)
Peptone	5.0
Yeast extract	1.0
Ferric citrate	0.1
NaCl	19.45
$MgCl_2 \cdot 6H_2O$	5.9
$MgSO_4 \cdot 7 H_2O$	3.24
$CaCl_2 \cdot 2H_2O$	1.8
KCl	0.55
$SrCl_2$	34.0 mg
$NaHCO_3$	0.16
KBr	0.08
H_3BO_3	22.0 mg
Na_2SiO_3	4.0 mg
NaF	2.4 mg
NH_4NO_3	1.6 mg
Na_2HPO_4	8.0 mg
pH	7.6

(CBRs) simultaneously until reactors reached operational volume (~350 mL each). Once the bioreactors reached steady state, culture inoculations began; 75 mL of *Desulfovibrio vulgaris* culture was injected directly into the reservoir (5% inoculant), then 10 hours after, 75 mL of *Thauera aromatica* was injected, followed by 75 mL of *Geoalkalibacter subterraneus* 2 hours later, and lastly 75 mL of *Pseudomonas putida* after another 8 hours. Injection of *P. putida* marked time zero for the experiment. On day 7, three (3) aliquots of 2 mL of media were collected from each bioreactors effluent line for planktonic testing. One coupon sleeve from each bioreactor was removed, and coupons were placed individually into 2 mL phosphate buffered solution (PBS) and sonicated for 2 times 5 minutes (5 minutes with each face up). Coupons were then removed from the PBS and used in corrosion weight loss measurements following NACE protocols while the PBS was used to assess sessile microorganisms. After sampling, the pump was paused for four hours and two reservoirs with fresh 1 L ASW each were connected to their respective bioreactors. After reconnecting the fresh reservoirs and restarting flow, biocide was added to the reservoir of one bioreactor (test reactor) for a final concentration of 37.5 ppm tetrakis hydroxymethyl phosphonium sulfate (THPS) or 1 ppm benzalkonium chloride (BAC). No biocide was added to the other bioreactor (control reactor). A single 2 mL aliquot of effluent media was removed from each bioreactor on days one and three post-biocide injection, along with a coupon sleeve for sessile testing (day 8 and 10, respectively). On day 14, sampling was done in the same manner as day 7 (three aliquots of effluent media and a coupon sleeve). After sampling and testing on day 14, each reservoir was disconnected from the bioreactors and drained, then 1 L fresh ASW was added to each, and reservoirs were reconnected to bioreactors. The fresh media was allowed to run through the bioreactors and drained into a waste beaker (not returned to reservoirs) to flush biocide from the test reactor. After 1 L of media was flushed through their respective bioreactors, pumps were stopped, and 1 L of fresh media was prepared and for each bioreactor and the system was reconnected in the normal closed-loop operating conditions (prior to media flushing). On day 21, planktonic and sessile samples were collected as described for day 7 for testing.

9.2.3 MICROBIAL GROWTH TESTING

9.2.3.1 Optical Density

1 mL of bioreactor effluent media (planktonic) or 1 mL PBS was added to clean 1 mL cuvettes and OD_{600} readings were taken using a Hitachi U-2000 Spectrophotometer, using fresh ASW as the reference solution. The 1 mL sample was collected from the cuvette, added to a sterile 1.7 mL microcentrifuge tube, and centrifuged at 10,000 × G for 10 minutes to pellet cells for further analysis.

9.2.3.2 ATP Activity

1 mL of bioreactor effluent media (planktonic) or 1 mL of PBS was collected into a 3-mL syringe and used for adenosine triphosphate (ATP) analysis according to manufacturers' protocol (LifeCheck, LuminUltra, and/or Water-Glo™ System, Promega). Amount of ATP (pg/mL) was calculated according to manufacturers' protocols as well.

9.2.3.3 DNA Extraction, Concentration, and Cleaning

After centrifugation, the 1 mL sample used in OD_{600} readings was collected, 800 μL of supernatant was discarded and the pellets were resuspended in the remaining 200 μL solution, and DNA extraction was performed according to manufacturer's protocol (FastDNA™ Spin Kit, MPBio). DNA was collected in 50 μL nuclease-free water. DNA concentrations were measured using Qubit™ fluorimeter and the dsDNA HS kits with a 2 μL aliquot for each sample. After measuring DNA concentrations, the DNA was cleaned using the OneStep™ PCR Inhibitor Removal Kit (Zymo Research) and collected in 50 μL nuclease-free water for use in qPCR.

9.2.3.4 Quantitative PCR

Quantitative PCR (qPCR) was done to target the genes listed in Table 9.3, along with their primer sequences and annealing temperatures. The primer design method and additional information on the primers used are available elsewhere [42]. Amplification was done using the following protocol: 50 °C for 2 minutes, 95 °C for 5 minutes, 40 cycles of 95 °C for 15 seconds then annealing temperature for 25 seconds (see Table 9.3), followed by a melt curve analysis from 60 °C to 95 °C. All reactions had a total volume of 20 μL and 1 μL template DNA with GoTaq® qPCR (Promega) reaction mixture. Reactions were performed on either a Bio-Rad CFX96 Real-Time PCR Detection System or a QuantStudio™ 3 Real-Time PCR System.

Quantification of each target gene was performed using a synthetic gene block from Integrated DNA Technologies (IDT) with known concentrations ranging from 10^8 to 10^4 copies/μL in tenfold dilutions. C_t values were then calculated for each

TABLE 9.3

qPCR Target Genes, Primer Sequences, and Annealing Temperatures

Gene Target*	Upstream Primer Sequence (5′-3′)	Downstream Primer Sequence (5′-3′)	Anneal. Temp. (°C)
16S rRNA	CAG CMG CCG CGG TAA	GGA CTA CHV GGG TWT CTA AT	58
T-RFLP	6-FAM- GTG CCA GCM GCC GCG GTA A	ACG GGC GGT GTG TRC	
acrB (P)	TGG YGG CGC WGT ACG AAA GC	TTG GCG AAC GCC ACC ATC AGG AT	57
acrB (G)	AGG AAC GCC TTT TGG ATG ACG C	CCC TGG CAG GTC AGA CCA AGA A	57
acrB1 (T)	CTA CAT CGT CGT ACC GTG GGC A	ATC AGC GAG ACC GTC ATC AGC A	55
acrB2 (T)	TGG CAG CGC AGT TCG AGA GC	TGG CGA ACT CCA CGA TCA GGA T	57
acrB3 (T)	ACC ARC AWG CCG AGC GCG AT	GGG CAT GGA GCT GAA CGT GGT	57
emrE (G)	TGC ACT GGT TTT TGA AAG CA	GGC GCT GCT TTC TAT TTA CTT TC	55
emrE (P)	GCC ATT GCC ATC TGC GCC GA	ATC CCC AGC CCC GAC CAG ATG G	55
norM (D)	ACG GCC TGC CCA GCG GCA TC	GCT GCC CTT GCC CAT GGC CT	57
norM (P)	TCG GCC TGC CGA TGG GGG TG	GTC CTG CGC GCC GGC CGA CT	57
norM (T)	TGG GCC TGC CGA TTG GCG GT	GTT GCC AGC GCC GTA GTA CA	57
qacA1 (P)	AGA ASA YCC AGC GCC ACG AM	TGC TGG CCC GTG TAC TGC AGG	55
qacA3 (P)	AGA ASA YCC AGC GCC ACG AM	TCG TAA TCC GGG TGA TCC AGG	55
mexD (P)	TGG YGG CGC WGT ACG AAA GC	TGG CGA ACT CCA CGA TCA GGA T	57

* D = *D. vulgaris*; G = *G. subterraneus*; P = *P. putida*; T = *T. aromatica*.
Degenerate nucleotide codes: **S** = C/G; **Y** = C/T; **W** = A/T; **H** = A/C/G; **M** = A/C; **R** = A/G.

unknown sample and the copy number calculated using the C_q values from the gBlocks. Melt curve analyses were performed for each run to ensure only the target amplicon was obtained.

9.2.4 COMMUNITY COMPOSITION VERIFICATION

Community composition was verified at each time point using terminal-restriction fragment length polymorphism (T-RFLP). Using a 6-FAM labeled upstream 16S rRNA primer (Table 9.3), labeled amplicons of 895 basepairs were created using the following amplification protocol: 95 °C for 5 minutes, followed by 35 cycles of 95 °C for 45 seconds, 54 °C for 60 seconds, 72 °C for 90 seconds, and a final elongation

TABLE 9.4
Community Composition Scores from T-RFLP Sequencing Data

Biocide	Reactor	Species (Fragment Length)	Time Point (Day) ->	7	8	10	14	21
THPS	Test	D. vulgaris (319)		1	1	0.5	0.5	1
		G. subterraneus (720)		0	0.5	0.5	0.5	0.5
		P. putida (467)		1	1	1	1	1
		T. aromatica (541)		0.5	0.5	0.5	0.5	0.5
	Control	D. vulgaris (319)		1	1	1	1	1
		G. subterraneus (720)		0	0.5	0.5	0.5	0.5
		P. putida (467)		1	1	1	1	1
		T. aromatica (541)		0.5	1	0.5	0.5	0.5
BAC	Test	D. vulgaris (319)		1	1	1	1	1
		G. subterraneus (720)		0.5	0.5	0.5	1	1
		P. putida (467)		1	1	1	1	1
		T. aromatica (541)		1	1	0.5	0.5	1
	Control	D. vulgaris (319)		1	1	1	1	1
		G. subterraneus (720)		1	1	1	1	1
		P. putida (467)		1	1	1	1	1
		T. aromatica (541)		1	0.5	0.5	1	1

Presence Score (1 = present, 0.5 weakly present, 0 = absent)

step of 72°C for 10 minutes. 1 µL of the amplicons from each coupon was then separately digested by the restriction enzymes (RE) *StuI* and *HaeIII* for 16 hours using manufacturer's protocols (Thermo Scientific). 5 µL of each RE digestion product as well as 5 µL of the undigested amplicons from the three coupons collected at each time point were then pooled and sent for fragment length analysis. The combination of these two RE digests and undigested amplicons allowed for detection of a unique fragment length originating from each species. The results of this analysis were scored as either 0 (absent), 0.5 (weakly present), or 1.0 (strongly present) and are shown in Table 9.4 for each biocide exposure.

9.2.5 MDREP RATIO CALCULATIONS AND STATISTICS

MDREP ratios were calculated using the results of qPCR for each gene target using equation 9.1:

$$\frac{\left(\text{MDREP R1 of Cx at Ty} + \text{MDREP R2 of Cx at Ty}\right) \div 2}{\left(\text{16S rRNA R1 of Cx at Ty} + \text{16S rRNA R2 of Cx at Ty}\right) \div 2} = \text{Ratio of Cx at Ty} \quad (9.1)$$

where R = technical replicate, C = coupon, and T = time

DNA extracted from each coupon was treated independently of the other coupons by averaging the technical replicates for each coupon. This resulted in three replicates (i.e., three coupons) available for determining ratios in each bioreactor at every time point. To account for potential variability in cell densities on each coupon, ratios were only calculated using the copy numbers within a coupon and never between coupons. Statistical significance was calculated for each ratio pairing between test and control bioreactors using an unpaired, two-tailed Student's t test. Values represent mean ratio +/– SEM; $n = 3$. *$P < 0.05$.

9.3 RESULTS AND DISCUSSION

This work was broken into biologically distinct trial runs for each biocide. Although two trials of each biocide were completed, only a single trial of each will be discussed at length for brevity and a statement regarding reproducibility between trials will be made. In each trial, growth of the mixed community was monitored five times over the course of 21 days with a focus following exposure to biocide. An additional time point to measure planktonic cells was done an hour after the final inoculation was performed. Technical replicates of the coupons sampled at each time point provide the error bars in the figures.

Our previous study compared the efficacy of different growth monitoring techniques and found that although all are comparable, each approach has its pros and cons (Brown and Turner, 2022, submitted), and it behooves environmental microbiologists to appreciate and be sure they are applying the best approach to their question. Although our study suggested that DNA concentration tended to be the most reliable, here we report our reactor data in all four of the different ways, DNA concentration, OD_{600}, ATP levels, and 16S rRNA qPCR levels as a means of monitoring a more complex community.

9.3.1 THPS TRIAL

9.3.1.1 Growth Monitoring

Figure 9.2 shows the four methods of monitoring growth used in this experiment for planktonic and sessile microbes for both the test and control bioreactors during the THPS trial. Comparing all four methods together allows for the observations of trends within and between each method despite not sharing the same scales or units. The most immediate observation is the difference between planktonic and sessile growth in both bioreactors. OD_{600} readings are similar in value, with the planktonic values being higher (Figure 9.2A). Planktonic ATP values are anywhere from 11 to 60 times higher than the sessile values in the test reactor (average of 39 times) and 11–20 times higher in the control reactor (average of 16 times) (Figure 9.2B). Unlike the other three lines of evidence, ATP shows an increase in planktonic values at day 21 compared to a decrease seen in DNA and 16S rRNA-targeted qPCR (readings were relatively stable in OD_{600}). The addition of THPS caused a decrease in DNA of the sessile test bioreactor not observed in the control bioreactor, after which the DNA concentrations from both bioreactors increased by 14 days, with the control bioreactor always remaining 2–3 times higher than the test bioreactor (Figure 9.2C). Unsurprisingly, 16S rRNA copy numbers followed a very similar trend as the DNA concentrations, with the greatest difference being the increased drop in planktonic values at day 8 in the test bioreactor (Figure 9.2D). These lines of evidence show a clear disruption in the microbial growth resulting from the addition of THPS, with a

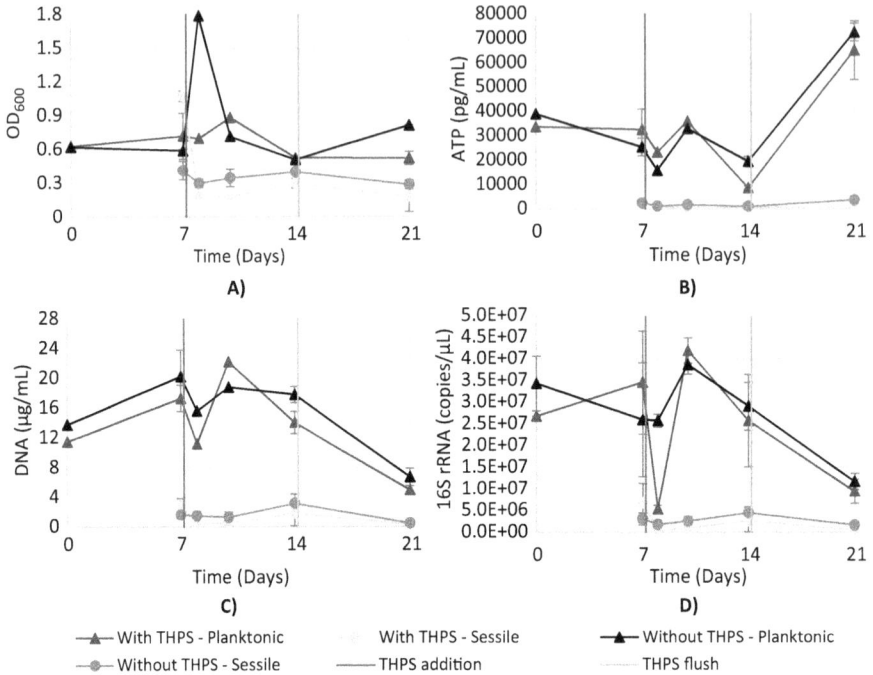

FIGURE 9.2 Growth monitoring of planktonic and sessile cells from parallel bioreactors treated with or without THPS (37.5 ppm) using **A)** OD_{600}, **B)** ATP activity (pg/mL), **C)** DNA concentration (µg/mL), and **D)** 16S rRNA-targeted qPCR (copies/µL) over 21 days. Vertical red lines indicate when fresh media was added to both reactors and THPS was added to one, vertical yellow lines indicate when media was flushed from both reactors and fresh media added.

greater affect being seen in the planktonic cells. This follows current understandings of biofilm-associated growth where tolerance of outside stresses is mitigated by the biofilm [43, 44].

9.3.1.2 MDREP Ratio

Figure 9.3 shows the calculated relative MDREP ratios as compared to 16S rRNA over three targeted days from the THPS trial. Two scales are shown for each time point to help illustrate the changes between different MDREP ratios. The days were chosen to reflect the time before biocide exposure, three days following biocide exposure and seven days following biocide exposure. Time points were selected following the findings of Vikram et al. (2015) who showed three days to be an intermediate point between susceptible and resistant phenotypes of *Pseudomonas fluorescens* exposed to glutaraldehyde [45]. MDREP ratios reflect the targeted MDREP copy number relative to the 16S rRNA copy number of the coupon at each time point; therefore, an increase in the MDREP ratio would indicate when a gene has increased in copies relative to cell counts. A change observed in the test bioreactor not seen in

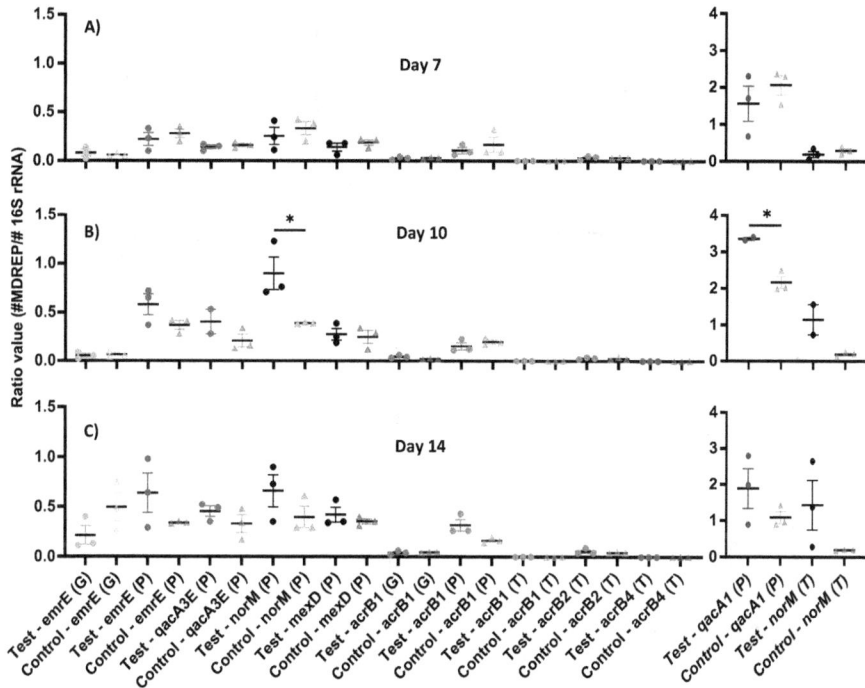

FIGURE 9.3 MDREP to 16S rRNA ratios for Test and Control bioreactors during the THPS trial as calculated using the copy numbers from qPCR. The ratios are shown for **A)** day 7, **B)** day 10, and **C)** day 14, to reflect pre-biocide, three days post-biocide exposure, and seven days post exposure, respectively. Stars indicate a significant difference according to two-tailed T test (p-value < 0.05).

the control bioreactor would indicate a gene is being selected for under the selective pressure of the biocide.

Two genes showed a significant difference between the test and control bioreactors: *norM* (P) and *qacA1* (P) (Figure 9.3B). This difference was only seen on day 10 (three days after biocide exposure), afterward the variance between the test bioreactor replicates increased. Though not statistically significant, the *norM* (T) ratio also showed an increased ratio in the test bioreactor following THPS exposure not seen in the control bioreactor. As with the *qacA1* (P) ratios, *norM* (T) showed greater variability in MDREP ratios of the test bioreactor following THPS exposure which was not seen in the control bioreactor ratios (Figure 9.3B and C). It is worth noting a replicate for the *qacA1* (P) and *norM* (T) test bioreactor both failed to amplify entirely at day 10, thus only two replicates are shown for each. Many of the *acrB* ratios were very low and showed no changes over time (Figure 9.3). The highest *acrB* ratios were observed in *acrB1* (P), where the ratios were stable between the two bioreactors but a slight increase in the test bioreactor was seen at day 14 (Figure 9.3C).

A detailed view of selected MDREP ratios is seen in Figure 9.4 to better highlight the changes over time. *mexD* (P) showed a trending increase between both bioreactors

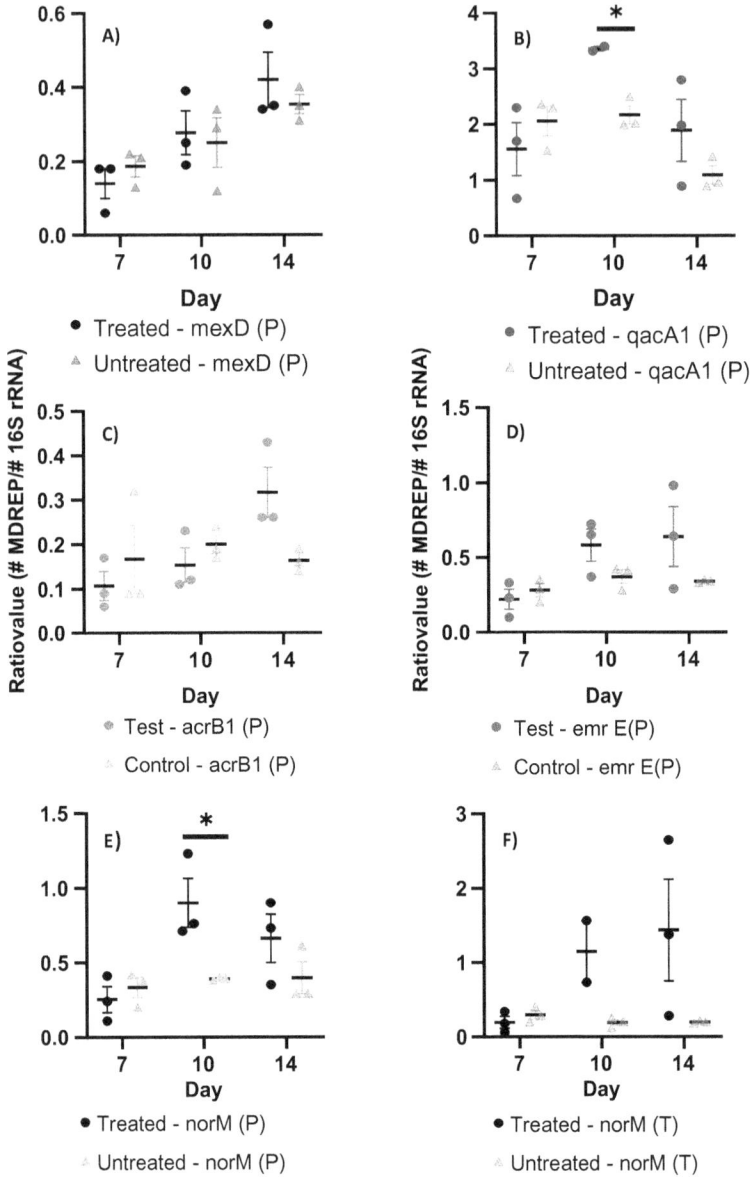

FIGURE 9.4 Selected MDREP ratios for **A)** *mexD* (P), **B)** *qacA1* (P), **C)** *acrB1* (P), **D)** *emrE* (P), **E)** *norM* (P), and **F)** *norM* (T) from the THPS trial of bioreactors to show changes in ratios closely over time. Stars indicate a significant difference according to two-tailed T test (*p*-value < 0.05).

(Figure 9.4A). *qacA1* (P) and *norM* (P) both show an increased MDREP ratio on day 10 followed by a decrease on day 14 (Figure 9.4B and E). MDREP ratios for *acrB1* (P), *emrE* (P), and *norM* (T) of the test bioreactor all increased with time but showed increasing variability between replicates (Figure 9.4B, C and F). This information suggests that *norM* and *qacA1* may be selected for by an exposure of THPS.

The growth monitoring methods from the second THPS trial showed a smaller difference between planktonic and sessile cells in both bioreactors and a lesser impact on sessile growth of the test bioreactor (data not shown).

9.3.2 BAC TRIAL

9.3.2.1 Growth Monitoring

The growth monitoring of the BAC trial bioreactors using OD_{600}, ATP, DNA, and 16S rRNA qPCR readings is shown in Figure 9.5. The planktonic cells were responsive to the BAC biocide, as seen in the decrease of the OD_{600}, ATP, and DNA values of the test bioreactor on day 8 (Figure 9.5A, B, and C). 16S rRNA qPCR showed a slight increase in the test bioreactor on day 8, after which the counts decreased by day 10 and recovered by day 14 (Figure 9.5D). As with the THPS trial, sessile values were much lower than the planktonic readings except for day 21 readings when OD_{600} and 16S rRNA copy numbers became similar. Omitting day 21, the planktonic values for ATP, DNA, and 16S rRNA were 46, 41, and 38 times higher than their sessile counterparts in the test bioreactor and 27, 59, and 29 times higher in the control bioreactor, respectively. The largest discrepancies between the monitoring methods were observed in the planktonic cells of the control bioreactor where the OD_{600} and ATP increased between days 7 and 8 then decreased toward day 14 while the DNA and 16S rRNA dipped on day 8 and increased afterward (Figure 9.5B, C and D). To better illustrate the differences between sessile values of the test and control bioreactors, the DNA and 16S rRNA values were replotted separately (Figure 9.5E and F, respectively). From these isolated graphs, we can see the DNA and 16S rRNA copy numbers of the sessile cells from both bioreactors were very similar on days 7–10, after which the control bioreactor values increased on day 14. After the BAC was flushed, the test bioreactor values surpassed the control bioreactor by day 21 in both DNA and 16S rRNA copy numbers (Figure 9.5E and F).

9.3.2.2 MDREP Ratio

The MDREP to 16S rRNA ratio values for days 7, 10, and 14 from the BAC trial bioreactors are shown in Figure 9.6. Most ratios are near or below 1.0 with the exception of *acrB1* (P), which are greater than 2 on day 7 and 10 and plotted on their own axis (Figure 9.6A and B). All MDREP ratios decreased between days 10 and 14 (Figure 9.6B and C). *emrE* (G) showed a significant difference between the test and control bioreactors on day 7 (Figure 9.6A), while *acrB1* (G) showed a significant difference on day 10 (Figure 9.6B). No other MDREP ratios showed significant differences. To better see the changes in ratios from select MDREP primer targets, *emrE* (G), *emrE* (P), *acrB1* (G), *acrB2* (T), *mexD* (P), and *norM* (T) are shown in Figure 9.7. Despite there not being a significant difference between test and control bioreactor ratios on day 10, the *emrE* (G) values showed an increase in the ratio between days

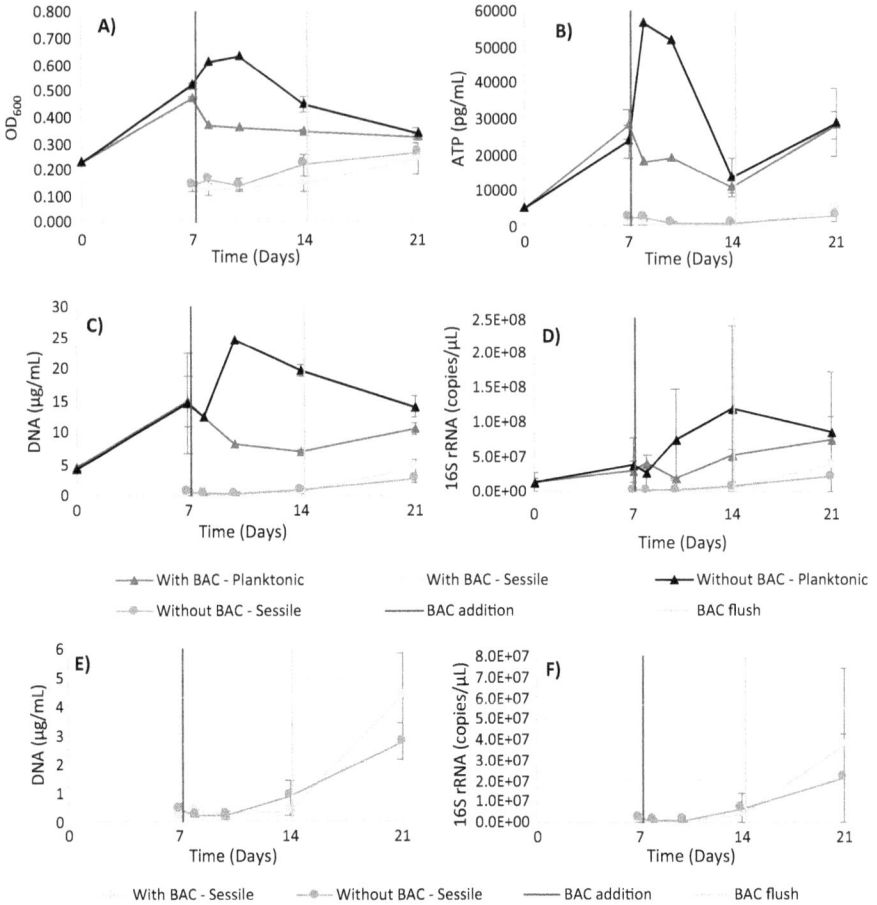

FIGURE 9.5 Growth monitoring of planktonic and sessile cells from parallel bioreactors treated with or without BAC (1 ppm) using **A)** OD_{600}, **B)** ATP activity (pg/mL), **C)** DNA concentration (µg/mL) and **D)** 16S rRNA targeted qPCR (copies/µL) over 21 days. Vertical red lines indicate when fresh media was added to both reactors and BAC was added to one, vertical yellow lines indicate when media was flushed from both reactors and fresh media added. To highlight the differences in DNA and 16S rRNA values between the sessile cells of the test and control bioreactors, the planktonic cells were removed, and plots redone in panels **E)** and **F)**, respectively.

7 and 10, which then decreased by day 14 (Figure 9.7A). This is also seen in *emrE* (P) ratios but to a lesser extent (Figure 9.7B), and in the *acrB1* (G) and *acrB2* (T) MDREP ratios (Figure 9.7 C and D, respectively). The MDREP ratios from the test bioreactors for *acrB1* (G) and *acrB2* (T) began higher than the control ratios on day 7 and only increased slightly on day 10. *mexD* (P) showed a slight increase in both reactors on day 10, but the variability between replicates from both bioreactors makes drawing conclusions difficult (Figure 9.7E). *norM* (T) ratios increased in both bioreactors on day 10, with both ratios having greater variability before dropping on day 14 (Figure 9.7F).

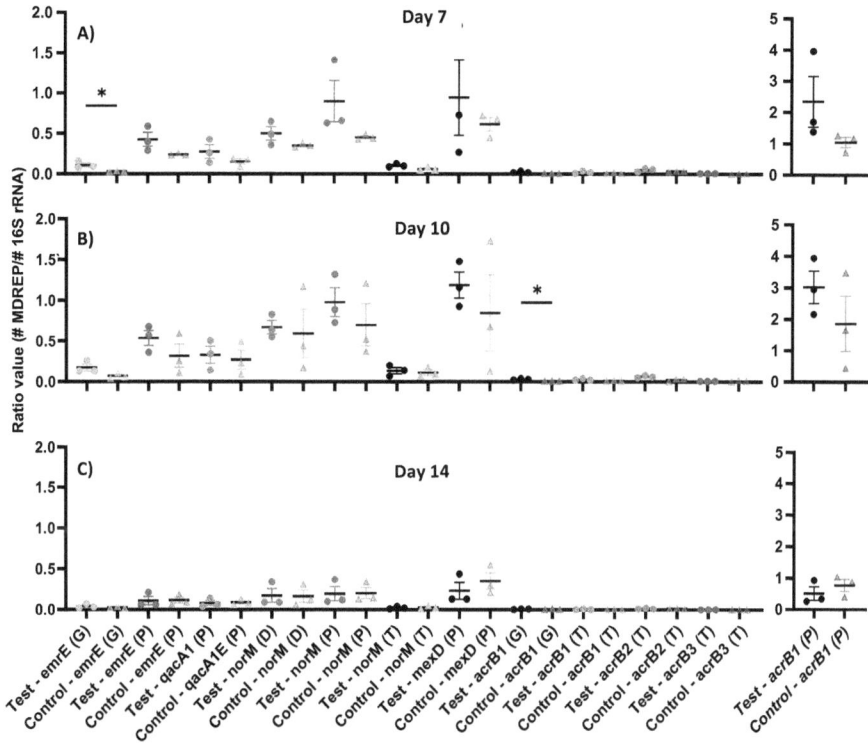

FIGURE 9.6 MDREP to 16S rRNA ratios for Test and Control bioreactors during the BAC trial as calculated using the copy numbers from qPCR. The ratios are shown for **A)** day 7, **B)** day 10, and **C)** day 14, to reflect pre-biocide, three days post-biocide exposure and seven days post exposure respectively. Stars indicate a significant difference according to two-tailed T test (p-value < 0.05).

These results suggest *emrE*, regardless of the source species, is being selected by BAC for improved community fitness. The response in terms of the actual ratio value was different between the two *emrE* genes (the *emrE* (P) ratio was at least three times higher than *emrE* (G) ratios), but this may be a result of community dynamics and the relative abundance of the different species.

The second trial of BAC showed very similar trends in growth methods, but the difference between planktonic and sessile cells was smaller. Sessile cells had similar values between test and control bioreactors until day 14 when the control bioreactor increased, but the values were similar again on day 21.

9.3.2.3 Community Composition

Comparing community fragment analyses to pure culture fragment analyses using T-RFLP, we were able to identify unique fragments which corresponded to each species. The results from the T-RFLP analysis are shown in Table 9.4. Fragment analyses of the pooled DNA from every bioreactor successfully identified all community members (Table 9.4). Unfortunately, we were unable to quantitatively use T-RFLP to

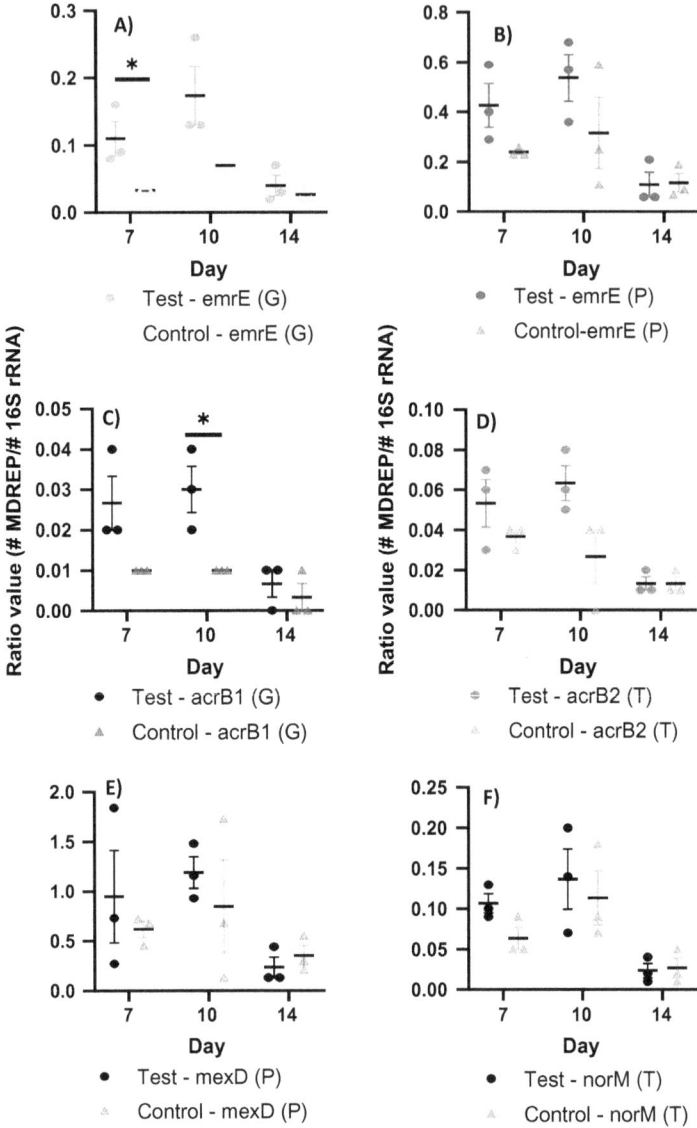

FIGURE 9.7 Selected MDREP ratios for **A)** *emrE* (G), **B)** *emrE* (P), **C)** *acrB1* (G), **D)** *acrB2* (T), **E)** *mexD* (P), and **F)** *norM* (T) from the BAC trial bioreactors to show changes in ratios closely over time. Stars indicate a significant difference according to two-tailed T test (*p*-value < 0.05).

assess the relative abundances of the community and thus were limited to the presence or absence of each species. The presence of each species was scored manually as either a 0 to represent absence, 0.5 for weakly present (peak was observed but either labeled and very small, or too small to be labeled but still clearly present), and 1 for clearly present. From these results, we can see that *D. vulgaris* is always present in all reactors though at reduced levels on days 10 and 14 in the THPS test bioreactor. *P. putida* was always clearly present in all bioreactors. *T. aromatica* was weakly or strongly present in all four bioreactors, but these scores were more random over the time course of the reactors. *G. subterraneus* was only clearly present in the BAC bioreactors and was undetectable at the start of both THPS bioreactors but was then weakly present for the remainder of time points (Table 9.4).

Though not quantitative, these results indicate that the robust biofilm former *P. putida* was a dominant species in all bioreactors, *G. subterraneus* was the weakest community member while *T. aromatica* and *D. vulgaris* were in between (Table 9.4). These community composition scores suggest a possible reasoning for the calculated low MDREP ratios seen in BAC and THPS bioreactors (Figures 9.3 and 9.6).

9.4   CONCLUSIONS

Here, we hypothesized that MDREP genes would contribute to the increased tolerance toward specific biocides and that this could be monitored using qPCR. Although we have not yet applied this approach to field samples, we believe this work demonstrates the proof of principle for measuring a community's potential tolerance to either BAC of THPS based upon the abundance of the MDREP genes *emrE* or *norM/ qacA*, respectively. This work demonstrated the ability to track a community's genetic response to low concentrations of biocides through targeted qPCR of MDREP genes. It also demonstrated that the genetic response is unique for each biocide, indicating the potential for this work to be built upon to identify additional genetic markers of tolerance for each biocide in addition to other biocides.

With this approach, the ground has been laid to demonstrate a cost-effective method for assessing the potential for biocide tolerance, either as a pre-existing condition or acquired through insufficient biocide killing efficacy. Observing trends across all four methods, we see that again the four methods generally followed the same trends, supporting the conclusions of Brown and Turner (2022) (submitted) that the methods agree with each other even when applied to a mixed community. With further development, the work stands to show that a community's genetic response may be tracked simply with qPCR to such an extent that tolerance toward different biocides may be predicted and prevented with proactive modifications to biocide treatment programs. This will involve the development of more robust primer sets, aiming toward "universal" primers of the desired MDREP target. As demonstrated by Brown et al., (2021), this can be simply obtained and the library of primers expanded upon as needed with significant changes in the microbial community [42]. Furthermore, as more work is done in this field, the library of relevant MDREP genes will grow, providing more robust databases to draw upon.

REFERENCES

1. Sharma M, Liu H, Chen S, Cheng F, Voordouw G, Gieg L. Effect of selected biocides on microbiologically influenced corrosion caused by Desulfovibrio ferrophilus IS5. *Sci Rep*. 2018;8(1):16620. DOI: 10.1038/s41598-018-34789-7
2. Jones IA, Joshi LT. Biocide use in the antimicrobial era: A review. *Molecules*. 2021; 26(8):2276. DOI: 10.3390/MOLECULES26082276

3. Poole K. Mechanisms of bacterial biocide and antibiotic resistance. *J Appl Microbiol.* 2002;92(s1):55S–64S. DOI: 10.1046/j.1365-2672.92.5s1.8.x

4. Mcdonnell G, Russell AD. Antiseptics and disinfectants: Activity, action, and resistance. *Clin Microbiol Rev.* 1999;12(1):147–179. DOI: 10.1128/CMR.12.1.147

5. Russell AD. Mechanisms of bacterial resistance to antibiotics and biocides. *Prog Med Chem.* 1998;35(C):133–197. DOI: 10.1016/S0079-6468(08)70036-5

6. Russell AD. Plasmids and bacterial resistance to biocides. *J Appl Microbiol.* 1997;83(2):155–165. DOI: 10.1046/J.1365-2672.1997.00198.X

7. Lyont BR, Skurray R. Antimicrobial resistance of Staphylococcus aureus: Genetic basis. *Microbiol Rev.* 1987;51(1):88–134.

8. Leelaporn A, Paulsen IT, Tennent JM, Littlejohn TG, Skurray RA. Multidrug resistance to antiseptics and disinfectants in coagulase-negative staphylococci. *J Med Microbiol.* 1994;40(3):214–220. DOI: 10.1099/00222615-40-3-214

9. Kucken D, Feucht H-H, Kaulfers P-M. Association of qacE and qacE Δ1 with multiple resistance to antibiotics and antiseptics in clinical isolates of Gram-negative bacteria. *FEMS Microbiol Lett.* 2000;183(1):95–98. DOI: 10.1111/J.1574-6968.2000. TB08939.X

10. Rouch DA, Cram DS, Di Berardino D, Littlejohn TG, Skurray RA. Efflux-mediated antiseptic resistance gene qacA from Staphylococcus aureus: Common ancestry with tetracycline- and sugar-transport proteins. *Mol Microbiol.* 1990;4(12):2051–2062. DOI: 10.1111/j.1365-2958.1990.tb00565.x

11. Neyfakh AA, Bidnenko VE, Chen LB. Efflux-mediated multidrug resistance in Bacillus subtilis: Similarities and dissimilarities with the mammalian system. *Proc Natl Acad Sci U S A.* 1991;88(11):4781–4785. DOI: 10.1073/PNAS.88.11.4781

12. Tennent JM, Lyon BR, Gillespie MT, May JW, Skurray RA. Cloning and expression of Staphylococcus aureus plasmid-mediated quaternary ammonium resistance in Escherichia coli. *Antimicrob Agents Chemother.* 1985;27(1):79–83. DOI: 10.1128/ AAC.27.1.79

13. Nikaido H, Pagès JM. Broad-specificity efflux pumps and their role in multidrug resistance of Gram-negative bacteria. *FEMS Microbiol Rev.* 2012;36(2):340–363. DOI: 10.1111/j.1574-6976.2011.00290.x

14. Jack DL, Storms ML, Tchieu JH, Paulsen IT, Saier MH. A broad-specificity multidrug efflux pump requiring a pair of homologous SMR-type proteins. *J Bacteriol.* 2000;182(8):2311–2313. DOI: 10.1128/JB.182.8.2311-2313.2000

15. Lewis K. Multidrug resistance pumps in bacteria: Variations on a theme. *Trends Biochem Sci.* 1994;19(3):119–123. DOI: 10.1016/0968-0004(94)90204-6

16. Nikaido H. Prevention of drug access to bacterial targets: Permeability barriers and active efflux. *Science.* 1994;264(5157):382–388. DOI: 10.1126/SCIENCE.8153625

17. Xue Y, Voordouw G. Control of microbial sulfide production with biocides and nitrate in oil reservoir simulating bioreactors. *Front Microbiol.* 2015;6(DEC):1387. DOI: 10.3389/FMICB.2015.01387

18. Davidova I, Hicks MS, Fedorak PM, Suflita JM. The influence of nitrate on microbial processes in oil industry production waters. *J Ind Microbiol Biotechnol.* 2001;27(2):80–86. DOI: 10.1038/SJ.JIM.7000166

19. Xu D, Jia R, Li Y, Gu T. Advances in the treatment of problematic industrial biofilms. *World J Microbiol Biotechnol.* 2017;33:97. DOI: 10.1007/s11274-016-2203-4

20. Fernando Bautista L, Vargas C, González N, Molina MC, Simarro R, Salmerón A, et al. Assessment of biocides and ultrasound treatment to avoid bacterial growth in diesel fuel. *Fuel Process Technol.* 2016;56–63. DOI: 10.1016/j.fuproc.2016.06.002

21. Elumalai P, Parthipan P, Narenkumar J, Sarankumar RK, Karthikeyan OP, Rajasekar A. Influence of thermophilic bacteria on corrosion of carbon steel in hyper chloride environment. *Int J Environ Res.* 2017;11:339–347. DOI: 10.1007/s41742-017-0031-5

22. Enning D, Venzlaff H, Garrelfs J, Dinh HT, Meyer V, Mayrhofer K, et al. Marine sulfate-reducing bacteria cause serious corrosion of iron under electroconductive biogenic mineral crust. *Environ Microbiol.* 2012;14(7):1772–1787. DOI: 10.1111/j.1462-2920.2012.02778.x

23. Venzlaff H, Enning D, Srinivasan J, Mayrhofer KJJ, Hassel AW, Widdel F, et al. Accelerated cathodic reaction in microbial corrosion of iron due to direct electron uptake by sulfate-reducing bacteria. *Corros Sci.* 2013;66:88–96. DOI: 10.1016/J. CORSCI.2012.09.006

24. Little BJ, Lee JS. Microbiologically influenced corrosion: An update. *Int Mater Rev.* 2014;59(7):384–393. DOI: 10.1179/1743280414Y.0000000035

25. Hostacka A, Ciznar I, Steflovicova M. Temperature and pH affect the production of bacterial biofilm. *Folia Microbiol.* 2010;55(1):75–78.

26. Marsden AE, Grudzinski K, Ondrey JM, DeLoney-Marino CR, Visick KL. Impact of salt and nutrient content on biofilm formation by Vibrio fischeri. *PLoS One.* 2017;12(1):e0169521. DOI: 10.1371/JOURNAL.PONE.0169521

27. Seiler C, van Velzen E, Neu TR, Gaedke U, Berendonk TU, Weitere M. Grazing resistance of bacterial biofilms: A matter of predators' feeding trait. *FEMS Microbiol Ecol.* 2017;93(9):112. DOI: 10.1093/FEMSEC/FIX112

28. Hathroubi S, Mekni MA, Domenico P, Nguyen D, Jacques M. Biofilms: Microbial shelters against antibiotics. *Microbial Drug Resist.* 2017;23(2):147–156. DOI: 10.1089/ MDR.2016.0087

29. Mah T-FC, O'Toole GA. Mechanisms of biofilm resistance to antimicrobial agents. *Trends Microbiol.* 2001;9(1):34–39. DOI: 10.1016/S0966-842X(00)01913-2

30. Gilbert P, McBain AJ. Biofilms: Their impact on health and their recalcitrance toward biocides. *Am J Infect Control.* 2001;29(4):252–255. DOI: 10.1067/MIC.2001.115673

31. Stalder T, Top E. Plasmid transfer in biofilms: A perspective on limitations and opportunities. *NPJ Biofilms Microbiomes.* 2016;2:16022. DOI: 10.1038/npjbiofilms.2016.22

32. Abe K, Nomura N, Suzuki S. Biofilms: Hot spots of horizontal gene transfer (HGT) in aquatic environments, with a focus on a new HGT mechanism. *FEMS Microbiol Ecol.* 2020;96(5):31. DOI: 10.1093/FEMSEC/FIAA031

33. Aminov RI. Horizontal gene exchange in environmental microbiota. *Front Microbiol.* 2011;2(JULY):158. DOI: 10.3389/FMICB.2011.00158/BIBTEX

34. Henriques Normark B, Normark S. Evolution and spread of antibiotic resistance. *J Intern Med.* 2002;252(2):91–106. DOI: 10.1046/J.1365-2796.2002.01026.X

35. Bengtsson-Palme J, Kristiansson E, Larsson DGJ. Environmental factors influencing the development and spread of antibiotic resistance. *FEMS Microbiol Rev.* 2018;42(1):68–80. DOI: 10.1093/FEMSRE/FUX053

36. Guo J, Li J, Chen H, Bond PL, Yuan Z. Metagenomic analysis reveals wastewater treatment plants as hotspots of antibiotic resistance genes and mobile genetic elements. *Wat Res.* 2017;123:468–478. DOI: 10.1016/j.watres.2017.07.002

37. Yu Z, Gunn L, Wall P, Fanning S. Antimicrobial resistance and its association with tolerance to heavy metals in agriculture production. *Food Microbiol.* 2017;64:23–32. DOI: 10.1016/J.FM.2016.12.009

38. Chen S, Li X, Sun G, Zhang Y, Su J, Ye J. Heavy metal induced antibiotic resistance in bacterium LSJC7. *Int J Mol Sci.* 2015;16(10):23390–23404. DOI: 10.3390/ IJMS161023390

39. Ortega Morente E, Fernández-Fuentes MA, Grande Burgos MJ, Abriouel H, Pérez Pulido R, Gálvez A. Biocide tolerance in bacteria. *Int J Food Microbiol.* 2013;162(1):13–25. DOI: 10.1016/J.IJFOODMICRO.2012.12.028

40. Bock LJ. Bacterial biocide resistance: A new scourge of the infectious disease world? *Arch Dis Child.* 2019;104(11):1029–1033. DOI: 10.1136/ARCHDISCHILD-2018-315090

41. Condell O, Iversen C, Cooney S, Power KA, Walsh C, Burgess C, et al. Efficacy of biocides used in the modern food industry to control salmonella enterica, and links between biocide tolerance and resistance to clinically relevant antimicrobial compounds. *Appl Environ Microbiol.* 2012;78(9):3087–3097. DOI: 10.1128/AEM.07534-11

42. Brown DC, Turner RJ. Creation of universal primers targeting nonconser horizontally mobile genes: Lessons and considerat. *Appl Environ Microbiol.* 2021;87(4):1–18. DOI: 10.1128/AEM.02181-20

43. Ceri H, Olson ME, Stremick C, Read RR, Morck D, Buret A. The Calgary Biofilm Device: New technology for rapid determination of antibiotic susceptibilities of bacterial biofilms. *J Clin Microbiol.* 1999;37(6):1771–1776.

44. Yan J, Bassler BL. Surviving as a community: Antibiotic tolerance and persistence in bacterial biofilms. *Cell Host Microbe.* 2019;26(1):15–21. DOI: 10.1016/J.CHOM.2019.06.002

45. Vikram A, Bomberger JM, Bibby KJ. Efflux as a glutaraldehyde resistance mechanism in Pseudomonas fluorescens and Pseudomonas aeruginosa biofilms. *Antimicrob Agents Chemother.* 2015;59(6):3433. DOI: 10.1128/AAC.05152-14

46. Nikaido H, Zgurskaya HI. AcrAB and related multidrug efflux pumps of Escherichia coli. *J Mol Microbiol Biotechnol.* 2001;3(2):215–218.

47. Nikaido H. Multidrug efflux pumps of gram-negative bacteria. *J Bacteriol.* 1996;178(20):5853–5859.

48. Nikaido H, Basina M, Nguyen VY, Rosenberg EY. Multidrug efflux pump AcrAB of Salmonella typhimurium excretes only those beta-lactam antibiotics containing lipophilic side chains. *J Bacteriol.* 1998;180(17):4686–4692.

49. Bay DC, Turner RJ. Small multidrug resistance protein emre reduces host ph and osmotic tolerance to metabolic quaternary cation osmoprotectants. *J Bacteriol.* 2012;194(21):5941–5948. DOI: 10.1128/JB.00666-12

50. Bay DC, Rommens KL, Turner RJ. Small multidrug resistance proteins: A multidrug transporter family that continues to grow. *Biochim Biophys Acta - Biomembr.* 2008;1778(9):1814–1838. DOI: 10.1016/j.bbamem.2007.08.015

51. Lomovskaya O, Lewis K. Emr, an Escherichia coli locus for multidrug resistance. *Proc Natl Acad Sci U S A.* 1992;89(19):8938–8942. DOI: 10.1073/pnas.89.19.8938

52. Lomovskaya O, Lewis K, Matin A. EmrR is a negative regulator of the Escherichia coli multidrug resistance pump emrAB. *J Bacteriol.* 1995;177(9):2328–2334. DOI: 10.1128/jb.177.9.2328-2334.1995

53. Nishino K, Latifi T, Groisman EA. Virulence and drug resistance roles of multidrug efflux systems of Salmonella enterica serovar Typhimurium. *Mol Microbiol.* 2006;59(1):126–141. DOI: 10.1111/j.1365-2958.2005.04940.x

54. Paulsen IT, Skurray RA, Tam R, Saier MH, Turner RJ, Weiner JH, et al. The SMR family: A novel family of multidrug efflux proteins involved with the efflux of lipophilic drugs. *Mol Microbiol.* 1996;19(6):1167–1175. DOI: 10.1111/j.1365-2958.1996.tb02462.x

55. Paulsen IT, Littlejohn TG, Radstrom P, Sundstrom L, Skold O, Swedberg G, et al. The 3' conserved segment of integrons contains a gene associated with multidrug resistance to antiseptics and disinfectants. *Antimicrob Agents Chemother.* 1993;37(4):761–768. DOI: 10.1128/AAC.37.4.761

56. Masuda N, Sakagawa E, Ohya S, Gotoh N, Tsujimoto H, Nishino T. Substrate specificities of MexAB-OprM, MexCD-OprJ, and MexXY-OprM efflux pumps in Pseudomonas aeruginosa. *Antimicrob Agents Chemother.* 2000;44(12):3322–3327.

57. Welch A, Awah CU, Jing S, Van Veen HW, Venter H. Promiscuous partnering and independent activity of MexB, the multidrug transporter protein from Pseudomonas aeruginosa. *Biochem J.* 2010;430(2):355–364. DOI: 10.1042/BJ20091860

Section VI

Future Perspectives on Microorganisms in the Energy Transition

10 Future Perspectives
Where Do We Go from Here?

Andrea Koerdt
Bundesanstalt für Materialforschung und -prüfung (BAM), Berlin, Germany

Jerzy Samojluk
AGH University of Science and Technology, Kraków, Poland

Biwen Annie An Stepec
Norwegian Research Centre, Bergen, Norway

10.1 INTRODUCTION: BACKGROUND AND DRIVING FORCES

In recent years, the consequences of climate change have led to increased political tensions and public protests. Both the government and industry are demanded to take immediate actions by increasing the use of alternative energy sources. In fact, the energy sector is experiencing a drastic change at this moment, which unfortunately is developing at an inadequate rate. By the end of 2018, 79.9% of the energy used worldwide was still generated from fossil fuels (Figure 10.1). Thus, rapid actions are required to decrease the proportions of fossil fuels by promoting renewable energies.

In the midst of this energy transition, microbiology plays an important role. Not only are microorganisms found in virtually all environments, the vast amount of knowledge generated from traditional energy sectors is also applicable in emerging energy systems. In this chapter, we will examine and compare the different types of renewable energy resources and briefly discuss how microorganisms may impact individual systems, specifically:

- Which countries are leading in the production of renewable energy, and what kinds of options do they have/use?
- What kinds of options exist for renewable energy production, and how developed they are?
- What are the advantages and disadvantages?
- How do we store the produced energy?
- How would the system behave under the influence of microorganisms, for example, microbiologically influenced corrosion (MIC)?
- How to transfer knowledge from the petroleum sector to the renewables?

DOI: 10.1201/9781003287056-16

191

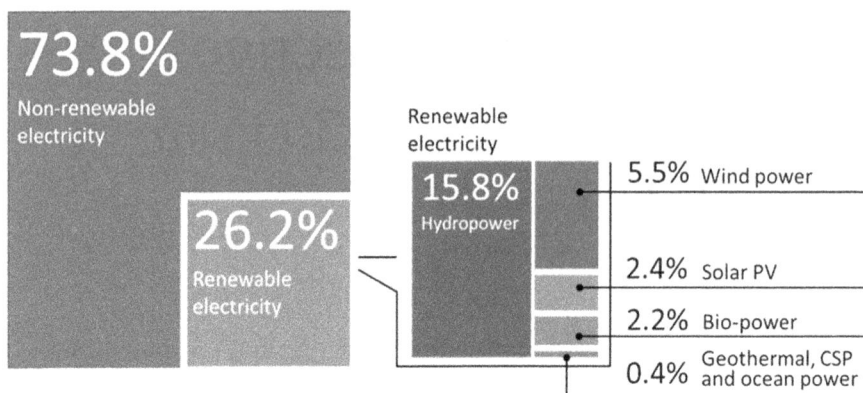

FIGURE 10.1 Estimated renewable energy share of global electricity production, end 2018. (Adapted from Ren21: Renewables 2019 global status report/Source based on OECD/IEA and IEA SHC [1].)

There are generally five different types of renewable energy sources: Solar, Wind, Hydro, Biomass, and Geothermal. However, these options are not equally available to all countries to the same extent. For instance, countries that have more land, such as China or the USA, are leading in wind power, with 210 MW and 130 MW, respectively[2]. Countries that receive more hours of sunshine will also produce larger amounts of solar energy than countries with shorter daylight hours. In this context, it is interesting to note that a rather northern country like Germany is a leading country in solar power generation (Figure 10.2).

The Renewables 2019 Global Status Report published in 2019 by REN21[1] (Renewable Energy Policy Network for the 21st Century) suggests that in 2018, China (404 gigawatts), the USA (180 gigawatts), and Germany (113 gigawatts) were leading in renewable energy production. However, the high values from China and the USA were accompanied with a much larger land area and population. Thus, if this is converted to a single inhabitant, the average per capita value would be 0.3 kW (China), 0.6 kW (USA), and 1.4 kW (Germany).

10.2 RENEWABLE ENERGY SOURCES: SOLAR ENERGY AND PHOTOVOLTAICS

In 1958, solar energy was used for the first time on the Vanguard 1 satellite. The efficiency at that time was only 6% and the production costs were immensely high, but this marked its first major application. Due to the energy crisis in the 1970s, increased urgency to search for an alternative terrestrial energy source began, which led to the expansion of the solar energy market. The use of solar energy makes absolute sense, if one considers how much energy the sun theoretically, and in principle inexhaustibly, supplies. The sun radiates 10,000 times more than the current energy consumption, which is equivalent to around 1.2×10^{14} kW. In other terms, the energy received within one hour from the sun equates to the total annual energy consumption[3].

Gigawatts

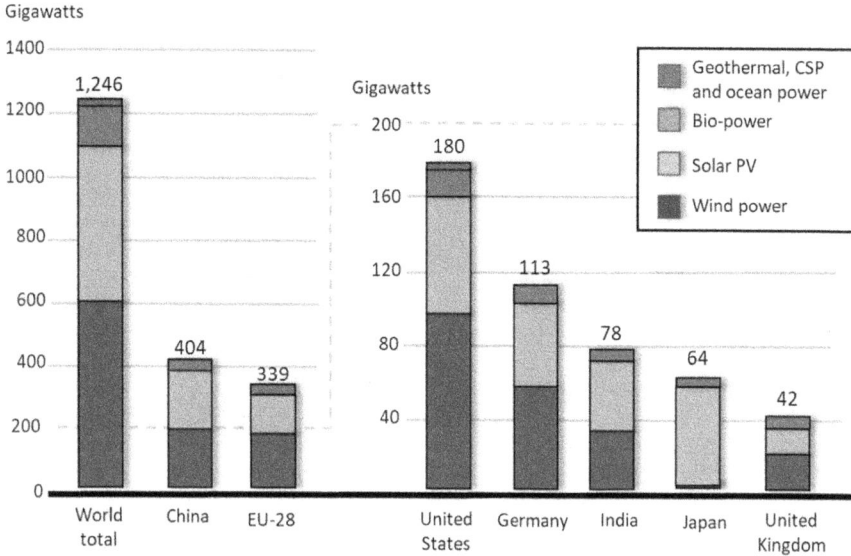

FIGURE 10.2 Renewable power capacities in the world, China, EU-28, and top 6 countries, 2018.

(Adapted from Ren21: Renewables 2019 global status report/ Source based on OECD/ IEA and IEA SHC[1].)

The energy from solar radiation can be used by solar cells (direct conversion into electricity by semiconductor devices) or solar collectors (accumulation of heat). Solar cells should not be confused with solar collectors. Solar cells convert solar radiation directly into electricity, which is often referred to as photovoltaic (PV) energy conversion and is based on the photovoltaic effect. Solar collectors use the sun's energy to generate heat, which is largely used to heat water. Because of this limited use, the further focus will be on solar cells and photovoltaics (Table 10.1).

Solar cells consist of semiconductors that become electrically conductive when exposed to light or heat and have an insulating effect at low temperatures. Most of the solar cells produced worldwide are made of silicon (Si). This is a great advantage because silicon is the second-most abundant element in the Earth's crust, so the raw materials are available in sufficient quantities, and it is also environmentally friendly. In the production of solar cells, the semiconductor material is "doped." This means the defined introduction of chemical elements with which one can achieve either positive charge carrier surpluses (p-conducting semiconductor layer) or negative charge carrier surpluses (n-conducting semiconductor layer) in the semiconductor material. The formation of two differently doped semiconductor materials creates a pn-junction at the interface where an electric field is built up. This then leads to a charge separation of the charge carriers released by the incidence of light. The electrical voltage can be tapped via metal contacts. As soon as a consumer is connected, the circuit closes, and direct current can flow.

TABLE 10.1
Advantages and Disadvantages of Solar Energy and Photovoltaic Systems

Advantage	Drawbacks
• environmentally friendly • no noise, no moving parts • no emissions • no use of fuel and water • minimal maintenance requirements • long lifetime, up to 30 years • electricity is generated wherever there is light, solar, or artificial light sources • PV operates even in cloudy weather conditions • modular or "custom-made" energy, can be designed for any application from watch to a multi-megawatt power plant	• PV cannot operate without light • high initial costs that overshadow the low maintenance costs and lack of fuel costs • large area needed for large-scale applications • PV generates direct current: special DC appliances or inverters are needed in off-grid • applications energy storage is needed, such as batteries

TABLE 10.2
Four General Types of Solar Cells

	Monocrystalline Solar Cell	Polycrystalline Solar Cell	Thin-film Modules	CIGS Modules
Advantage	• high efficiency (~21%) • low area demand	• medium efficiency (~16%) • low production costs	• low weight • low production costs	• medium efficiency (~17%) • low wight
Drawbacks	• high production costs	• middle area demand	• low efficiency (~7%) • high area demand	• high production costs

There are currently four different types of solar cells, each with its own advantages and disadvantages (Table 10.2).

Currently, solar cells cannot convert 100% of the sun's energy. The efficiency (η) indicates the effectiveness with which the solar cell works and is calculated from the ratio of electrical energy emitted to the incident light energy (max. 1 = 100%). The efficiency describes how well the solar cell utilizes the energy available as light. The solar systems available on the market have an efficiency of ~21%. There are several factors that influence their efficiency, including electrical current losses, the temperature of the modules, and the processed frequencies of the light spectrum. The latter determines the physically possible limit of efficiency from monocrystalline silicon. Based on thermodynamic calculations, the maximum conversion for a single solar cell is up to 33%, but this is not achieved in practice yet. However, a solar cell performs better than a photovoltaic system because the losses of all components of a system are included in the total efficiency. In the inverter, direct current is converted into alternating current, which is also associated with energy losses. In addition, the length and cross-sections of connecting lines, including the long lines for alternating current, further reduce their effectiveness[4].

A major barrier to the expansion of the use of solar energy, as well as other renewable energies, has long been the relatively high cost of installing and operating the systems. In the last 10 years, politically supported developments have significantly reduced the cost of grid infrastructure, capital costs, and operating costs. In fact, they are currently at a level like that of fossil fuels, with the distinct advantage of a very low carbon footprint (Figure 10.3)[5].

10.2.1 FUTURE PERSPECTIVES OF SOLAR ENERGY AND PHOTOVOLTAIC

The Intergovernmental Panel on Climate Change (IPCC) assumes that humanity must be climate neutral by 2050 to limit global warming to 1.5°C. In a special report, the IPCC published that photovoltaics would provide about 12.5 trillion kWh

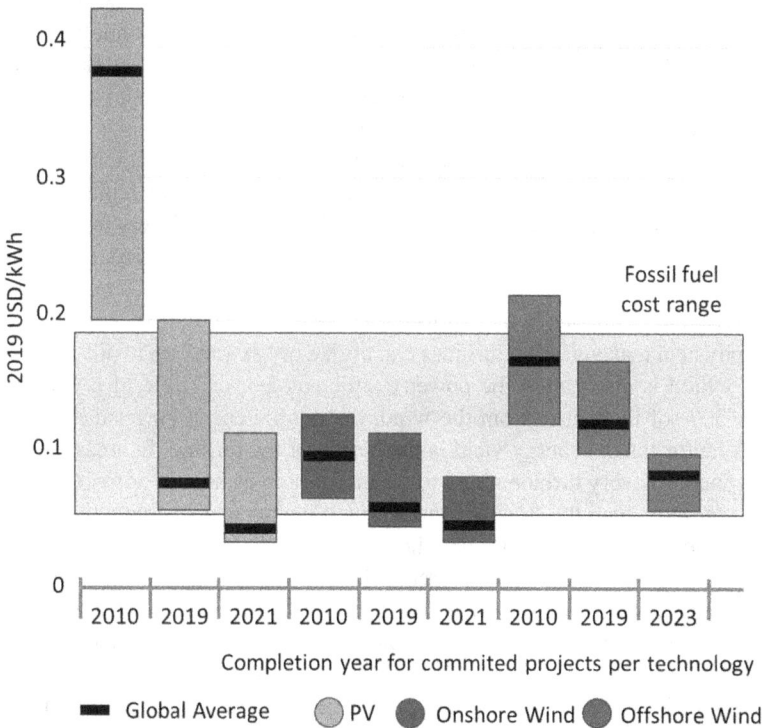

FIGURE 10.3 Overview of costs in US dollars/KWh for grid infrastructure, capital costs, and operating costs of renewable energy in 2010, 2019, and 2021/2023 compared to fossil fuels (light blue box represents the averages for fossil fuel-fired power generation usually between $0.05/kWh and $0.18/kWh). Photovoltaics (PV), wind onshore, and offshore are shown as examples.

(Figure adapted from Douglas Broom and based on the IRENA Report 2019[6].)

annually by then. Experts from the solar industry are now accusing the IPCC for underestimating this figure, since they are basing their estimates on outdated assumptions about the cost development of photovoltaics. Rather, they assume 41–96 trillion kWh, citing an international team of researchers led by Aarhus University in Denmark. The price of electricity from solar cells assumed by IPCC is too high and the models are too conservative for the potential of renewable energy, Marta Victoria (lead author of the study) informed. She predicts that photovoltaics will be one of the most important sources of electricity in the world, as continuous development will not only increase efficiency but also reduce production costs. In fact, experiments are currently underway with so-called tandem solar cells, which should reach market maturity in 2022/2023. They consist of silicon and perovskite and have achieved efficiencies of almost 30% [7]. Like pure silicon solar cells, tandem solar cells are expected to become more affordable over time.

10.3   RENEWABLE ENERGY SOURCES: WIND ENERGY

Wind energy is one of the oldest forms of energy used by humankind, first for transport (sailing, later balloons), then for performing mechanical work (windmills/pumps). The importance of wind as a source of energy was much greater in the past than it is today, and until the 19th century it was used almost exclusively for shipping. The discovery of America would not have been possible without wind power. Ironically, the importance of wind power was diminished and replaced by fossil fuels, and the resulting consequences are well known today. It was not until the oil crisis in the 1970s that wind power regained its importance as alternative sources of energy were needed. Since the early 1990s, the wind industry has been one of the fastest-growing industries in the world[8].

For physical reasons, wind turbines can utilize or "extract" up to 50% of the wind energy, which is also called the power coefficient. For example, if a wind turbine extracts 50% of its energy from the wind, the coefficient of performance is 0.5. A decisive factor for the energy yield is the height of the turbine. In areas close to the ground, the air is very turbulent due to obstacles such as houses or trees. The higher you go the more even the flow becomes, which makes power production more efficient. Furthermore, at high altitudes the wind speed is higher than at ground level, which is called wind shear. On average, the electricity yield increases by 1% for every meter a wind turbine is built higher. Also important for electricity production is the length and number of rotor blades. Over time, it has been found that turbines with three blades are the most efficient. Doubling the rotor length multiplies the electricity yield – doubling the wind speed results in eight times the yield. For this reason, wind turbines have become taller over time with correspondingly longer rotor blades. In the first modern turbines in the 1980s and early 1990s, the rotor diameter was 30–40 meters, and the center of the rotor hub was 40–60 meters above the ground. In 2018, German wind turbines offered on the market already had an average rotor diameter of 118 meters and a hub height of 132 meters (Table 10.3).

TABLE 10.3
Estimated Wind Power Output and Annual Energy Yields from 1980 to 2010[9,10]

	1980	1985	1990	1995	2000	2005	2010
Power output (kWh)	30	80	250	600	1500	3000	7500
Rotor diameter	15	20	30	46	70	90	126
Hub height	30	40	50	78	100	105	135
Annual energy yield (MWh)	35	95	400	1250	3500	6900	20000

10.3.1 PHYSICS OF WIND TURBINES

The moving mass (air) contains kinetic energy, which increases as a function of the square of the wind speed.

$$E = 1/2\,m^*v^2$$

(E = energy, m = mass, v = speed)

The mass flow (air flow rate) passing through the rotor surface of the wind turbine in a given time increases proportionally with the wind speed.

$$\dot{m} = A^*\rho^*v$$

($\dot{m}$ = mass flow, A = area of rotor, ρ = density of air, v = velocity).

The power P is equal to the energy E per unit time. Thus, the result for the power of the wind is:

$$P = \dot{E} = 1/2\,A^*\rho^*v3$$

Thus, the power of the wind depends on the third power of the wind speed. From this it can be concluded that if the wind speed is doubled, the energy supply, which is converted into rotational energy by the wind turbine, causes an eightfold increase.

When the kinetic energy of moving air is converted into electrical energy, the energy is first converted into mechanical rotational energy via the rotor blades. A generator then supplies the electric current. This conversion is subject to energy losses. In purely physical terms, no more than 59% of the power can be extracted from the wind. In addition, there are aerodynamic losses due to friction and turbulence on the rotor blade. It is estimated that a further 10% is caused by friction in the bearings and the gearbox, as well as the generator itself. According to the IEA-report 2021, 1591 TWh of energy were generated globally by wind energy in this way in 2020[10]. The leading countries worldwide are China (288 GW), the USA (122 GW), Germany (62 GW), India (38 GW), and Spain (27 GW)[9].

10.3.2 Types of Wind Turbines

Since the three-bladed rotor is the most used, this chapter focuses on this type of wind turbine. However, there are also other types of wind turbines, which will be presented in the following.

10.3.2.1 Three Blade Horizontal Wind Turbine

This type of wind turbine is the most efficient and therefore the most widely used. This turbine can rotate at a wind speed of 4 m/s and reaches its full efficiency at a wind speed of 11 m/s. Due to the three-blade rotor, the turbine is particularly running smoothly and has a lifetime of about 25 years.

10.3.2.2 Vertical Wind Turbine

In this type, the rotor axis is in vertical position (stationary axis) and was used especially in the first wind turbines. This type of turbine is based on the principle of the Persian windmill, which dates back to the 7th century. The vertical wind turbine exists in different designs like the Darrieus rotor and the Savonius rotor as well as mixed forms.

In the Darrieus rotor, the vertical axis allows the gearbox and generator to be placed on the ground. In addition, the Darrieus rotor does not have a wind alignment position, which means it rotates independently of the wind direction but only at 4 m/s. A disadvantage of this type is its large space requirement.

The Savonius rotor, also with a vertical axis, has two semi-circular loops as rotor blades, which are offset against each other at the top and bottom. This type of rotor can rotate at wind speeds as low as 2 m/s. However, its performance is significantly lower than that of the Darrieus rotor and far below that of the three-bladed rotor.

10.3.2.3 Bladeless Wind Turbines

This type of wind turbine was developed by the Spanish startup Vortex Bladeless, and 100 turbines were produced in 2021. These turbines take advantage of aeroelastic resonance, generating electricity from the vibration of the machines in the wind. When the machines are swayed back and forth, vibrations occur that are converted into electricity by a generator. Since there are no rotor blades, these turbines have no noise emission. They have a lightweight design, are only 85 cm high, and require a significantly smaller foundation. This saves resources and therefore results in lower construction costs. A disadvantage, however, is the lower energy efficiency compared to classic wind turbines.

10.3.2.4 Hybrid Wind Turbines (Wind and Solar Energy)

Here, manufacturers and developers have thought about the non-constant supply of energy – namely, that the two renewable energy sources, wind and sun, complement each other (they often appear complementary to each other). The wind turbine is covered with flexible photovoltaic cells, enabling hybrid energy production.

10.3.3 The Tower of Wind Turbines

The type of tower depends on the wind turbine, as each must meet different conditions (e.g., height, strength of the wind). If the tower has no foundation, it is called a mast.

Tensioned masts are slender tubular constructions held by steel cables. They are very inexpensive and light but can only be used for small wind turbines (up to 250 kW). Installation is simple and can be done without a crane. However, they require a large ground area and are therefore rather unsuitable for wind farms.

Lattice towers were often used in first-generation wind turbines and require less material (half as much as tubular steel towers). Consequently, they are lighter and easier to assemble. However, they are more expensive than cylindrical towers because of the time required for assembly and, consequently, higher labor costs.

Nevertheless, in Europe they are more expensive than cylindrical towers, as a lot of labor time has to be used in their manufacture or assembly, resulting in significantly higher labor costs. They are therefore more common in countries with low labor costs. In Europe, they are rarely seen and are only used for very tall towers (160 meters). Tubular steel towers are the most common and widespread type of tower today. They are available in different variants, for example, cylindrical, conical, or sub-conical. They are divided into two to five segments, each 20 to 30 meters long. Towers with a length of 60–120 meters reach a weight of 60–250 tons. They are made from steel plates, which are then rolled and then welded. Transporting the segments can be a challenge, especially for installations larger than 2 mW.

Concrete towers are made of reinforced concrete; they are much thicker and heavier than steel towers (5–6 times heavier than a tubular steel tower of the same height). They have more favorable vibration characteristics and thus cause reduced noise emissions. They are usually built on site. Hybrid towers (not to be confused with the hybrid photovoltaic tower) are made of reinforced concrete in the lower part while the upper part is made of steel. They are mostly used for high towers, since the large diameter of the lower part, consisting only of steel, would cause transport problems.

The tower of wind turbines must withstand high loads at times, which makes it a key component. In addition to the weight of the rotor and machine nacelle, whose mass can amount to several hundred tons, vibrations also act on the tower. Furthermore, high bending moments occur at the base of the tower. The higher the tower (the higher the yield), the wider the tower base. Most wind turbines, to be precise the rotor blades and the nacelle, have a service life of 20–25 years. The old towers that are currently being dismantled will not be re-equipped with a nacelle due to the low height that the towers had at the time. Because of the too low energy balance, it is more attractive for most operators to dismantle the tower and build a new one. This is a major criticism in the context of the urgent need to save resources. It is indeed the case that the recovery of the built-in resources of wind turbines has not/barely been considered in the selection of materials, the design, and installation of the wind turbine in the past. It is criticized that the environmentally friendly dismantling is not legally regulated. It is possible for the operators to simply dispose of the waste without recovering the important and rare resources (e.g., metals). This is even more problematic for the comparatively new offshore turbines[11,12]. The dismantling of the foundations is still the biggest problem, which will be considered in more detail in the next section.

10.3.4 THE FOUNDATION OF THE WIND TURBINES

The basis of the wind turbine is the foundation. A high level of stability must be ensured. The most common variant on land is the shallow foundation. However, if

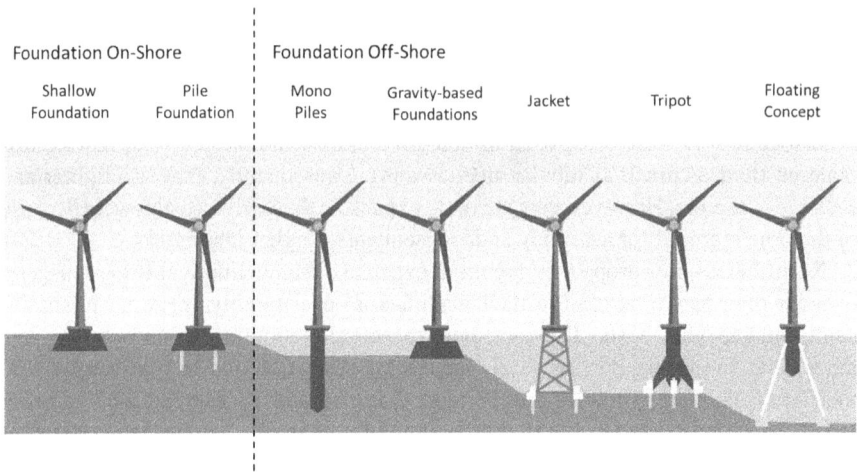

FIGURE 10.4   Different types of foundations used for onshore and offshore wind turbines.

only very soft soils are available, it is also necessary to drill/drive into more stable layers with piles. The capped heads are then interlaced with the foundation. This is then called pile foundation or deep foundation (Figure 10.4). The advantage is that the additional stability of the piles means that the associated foundations are usually smaller[13]. Negatively, besides the additional costs, are the problems of environmentally friendly dismantling, which is unfortunately not well regulated by law. In many cases, operators are not legally obliged to remove the foundation completely. In Germany, for example, the operator is only obliged to remove the foundation up to 10 meters below ground (even in the case of a shallow foundation). This means that with the intended expansion of wind energy, the soil is literally sealed. Environmentalists criticize that this means that not enough rainwater will reach the groundwater. If operators decide to completely remove the pile foundation, they face another problem, which also affects the groundwater. Pile foundations extend up to 40 meters into the ground and penetrate various water-bearing strata. If the piles were now removed, there would not only be a risk of saltwater getting into the drinking water but also pesticides from agriculture.

The first offshore wind farm started in 1991 in Denmark as a pioneer project. Offshore plants are therefore rather one of the younger variants of renewable energy, and many aspects are therefore not yet exactly clarified. For the foundation of offshore plants, several methods exist meanwhile. Hollow steel piles are often driven into the ground. Small wind turbines can be mounted on single piles and are called monopiles. Larger turbines require more stability and are usually mounted on three (tripot/tripile) or four (jacket). Increasingly, so-called bucket foundations are being used. These are placed under pressure instead of noisy pile driving. Also possible is the gravity foundation, which is like onshore installations, a shallow foundation with a precast concrete element. In the meantime, there are also concepts for wind turbines with floats anchored to the seabed. Here, the measures to counteract heeling (list) differ. The first prototypes of these turbines were installed in 2018 and are currently being tested. They have the advantage that they can be used on steeper sloping coasts. However, they are very costly. After the offshore wind turbine is installed, it must be connected to the power grid. Here, an inner park cabling as well as the external grid connection to the mainland is required.

10.3.5 OFFSHORE EQUIPMENT

Offshore wind turbines, which are largely made of steel, are subject to harsh environmental conditions throughout their operating life and, like all offshore installations, require special protection. The chemical properties of the submerged medium, that is, seawater or brackish water, influence the corrosion processes of metals[14]. Saltwater is extremely corrosive compared to drinking water, as corrosion increases with increasing salinity, but parameters such as pH (seawater 7.9–8.3), temperature, or oxygen content also have an influence. Different types of corrosion can occur, for example, uniform corrosion (or general corrosion), pitting corrosion, crevice corrosion, galvanic corrosion, erosion corrosion, or MIC[15,16]. In addition, waves and wind can initiate stress corrosion cracking and corrosion fatigue. Offshore wind turbines therefore require additional protective measures to ensure the planned operating time of 25 years or more. This is a particular challenge because different challenges may exist depending on the zone/part of the wind turbine (Figure 10.5). The different parts of the wind turbines are therefore protected against corrosion with different strategies; often different strategies/techniques are combined. Offshore wind turbines (Figure 10.5; mono pile as an example) can be divided into four parts: the turbine, the tower, the transition piece, and the foundation. The foundation and transition piece are further divided into "sub" zones. The underwater zone (UWZ), which is constantly underwater, and the "tidal" water zone (TWZ), which includes the area of ebb and flow (changing influence of waves, sun, biological growth, or other floating objects). This is followed by the "splash" water zone (SWZ), which is influenced by waves and seawater (Figure 10.5). Corrosion on this zone is most often caused by salt remaining on the surface. Below sea level, the calculated corrosion rates are between 0.08 and 0.14 mm/year for uniform corrosion and 0.07–0.21 mm/year in the splash water and tidal zones[17]. Corrosion in the surge and tidal zones is very high due to high chloride concentrations, humidity, and changes in pH during wet and dry cycles[18]. In the submerged zone, combinations of different techniques are often used. These can include (organic) coatings, corrosion surcharges to compensate for corrosion losses, and galvanic anode cathodic protection (GACP) or impressed current cathodic protection (ICCP) systems for the steel exposed to water inside and outside the foundation structures.

Protection of these huge quantities of steel (e.g., the monopiles currently in use have a weight of up to 805 t[19] with a correspondingly large, exposed area inside and outside the monopiles e.g., 70 m length, 6.8 m diameter ≈ 1500 m² per side) requires a sufficiently high protective current provided by the different cathodic protection systems[19].

National and international standards regulate the technical requirements of the corrosion protection systems to ensure the use of the best available techniques (e.g., DNVGL RP-0416 (2016), DNVGL-RP-B401 (2017), NORSOK (e.g., M-501, 2012),

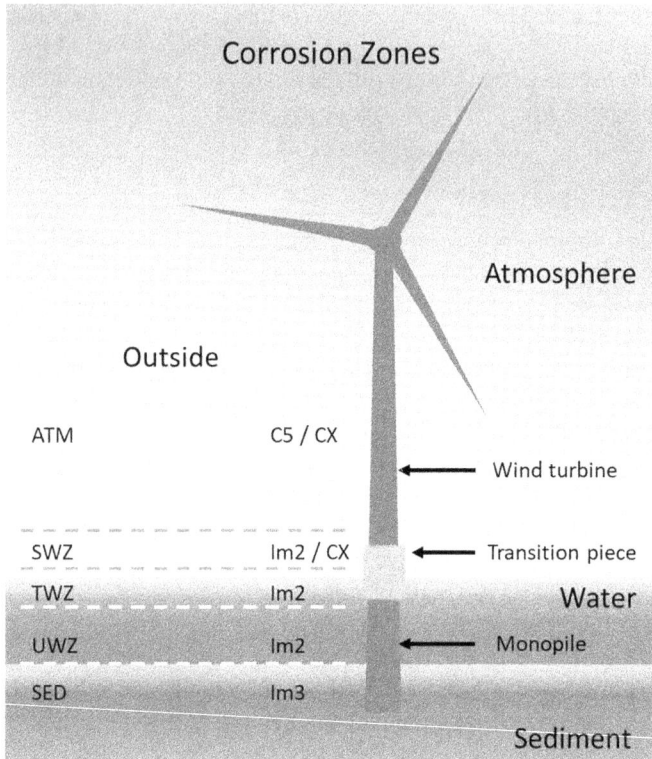

FIGURE 10.5 Illustration of the external corrosion zones of an offshore wind turbine: SED: Sediment, UWZ: Submerged Zone, TWZ: tidal water zone, SWZ: splash water zone, ATM: Atmosphere. The corrosion stress classes are based on DIN EN ISO 9223 (2012) and DIN EN ISO 12944-2 (2017)[20,21]; Corrosivity: C5 = severe; CX = extreme; lm2 = salt or brackish water; lm3 = terrestrial; lm2/lm3 with cathodic corrosion protection.

VGB/ BAW standards part 1-4 (VGB-S-021-01-2018-04 DE, 2018; VGB-S021-02-2018-04-DE, 2018; VGB-S-021-2018 03-04-DE, 2018; VGB-S021-04-2018-07-DE, 2018)). Unfortunately, environmental aspects are less regulated in these technical standards.

10.3.6 ADVANTAGES AND DISADVANTAGES OF WIND ENERGY

There are many advantages and, unfortunately, some disadvantages in the use of wind turbines. The first clear advantage is that it is a relatively clean energy production, which does not produce harmful emissions such as smog or greenhouse gases. Relatively because, although the turbines do not cause any emission during the operating time, the construction and the still not completely clarified dismantling (especially offshore) would also have to be taken into account. Another advantage is that wind is an inexhaustible energy source. This offers the advantage that even resource-poor countries have the possibility to be energy self-sufficient. Furthermore, wind turbines can be installed quickly, maintenance is simple, and operating costs are low. Today's wind turbines differ significantly in terms of productivity from the turbines of 20 years ago. In the future, the constant improvements will lead to the fact that more and more electricity can be generated on less and less space. Conversely, this means that wind turbines will also become more affordable because of greater efficiency.

At first glance, wind energy seems like the perfect way to generate energy: clean, renewable, effective. However, wind energy is not always constantly available. Since wind energy cannot be stored (another disadvantage), energy production will always fluctuate. Therefore, wind can never be the only source of energy, but always a backup. Although the installation costs are constantly decreasing, the locations where a high yield can be achieved are often in the ocean, on lakes, or in the mountains. Therefore, the costs are higher than for example in the countryside.

A much-discussed point of criticism of onshore wind turbines, especially among the population, is noise pollution. For this reason, there are certain requirements for the planning of new wind farms, which regulate certain limits and minimum distances to inhabited areas. Furthermore, the development of wind farms must be in harmony with nature. Further criticism of wind farms is the "destruction" of the landscape. For this reason, manufacturers try to make the turbines as inconspicuous as possible. Environmentalists state that the mortality rate of birds is said to be increased in the vicinity of wind turbines and, among other factors, shadow cast by wind turbines can have a negative impact on populations, as it can be mistaken for predatory birds, for example. Birds would notice the wind turbine rotors too late and would perish in high numbers (especially in flocks of birds). In Germany alone, 100,000 birds are said to die each year due to wind turbines. This seems high but it is estimated that about 18 million birds die annually in Germany due to glass panes[22]. A study from the USA stated that about 0.007% of all birds killed are by wind turbines and 70% by cats[23]. A study in Denmark performed by consulting companies for Vattenfall found that birds can avoid wind turbines better than previously thought. And a study from Norway found that when a wind turbine blade is black instead of white, birds are better able to detect and avoid wind turbines by 72%[24]. Wind power indeed has some disadvantages, but the advantages clearly outweigh the disadvantages, as further use of fossil fuels has more serious consequences.

10.4 RENEWABLE ENERGY SOURCES: HYDRO POWER

Hydropower is defined as the generation of energy from fast water flow[22]. Dams are used to create a physical barrier between two bodies of water and restricting the water flow. Massive electricity can be produced by sending the upstream water through a turbine to the downstream reservoir. By 2016, the projected global hydropower electricity generation is 52 PWh[22], but the actual energy produced (electricity only) was only 4.1 PWh. Unlike solar and wind energy, hydropower is highly dependent on the local environment. For example, if a region is experiencing severe drought or flooding, power production will be significantly reduced due to facility shutdowns. At the same time, the power demand of the region also dictates the size of the hydropower plant. Hydropower is generally harnessed through three main methods: storage, "run-of-river," and pumped storage. Storage-based hydropower plants drive electricity production by feeding water from a reservoir into the turbine generator. This form of powerplant requires medium to large storage space and can generate beyond 300 MW[25]. The "run-of-river" type power plants rely on the local river systems by changing their flow dynamics, which means only places with a sufficient flow can sustain long-term power production. Generally, the maximum power output from "run-of-river" plants is only 100 MW (medium sized) and down to less than 10 MW for smaller operations[25]. The last commonly known type of hydropower plants is pumped storage, which are designed to compensate energy demand during peak loads. These storage facilities increase power production by releasing the water from upper reservoir on-demand. During low-demand and low-cost times, water is pumped into the upper reservoir, which is a very cost-effective way to preserve energy. The cost of electricity production from hydropower plants is dependent on the local environment and legislative rules, but generally between US \$50 to 100/MWh[25]. The cost of operating and maintaining medium to large powerplants is between US \$5 to 20/MWh and approximately US \$10 to 40/MWh for small plants. By 2012, Asia had the higher amount of hydropower in operation (401,626 MW), followed by Europe (179,152 MW), North and Central America (169,105 MW), South America (139,424 MW), Africa (23,482 MW), and Oceania (13,370 MW)[25].

In comparison to the other renewable energy systems, hydropower has the greatest energy capacity. One of the most advantageous aspects of hydropower is its ability to combine with other systems, including agricultural irrigation and water usage[25]. Recently, hydropower is criticized for causing large-scale environmental impacts, such as ecological fragmentation, greenhouse gas production, and increase the impact of natural disasters. One of the key disadvantages of hydropower is the extent of influence it exerts on local habitats. For example, to build the Belo Monte Dam in Brazil, around 4000–5000 m^2 of forest is expected to be removed, this ecological disaster does not even include the amount of river flow that will be restricted due to the construction of this project[22]. Furthermore, once the powerplant is constructed, a sudden appearance of a large reservoir will cause local ecosystems to fragment, resulting in biodiversity loss[22].

10.4.1 MACRO- AND MICRO-ORGANISMS IN HYDROPOWER SYSTEMS

Microbiologically influenced corrosion is a long withstanding problem in the water systems due to the combination of physical, chemical, and biological changes. The extent of MIC in hydropower plants impacts both metallic and concrete infrastructures[26]. Formation of biofilm on the key infrastructures of the hydropower plants, including the pipeline system, can lower the energy efficiency of the overall facility by increasing resistance along the operating system, that is, decreasing flow rate[26]. In addition, corrosion damages through microbial reactions lead to the deposition of corrosion products and precipitate that further decreases the efficiency of the powerplant. Production of H_2S by sulfate-reducing and sulfur-cycling microorganisms causes additional stress and environmental damages to the system by propagating the risks of corrosion[27,28]. It was demonstrated in one case in South Africa that MIC was detected in nearly all hydropower plant infrastructures, which can potentially affect all power supply of the region[29]. Heat exchangers are susceptible to the attack of different organisms which can form a thick layer of slime on the surface, thus reducing the overall heat exchange efficiency[30].

In addition to microorganisms, macro-organisms such as rodents and invertebrates can also impact the overall infrastructure integrity. For example, rodents are known to cause hydraulic alterations by disrupting the flow-net of the hydropower system through burrowing[31]. They can also promote erosion by diverting the water through cracks and other voids within the underground dam system. On the surface, species such as beavers will cause blockage along the riverbanks as part of their natural survival instinct. However, their impact on the hydropower infrastructure should not be underestimated. Damages to the water dam due to beaver activity were reported in 32 of the 48 US states[31]. In Europe, burrowing activities of muskrats in the Netherlands have led to irreversible damages to some of the hydropower infrastructures[31]. It is to be noted that to control the biological activities typically both non-lethal and lethal methods are used, which range from live trapping, applying bio-repellant, habitat management, to deathtraps and rodenticides. Thus, the close association between biological activity, ecological footprint, and operations of the hydropower plants makes this energy source one of the most controversial "renewables."

10.5 RENEWABLE ENERGY SOURCES: BIOENERGY

Bioenergy is the production of energy from biomass, usually in the form of plants or plant residues. For the selected biomass, various options are available for bioenergy use. It can be specially cultivated crops (e.g., corn or rapeseed), fast-growing woody plants (e.g., pine), waste or residual materials (e.g., from agriculture, households, industry, or sewage sludge). The types of use are also diverse, as they can be "stored" in gas form (mainly methane), liquid form (vegetable oil/biodiesel), or solid (logs, wood pellets and flakes or straw pellets).

10.5.1 BIOGAS

About 1.5 billion people (more than 20% of the world's population) have no electricity, and about 3 billion people (about 45% of the world's population) rely on solid fuels such as firewood, crop residues, livestock manure, and coal to meet their food needs[32].

In these countries, with steadily growing populations, it has been observed that extreme problems with waste disposal have also arisen. In the case those countries continue to grow and urbanize, waste management becomes an important issue at the local and national levels[33]. Especially in developing and underdeveloped countries, the lack of effective and efficient waste and wastewater management systems poses a significant threat to human health and the environment. In Asia alone, waste generation has a value of 1 million dry tons per day[35]; up to 70% of municipal solid waste (MSW) consists of organic matter[36]. Biogas seems to be in this context an optimal source of energy as it addresses several aspects at once. The United Nations (UN) started in 2012 a special program with the goal to provide universal access to modern energy for all by 2030 to overcome these challenges[32]. In fact, after fossil fuels, biomass is the most widely used energy source in the world.

Currently, biogas is mainly produced by the microbial digestion of organic matter under anoxic conditions. This process consists of several stages carried out by different types of metabolism (microorganisms). First, polymeric components (celluloses, lignin, proteins) are usually converted to monomeric substances by extracellular enzymes. These monomers are then degraded by fermentative microorganisms to alcohols, organic acids, carbon dioxide (CO_2), and hydrogen (H_2). Alcohols and organic acids are then converted to acetic acid and hydrogen by acetogenic bacteria. The final step is carried out by methanogenic archaea, which form the energy carrier methane (CH_4) and water from carbon dioxide, hydrogen, and acetic acid[37]. However, it must be mentioned that the production of biogas is a very sensitive process that requires close monitoring. The composition of the feedstock is also very important and may vary only to a small extent; otherwise, the microorganisms will show limited activity.

The composition can vary depending on the feedstock. In general, biogas consists of 50–75% CH_4 and 25–50% CO_2. It may also contain traces of water vapor (H_2O), hydrogen sulfide (H_2S), and ammonia (NH_3)[40]. Before the biogas can be used, some particles and condensate must be removed – especially the hydrogen sulfide, which can otherwise lead to corrosion of engines and other plant components. In addition to desulfurization, the biogas is dehumidified, removing salts, minerals, and ammonia. Biogas can be used in a variety of ways. One option is decentralized-coupled electricity and heat production since heat is also generated when the gas is burned. Direct heat utilization or distribution via heat networks is also possible. Furthermore, the application in gas-powered household appliances, as well as the processing and feeding into the natural gas grid as a natural gas substitute in combined heat and power applications, for heat supply or as a fuel. Biogas can also be stored in gas grids, in decentralized gas storage facilities, or by means of heat storage systems, even over longer periods of time. New, but so far hardly used options are the use of biomethane as a natural gas substitute in the chemical industry or the integration of biogas in power-to-gas/power-to-heat concepts.

10.5.2 Biofuels

In the production of biofuels, different methods can be used to convert the biomass. There are numerous microorganisms that produce special enzymes that play a central role in the conversion of biomass as a substrate into various biofuels [41]. Biofuels are usually liquid, but sometimes also gaseous and used to power, for example, internal combustion engines (mobile and stationary). Feedstocks for biofuels are downstream raw materials such as oil crops, grain, sugar beets, sugar cane, forest and residual wood, special energy crops, and animal waste[42]. The prefix bio indicates the plant origin rather than that of organic farming. In contrast to other renewable energies, biofuels are unlimited as they are not dependent on fluctuating "drivers" such as wind or solar energy. However, the climate neutrality and environmental benefits of biofuels are highly controversial since land for agriculture can be involved and decrease the production of food. Although currently not applied and used in industry researchers just recently published a potential solution for this challenge. From lignin-containing organic material, like wood and plants, approximately 30% consist of lignin, which cannot be used in many industries because effective decomposition of lignin itself is too costly, complicated, and energy-intensive. In the paper industry, for example, up to 50 million tons of lignin waste are produced annually, which are burned in 98% of the cases. The utilization of this organic material, which is the most common polymer in the world besides cellulose and chitin, would be an outstanding source for sustainable energy production. However, from a biological point of view, its degradation is a major challenge, as there are no efficient, cost-neutral, and environmentally friendly ways to degrade it efficiently. An interesting solution was recently published by an international team of researchers. So-called "peroxidase mimetics" are used, which imitate the enzyme catalysis of a peroxidase. This was combined with nanoparticles and offers the possibility of a heat-resistant, pH-resistant, and long-term use to enable the decomposition of lignin. However, the long-term environmental impact of using these synthetic molecules remains to be determined[43].

Up to now, researchers have classified biofuels into different generations, although there is no clear demarcation. For this reason, some studies in the literature reference two, three, four, and others five generations. In the further course the following classification is used: edible biomass, non-edible biomass, algae biomass, and chemical processing (Table 10.4). Each generation has its own advantages and disadvantages and is listed in Table 10.4.

TABLE 10.4

Different Generations of Biofuels and the Used Feedstock Together with the Corresponding Advantages and Disadvantages[44,45]

1st Generation	2nd Generation	3rd Generation	4th Generation
Edible biomass	Non-edible biomass	Algae biomass	Chemical processing
Corn, sugar beet, Wheat, rice	Waste, straw, grass, wood	Macro and micro algae	Pyrolysis, solar fuels, genetic algae, gasification
Advantage:			
Emission of greenhouse gases is low	Reasonable use of non-edible food as feedstock	Simple algae cultivation	High production and biomass yield
For conversion, only a simple and inexpensive technology is needed	Use of non-agricultural land for limited crop cultivation	No edible plants are needed. Waste/sea water can be used	High CO_2 fixation
Disadvantage			
The yield is too small for the demand	Costly pre-treatment is required	Resource consumption for algae cultivation is higher	High costs for the bioreactor
Creates shortages of food	Sophisticated technology is required to convert the biomass into fuel	Lipid concentration and biomass accumulation in algae is lower	Investment costs for the early stage of research are high
High land requirement			

First-generation biofuels are bioethanol, biobutanol, and biodiesel. Each of these fuels requires either a different feedstock or different classes of microorganisms. Bioethanol is produced by the fermentation of carbohydrates such as starch (feedstock: wheat, barley, corn, rice grains, potatoes) or dual sugars (feedstock: sugar cane, sugar beet).

Biobutanol is produced using a similar principle but with different fermentative microorganisms[46,47]. Biodiesel is produced from vegetable (soybean, coconut, palm, sunflower, recycled waste cooking oil) or livestock lipids[48]. In some countries, such as Brazil or Germany, it is a legal requirement that fossil diesel must have a certain percentage of biodiesel. First-generation biofuels are useful up to a point. Since they are in direct competition with food supply and also threaten biodiversity, expanding their production is not a sustainable strategy[42,49]. However, to not compete with food production, it would be necessary to utilize whole plants, not just edible parts of food crops. In addition, production is dependent on subsidies and thus not competitive with existing fossil fuels. Some of the biofuels also lead to limited greenhouse emission savings when the resulting emissions from production as well as transportation are considered. For this reason, researchers focused on the further development of second-generation biofuels.

Second-generation biofuels produce bio-methanol, -butanol, -methane, -methanol, -hydrogen, DMF (2-methylfuran), lignocellulosic ethanol, and biomass-to-liquid (BTL). In contrast to first-generation biofuel, not only the mono- or disaccharide is used as feedstock, but also the lignocellulose and cellulose from which plants are largely made. Either organic waste (straw, wood residues, waste products from agriculture, waste wood, sawmill residues, and low-grade forest wood) or fast-growing, non-edible plants and wood varieties (jatropha, cassava, or miscanthus) serve as feedstock. These are converted to biofuels through various chemical, physical, and biological processes and have a positive carbon footprint[50]. The feedstock as well as the process used is crucial for the resulting product. Elaborate manufacturing processes provide particularly energy-rich fuels, such as biomethane and BTL (Figure 10.7). In summary, commercial production is not yet viable in many cases, as it requires expensive and sophisticated technology. In particular, the breakdown of all plant sugars into mono- and disaccharides is a costly and complex procedure that requires minimization of production costs[51]. An interesting approach that some researchers are pursuing is the biotechnological use of termites for the breakdown of, for example, lignocellulose[52].

Third-generation biofuel is produced from photosynthetic microalgae. This type of fuel is one of the most sustainable, environmentally friendly, and economical fuels of all existing variants for biofuel production. Methane, biodiesel, and biohydrogen can be produced from microalgae[54,55]. This type of production does not require agricultural land, fixes CO_2 from the atmosphere, and is therefore an optimal solution to achieve climate neutrality[56]. There are more than 300,000 different species of micro algae that can live in both fresh and saltwater[51]. Under optimal nutrient conditions, they have a doubling time of 24 hours. All microalgae produce lipids like triacylglycerols under stress conditions (e.g., nitrogen deficiency), but it depends on the strain of the microalgae and can vary between 2 and 58%[57]. This oil will subsequently be converted into a biofuel via a simpler transesterification process. However, the selection of suitable strains, cultivation, harvesting, and subsequent extraction of the oil is laborious and requires a high level of expertise. As this process is still very expensive, the production is not yet sustainable[58] as a lot of energy is required for production. For this reason, further development of the process is necessary, which would combine advanced methods of lipid metabolism with biotechnological tools[59].

For the fourth generation of biofuels, various methods are now being used to increase the lipid concentration of algae. In addition to the selection of suitable microalgae, various biotechnological (improvement of the production plant) and genetic tools are used. Genetic modifications are used, for example, to delete genes for lipid degradation or to introduce genes to increase lipid synthesis or photosynthesis while maintaining or even increasing the growth rate. This requires fine-tuning and time. Only when the necessary threshold is reached, and production is sustainable, will this type of fuel production be used on a large scale[57]. Biofuels were already used in the early days of the automotive industry, but it was not until the oil crisis in the 1970s that many countries showed renewed interest in producing commercial biofuels. However, only Brazil began large-scale production as part of a national ethanol program[60]. It was not until the 1990s, with the rise in crude oil prices and concerns about energy security, that policies in the USA and Europe changed. Climate change mitigation and strategies to reduce greenhouse gas emissions were now also part of the discussions that brought the further development of biofuels to the forefront[61]. Now, more than 60 countries worldwide produce biofuels led by Brazil, the USA, and Europe[62]. From 2008 to 2018, global bioethanol production increased 67% to 100.4 billion liters and biodiesel production tripled to 41 billion liters[63]. Biofuel production is controversial as it has both advantages and disadvantages. The life cycle assessment and global warming potential of biofuels depend on the type of biofuel, the feedstock, the location of production, the use of chemicals (e.g., fertilizers), the energy invested for transesterification or pyrolysis, the competition with food, the expansion of agricultural land (deforestation, monocultures), etc. In a recently published review, this issue was discussed in great detail which kind of aspects need to be considered to calculate the impact to the global warming (Figures 10.6 and 10.7). However, further research and development is needed to optimize yields and the production process. If the price of crude oil remains so favorable (compared to biofuel), this will continue to be a major challenge. For more information, please read the full review by Jeswani et al.[64].

10.5.3 Solid Bioenergy (Biomass)

Solid biomass, in addition to being used for the production of biogas and biofuels, can also be used without this type of "conversion" as a "direct" alternative energy source. According to the definition of the Joint Research Centre of the European Union, this type of energy source is raw or processed organic material of biological origin (but never fossil fuels) used for energy production, such as firewood, wood chips, wood pellets, tree pruning, stalks, or straw[65]. Compared to other regenerative energy sources, solid biomass (lignocellulosic containing biomass) has the advantage of being easily stored. It usually comes from forests, farms, and cities

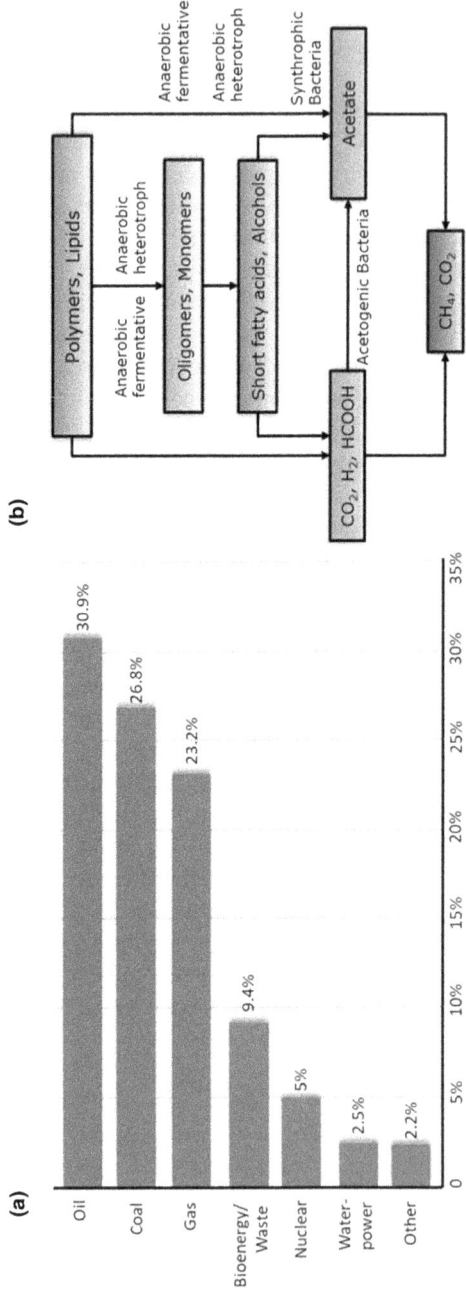

FIGURE 10.6 (a) Distribution of global energy production by energy carrier in 2019[38,39] (figure adapted from de.statista.com); (b) Schematic representation of the four-stage anaerobic digestion of complex polymers to biogas/methane.

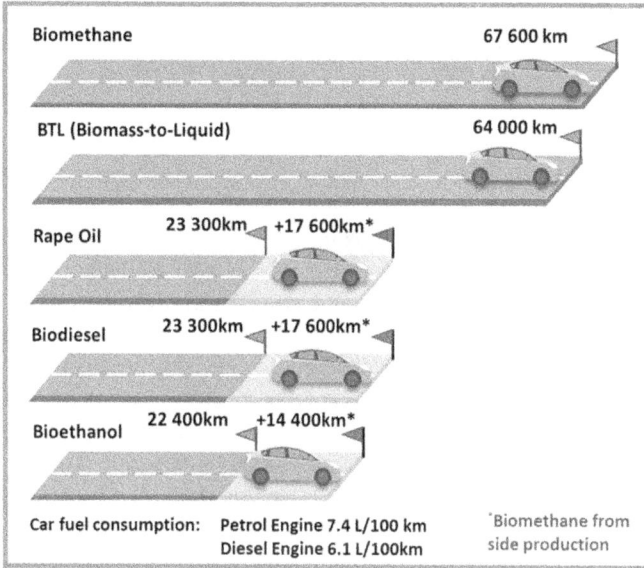

FIGURE 10.7   Comparison of biofuels. Achieved distance of the respective biofuel per hectare of cultivated area.

(Adapted from Agency of Renewable Resources e.V. (FNR).)[53]

and can therefore be used by countries that have fewer options for renewable energy. Depending on the combination of raw materials and conversion technology, solid biomass can be used to generate either heat or combined heat and power technologies (CHP) with both electricity and heat. With better development of energy production plants, this type of energy production could account for one-fifth of the world's energy consumption by 2050[62].

Basically, two main sources of lignocellulosic biomass can be distinguished: agricultural biomass (herbaceous biomass) and forestry biomass (woody biomass). Among the solid biogenic energy sources, wood is technically and environmentally the best fuel due to its low sulfur, nitrogen, and chlorine content and its high ash melting point. This type of energy source can be in the form of firewood, wood pellets, or wood chips. The firewood is often used in the private sector for cooking or heating. However, this type of conversion provides a maximum of 10–100 kW of energy. Even the use of special boilers provides only a comparatively small increase in output. Nevertheless, this type of use is widespread throughout the world and is practiced especially in third-world countries.

10.6   RENEWABLE ENERGY SOURCES: GEOTHERMAL ENERGY

Beneath the Earth surface lies an almost inexhaustible reserve of carbon-free energy that is independent from external weather conditions and available 24 hours a day all year long. Estimates are that around 47 TW of thermal energy is transferred to the Earth's surface from its interior[66]. This amount is six times more than the total global power capacity installed in 2022 (~8 TW) and more than double the world's energy demand in 2022, which reached ~160,000 TWh[67]. Hence, in the face of global climate change, environmental concerns, geopolitical tensions, and the growing demand for energy, further development and expansion of geothermal resources can play a critical role in solving some of the emerging challenges of the 21st century.

First, archeological evidence of human interest in the use of geothermal energy comes from the times reaching as far as the end of the last Ice Age. Around 10,000 years ago on the territories of nowadays Arkansas, people used hot springs to process food, to heal their bodies, or for spiritual reasons[68]. One of the earlier examples of wider-scale utilization of geothermal energy dates back to 1332 in the French village of Chaudes-Aigues, where 40 houses and a church were connected to district heating [69]. Modern, industrial history of the geothermal energy usage starts in 1904 in Larderello, Italy, where Prince Piero Ginori Conti constructed the first geothermal power plant[70]. As of present day, geothermal energy generation provides electricity in more than 30 countries and delivers heat and/or cooling in more than 80 countries[71,72]. In 2021, the global installed capacity for geothermal electricity generation and heat (including cooling) was approximately 16 GWe and 110 GWth, respectively.[72]. A ranking of the 10 countries with the highest geothermal energy production capacity is shown in Table 10.5. In the future, further increase of the geothermal energy generation capacity is anticipated, as the world geothermal technical potential is estimated at around 200 GWe and over 4000 GWth[73]. Based on recent findings presented by Augustine et al. (2023)[74], the geothermal power capacity in the USA is projected to reach 90 GWe by the year 2050. This projection is further supported by the successful completion of a horizontal doublet well system pilot project in Nevada[75]. As the industry progresses, we can anticipate many positive developments in the future.

TABLE 10.5
Top 10 Countries by Installed Power and Heat Generation Capacities[71,76]

Power		Heat	
Country	GWe	Country	GWth
USA	3.79	China	40.61
Indonesia	2.36	USA	20.71
Philippines	1.94	Sweden	6.68
Turkey	1.68	Germany	4.81
N. Zealand	1.04	Turkey	3.49
Mexico	0.96	France	2.60
Kenya	0.94	Japan	2.57
Italy	0.94	Iceland	2.37
Iceland	0.75	Finland	2.30
Japan	0.62	Switzerland	2.20

10.6.1 GEOLOGICAL FOUNDATIONS

Geothermal energy is intrinsically related with the Earth's interior. That is why in order to gain better understanding of the nature of geothermal resources, a few geological concepts require a brief explanation.

10.6.1.1 Earth's Internal Structure

Depending on the applied criteria, the Earth's structure can be divided according to its chemical composition or rheological properties that differentiate Earth's internal architecture into separate elements (Figure 10.8). Following the chemical composition classification of the Earth's internal structure, three main units can be distinguished: the core, mantle, and crust: core, mantle, and crust. The core, the innermost layer, is primarily composed of a mixture of solid and molten iron and nickel. It has a radius of approximately 3500 km, with temperatures near its boundary reaching close to 6000°C. The core is wrapped by the mantle which comprises dense iron- and magnesium-rich rocks. It is approximately 2800 km in thickness and its temperature varies from 5000°C in the bottom part to 1500°C in the upper portion. The crust is the outermost layer composed of oceanic and continental elements. The oceanic crust is formed of basalts, and its thickness is in the range of 5–7 km. In contrast the continental crust is mostly made from igneous granites and metamorphic gneiss that are locally overlaid by the sedimentary rocks. Its thickness may be between 10 and 70 km. Based on the rheological properties, the Earth's internal structure may be divided into inner core, outer core, mesosphere, asthenosphere, and lithosphere. In the center of the Earth is located the inner core which is

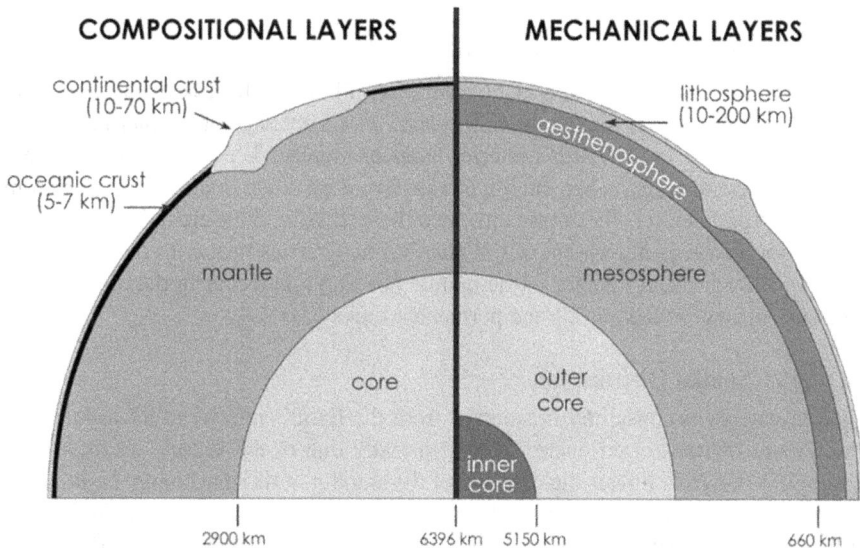

FIGURE 10.8 Cross-sectional view of Earth's compositional and mechanical layers.

(Figure adapted from Boden, 2016[77].)

characterized by the solid state and radius of around 1300 km. Above is a layer of outer core that is around 2200 km in thickness and is in the liquid state. The mesosphere, situated above the outer core and below the asthenosphere, also has a thickness of about 2,200 kilometers. Rocks in this layer are solid, yet ductile, thereby exhibiting the ability to flow. In the asthenosphere, which overlies the mesosphere and underlies the lithosphere, rocks are mainly in solid state, but in comparison to the mesosphere they have weaker mechanical properties and thus exhibit higher flow rate. This layer has around 200 km in thickness. The uppermost layer, the lithosphere, is mechanically strong and exhibits brittle behavior. On average, it has a thickness of about 100 km [77].

10.6.2 HEAT (SOURCE, MECHANISM OF TRANSFER, SPATIAL DISTRIBUTION)

10.6.2.1 Source

Earth's heat comes from three main sources: *primordial heat*, *radiogenic decay*, and gravitational compression. Primordial heat is the internal, residual heat generated at the stage of the Earth formation when space dust and debris accreted due to gravitational forces. With every collision of a celestial body, kinetic energy was transformed into thermal energy and as a result, the temperature of the proto-Earth rose. Heat generated due to the transformation of unstable elements, primarily uranium, thorium, rubidium, and potassium, is known as radioactive decay. This provides around 60% of the heat energy in the continental crust[78]. The third source of thermal energy originates from the gravitational pressure, which causes mechanical compression and expansion of the rock structure[79]. Of the sources, primordial heat and radiogenic decay equally contribute to the Earth's internal heat budget[77,80].

10.6.2.2 Mechanism of Transfer

As Earth's interior is in a state of thermal disequilibrium, heat is transferred from areas of higher to lower temperatures (Figure 10.9). Exchange of the thermal energy between materials in the Earth's interior happens chiefly via conduction or convection. Conduction occurs when energy is transferred by direct contact from one atom or object to another. It is the dominant way of heat transfer in the crust and inner core. Convection involves movement of the material that carries heat with it. Convection is responsible for heat transfer in the mantle but also may occur in the crust if only fluids are present in the porous and permeable rocks.

10.6.2.3 Spatial Distribution

Thermal energy is constantly transferred from the Earth's interior to its surface, but the amount of transferred energy varies spatially due to the Earth's heterogeneity (Figure 10.10). Heat flow is the measure of the amount of thermal energy emitted per unit area per unit time. The average heat flow of the Earth is about 87 mW/m^2, but it differs between oceanic and continental crusts, which have heat flows of 101 mW/m^2 and 65 mW/m^2, respectively. Moreover, areas of higher heat flow are associated with boundaries of the tectonic plates and hot spots, where it can reach 2000 mW/m^2 and more as it was measured in the Yellowstone National Park[77,81].

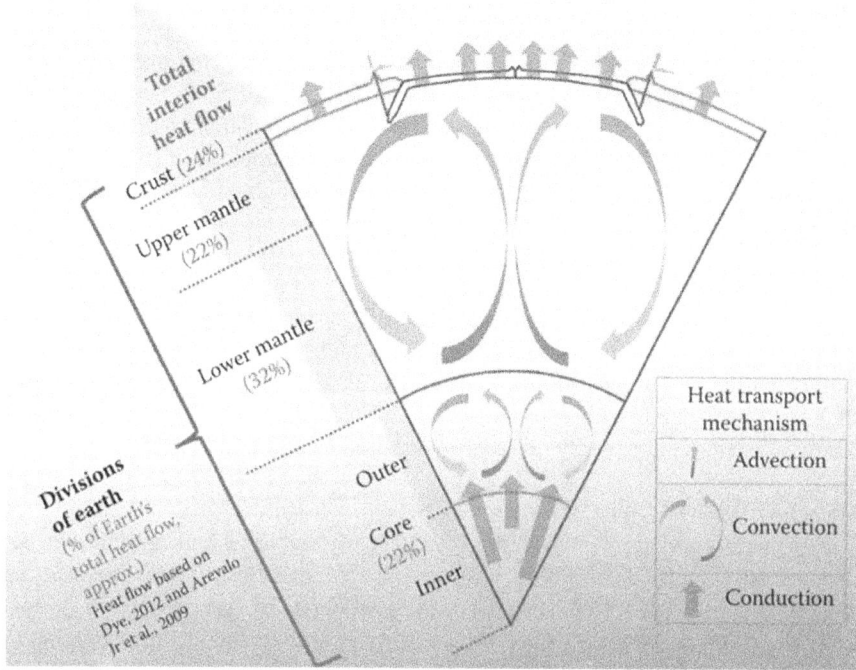

FIGURE 10.9 Pie slice through Earth's interior showing major compositional and rheological divisions and the relative proportion and type of heat flow from each division.

(Figure adapted from Boden, 2016 [77].)

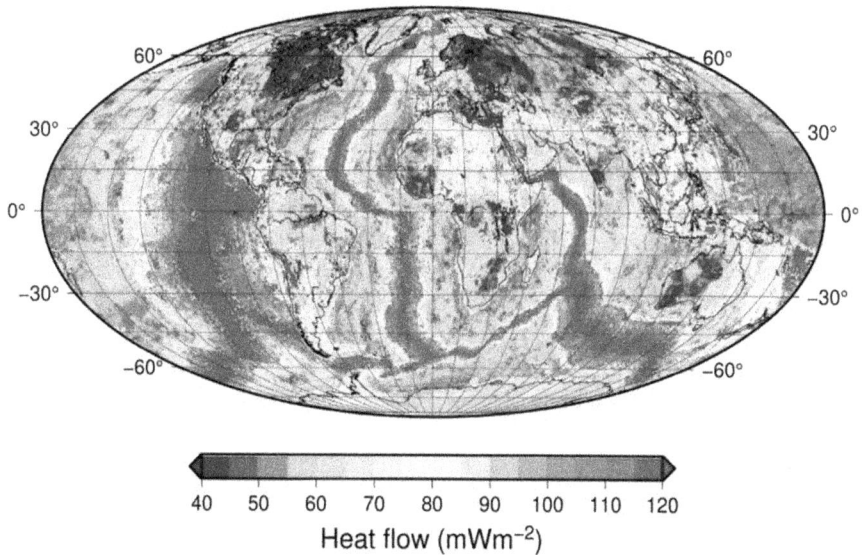

FIGURE 10.10 Global heat flow map.

(Figure adapted from Lucazeau, 2019[82].)

10.6.3   GEOTHERMAL SYSTEMS AND THEIR UTILIZATION

Geothermal systems can be classified in numerous ways, reflecting their geological position, predominant mechanism of heat transfer, state of fluid, and other criteria[77]. For the sake of brevity, we will follow the classification based on the heat content (enthalpy), which is often used in engineering applications. According to this classification, these systems may be divided into high-enthalpy (>150°C) and low-enthalpy (<150°C) categories[83].

10.6.3.1   High-Enthalpy Systems

Most of the geothermal resources related to these systems are in volcanically active areas or near the margins of tectonic plates, which means that subsurface fluids mainly consist of a mixture of hot water and steam, but they can also contain gases such as CO_2, H_2S, HCl, HF, and others[84]. Fluids produced from high-enthalpy geothermal systems are primarily utilized for power generation.

10.6.3.2   Low-Enthalpy Systems

These systems are commonly found in tectonically stable regions with thick sedimentary cover, such as the Polish Lowlands or the Paris Basin[85,86]. Moreover, the utilization of ground-source heat pumps in the shallow subsurface opens vast possibilities to harness geothermal energy across the globe. Geothermal resources of the low-enthalpy systems are mainly applied in direct-uses, including bathing and swimming, local and district heating and cooling, thermal storage, greenhouse heating, and aquaculture, as well as agricultural and industrial applications[71].

10.6.3.3   Power Generation

There are three main types of power plants: dry steam, flash steam, and binary cycles power plants (Figure 10.11). Dry steam power plants use steam from a geothermal reservoir as a processing medium. After passing through the turbine, the steam loses its pressure and is condensed back into water, which is then injected into the subsurface to replenish the geothermal resources and maintain reservoir pressure (Figure 10.11a). This type of power plant is known for being the most energy-efficient among all types of geothermal power plants. However, mostly due to the scarcity of vapor-dominated reservoirs, by 2015, they accounted for 22% (2863 MWe) of total global geothermal power capacity[77,87]. In flash steam power plants, hot pressurized water (usually >175°C) produced from the well is separated into steam and water through a process called depressurization. This separation begins early at the wellhead and continues further in the separator. The depressurization causes the geothermal fluid to "flash" into steam, which is then directed to drive a turbine, while the remaining water is reused within the system (Figure 10.11b). The water can be reinjected into the subsurface or sent for another cycle of separation. As of 2015, the installed world power capacity of combined flash steam power plants (single-, double-, triple flash) accounted for 63% (8038 MWe)[77,87]. Since binary cycle power plants (Figure 10.11c) operate with water temperatures in the range of 75–200°C, they can be developed in both high- and low-enthalpy geothermal systems. Power generation is based on the Organic Rankine Cycle or Kalina Cycle, wherein geothermal fluids transfer energy through the heat exchanger to the working fluid, which has a lower

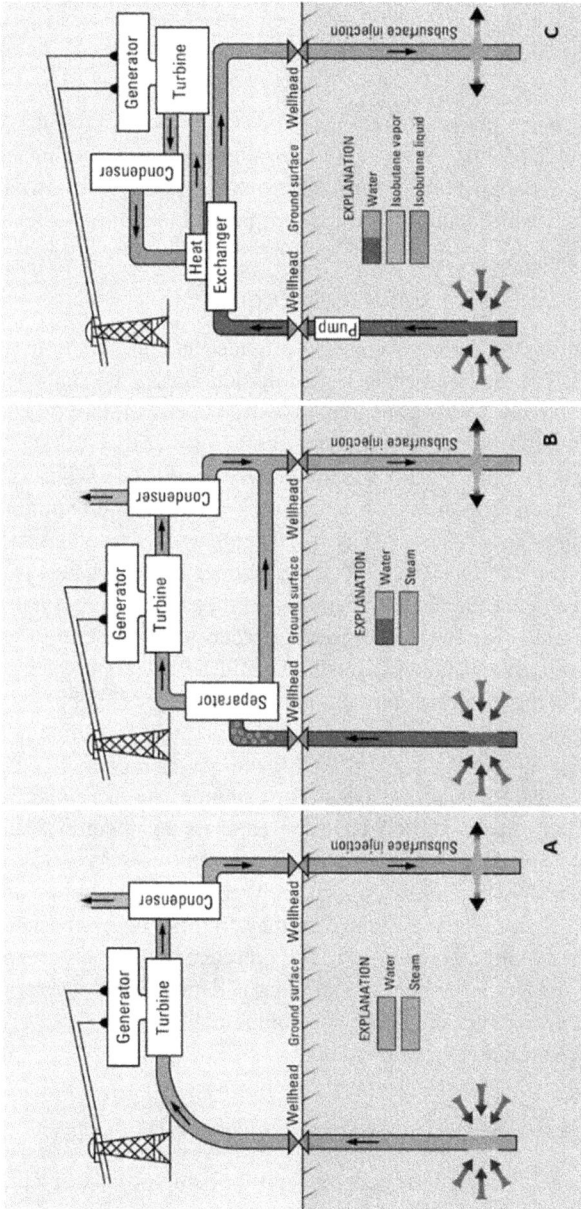

FIGURE 10.11 Diagrams showing three basic types of geothermal power plants: (a) dry steam, (b) flash steam, and (c) binary cycle. (Figure adapted from Duffield and Sass, 2003[88].)

boiling point than water. The resulting steam released from the working fluid drives the power-generating turbine and undergoes condensation to form a closed cycle. Meanwhile, water is reinjected into the reservoir (Engineers VC 2014). Although they are the most common type of power plants, their contribution to the overall power capacity in 2015 was only 14% (1726 MWe)[87].

10.6.3.3.1 Direct Use

Geothermal resources are predominantly used for direct utilization of the heat from the subsurface[72]. Out of 110 GWth of the installed heat and cooling generation capacity, around 71% is contributed by ground-source heat pumps, followed by space heating and bathing, each accounting for approximately 11% of the total share[71].

10.6.4 GEOTHERMAL ENERGY AND MICROBIOLOGY

The use of water during the extraction process provides an opportunity for microbial growth[89]. Similar to petroleum reservoirs, the microorganisms are challenged with a large shift in environmental conditions, that is, extreme temperature fluctuations[89]. For example, temperature within the shallow geothermal energy extraction fluctuates with the season, resulting in a variation between ± 6°C[89]. Whereas the deep geothermal energy extraction process is faced with a microbial community that is much more accustomed to harsh environments, that is, high temperature and salinity, than shallow processes[89]. It is expected that mesophiles (microorganisms that grow up to 45°C) will dominant the shallow extraction processes, whereas thermophiles and extreme thermophiles are found in the deeper processes[89]. Similar to other high-temperature reservoirs, these microorganisms should be anaerobic such as methanogens, sulfate-reducing bacteria, fermenters, and acetogens[89], but with higher probability to be spore-formers than reservoirs that are more moderate in temperature ranges. In addition, risks for corrosion, clogging, H_2S production, etc., are still prominent within the geothermal reservoirs. For example, the thermophilic archaea *Thermococcus* (growth up to 100°C) have the capacity to undergo fermentation, sulfur-reduction, and iron dissolution, thus increasing the risks for acid production and reservoir corrosion[90]. Several cases have been reported that microorganisms can have a detrimental effect on heat exchange as they foul the heat exchangers, cause MIC, and reduce the overall efficiency[91–94]. The presence of the microorganisms and the extent of their impact on the geothermal plant is dependent on the temperature, the fluid dynamics, availability of the organic compounds, and the types of gas present, that is, hydrogen stimulates microbial activity[91].

10.6.5 ADVANTAGES AND DISADVANTAGES OF GEOTHERMAL ENERGY

The nature of the energy transition goes beyond the simple replacement of fossil fuels with low-carbon sources. It involves a transformative shift in methods of production, distribution, and consumption. Thus, when designing decarbonized energy systems that cater to societal and environmental concerns, it's crucial to recognize the strengths and limitations associated with specific technologies. Knowledge of these allows for flexible adaptation to local conditions and helps maximize the utilization of their energy potential.

10.7   HOW TO STORE RENEWABLE ENERGY?

Energy storage is essential for peak power demand. Resources with high storage stability, such as fossil fuels, are the most ideal candidate for these purposes. However, with the increased demand for renewable energy, how to efficiently harvest and store them for mid- to long-term usage is essential. Short-term energy storage ranges between a few hours to days, while long-term storage can store for an entire season. For example, it is often the case to store energy during the summertime for later use in the winter season. There are several types of energy storage (Figure 10.12), including electrochemical, electrical, mechanical, thermal, and chemical[95]. There are numerous in-depth reviews on the different characteristics, applications, and technical features of the various types of energy storage. Koohi-Fayegh and Rosen[95] discussed and compared in detail the different technical specifications of energy storage systems, including their advantages and disadvantages. Guney and Tepe provided a detailed description to classify the various energy storage systems, such as their features and applications[96]. Luo et al. compared the different technical and economic performance capabilities of the different electrical energy storage technologies and provided general guidance for future developments [97]. For this chapter, we will briefly discuss the advantages and disadvantages of the electric power storage system compared to natural storage systems, particularly within geological formations.

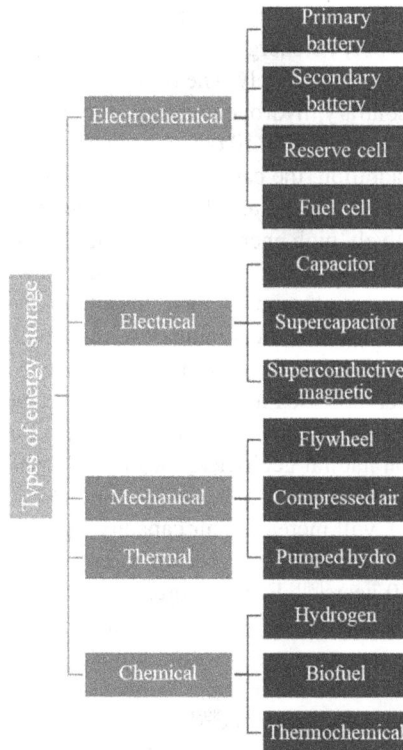

FIGURE 10.12   Types of energy storage.

10.7.1 Electric Power Storage

Electricity is one of the most utilized forms of energy in our society[98]. Electrical energy storage technologies are important to balance supply and demand during different seasons. As electricity is not a storable energy, efficiently converting it into a stable form is important to meet the energy demand[96,98]. Electricity is one of the most utilized forms of energy in our society[98]. Electrical energy storage technologies are important to balance supply and demand during different seasons. As electricity is not a storable energy, efficiently converting it into a stable form is important to meet the energy demand[96,98]. There are many different methods to convert electricity, that is, generally categorized into mechanical, chemical, electrochemical, superconducting magnetic energy, and cryogenic energy[96,98]. There are several detailed reviews that define and compare the different types of storage systems[96,98]. This work will focus on the advantages and disadvantages of the different categories of electric power storage systems. An efficient electricity storage system is evaluated based on its energy generation capacity, transmission, lifetime, distribution, response time, and technology availability[96,98]. For example, the energy density is closely related to the total volume capacity of the storage system, which is directly influenced by the round-trip efficiency of the system[98]. Flywheel energy storage is a mechanical energy storage system that stores energy in the form of kinetic energy by spinning a heavy rotor (the flywheel) at very high speeds[96,98]. It is a mature technology that was developed in the 1950s, its energy and power density are between 10–30 Wh/kg and 400–5000 W/kg[98], respectively. The round-trip efficiency of this technology is up to 95%. On the contrary, hydrogen energy storage, in the form of fuel cells, has an energy density between 800–10,000 Wh/kg[98], with a round-trip efficiency of only 20–35%[98]. In addition, the capital cost of flywheel is only 2.5% of that of fuel cells[98]. On the other hand, battery-based systems, such as lithium-ion, NaS, and NaNiCl$_2$, have relatively high energy densities of 70–200 Wh/kg, 150–240 Wh/kg, and 100–120 Wh/kg, respectively. However, their relative capital costs are relatively high with a lifetime much shorter than that of other technologies[98]. This is mainly due to mechanical and chemical failure during operation. Another key factor to consider for storage systems is their storage capacity and duration. For example, energy storage such as compressed air and pumped hydroelectric have longer storage duration ranging between hours to months. A limitation to their operations is that they both depend on natural geological locations. Novel gravity power module and advanced rail energy storage are expected to be able to be applied for grid-scale applications, which will increase their capacity significantly as the technology matures[98]. Their storage duration is expected to be up to several months with a lifetime of more than 30 years[98]. Furthermore, different technologies have various response time, which is defined to be the release time of stored energy to meet energy demand[98]. The fastest response time is within milliseconds and seconds for supercapacitors, batteries, and flywheel. Whereas pumped hydroelectric storage systems can take up to 24 hours for response time[98].

In addition to variations between the technologies in terms of capacity, response time, energy density, etc., there are other factors that need to be considered when constructing a storage system. Pumped hydroelectric systems are highly dependent on geological features, and cause several environmental issues such as erosion, ecological disruption, and flooding[96]. Battery-based systems are relatively reliable but require high investment costs and more importantly are limited by raw materials. In addition, disposal of lead acid batteries is an environmental concern due to the potential leakage of sulfuric acid electrolytes, whereas nickel-metal-based batteries require high raw material demand and emit greater greenhouse gas during the manufacturing process compared to other technologies[96]. One of the key operation limitations with electric energy storage systems is infrastructure and space requirements. As volume is directly related to energy and power density, the size of the storage system needs to be constructed carefully to meet the necessary demand while considering the associated costs related to disposal, charging, maintenance, labor, and decommissioning[96]. Currently, more discussions and research related to smart grids are under development to increase energy efficiency. Lastly, as more processes in our society are becoming reliant on electricity, increasing renewable energy production and developing effective energy storage systems will ultimately drive our society into a more sustainable future.

10.7.2 Underground Natural Geological Storage Systems

A key challenge with the current renewable energy landscape is storage capacity. Large-scale above-ground storage facilities are limited by infrastructure need, space, environmental impact, and social acceptance[99–102]. Such issues are less problematic for underground formations, as exemplified by the decades-old petroleum industries. As of 2010, there are 76 salt caverns, 82 aquifers, and 476 hydrocarbon reservoirs active worldwide for underground energy storage (UES). In addition, more UES is expected to be developed to increase the current capacity for renewable energy storage, such as hydrogen storage within salt caverns[99,103–105]. In addition, since the 1990s, carbon capture to reduce greenhouse emissions has been gaining more traction and is expanding in operation capacity[103,104]. The storage capacity within the underground formations ranges greatly depending on the type of the formation. For example, the bedded salt caverns at Teeside, UK, have a storage volume of around $3 \times 70,000$ m^3 while the saline aquifer in Beynes, France, has a storage volume of 3.3×10^8 m^3 [99,101–106]. In addition, since the 1990s, carbon capture to reduce greenhouse emissions has been gaining more traction and is expanding in operation capacity recently[103,104]. The storage capacity within the underground formations ranges greatly depending on the type of the formation. For example, the bedded salt caverns at Teeside, UK, have a storage volume of around $3 \times 70,000$ m^3 while the saline aquifer in Beynes, France, has a storage volume of 3.3×10^8 m$^{3[101,102,106]}$.

Currently, porous media and engineered cavities are the two most common types of UES. Porous media includes depleted hydrocarbon reservoirs and, in some cases, saline aquifers[101,102,107]. The storage capacity depends on the porosity and permeability of the sedimentary rocks (i.e., sandstones), but their large volume and potentially high contact surface with fluids increase potential concerns surrounding chemical and biological reactions[101]. Engineered cavities on the other hand are artificially constructed with well-defined characteristics. Through continuous mining and leaching, a large cavity is created within a salt bed or salt dome[107]. One key aspect is that such salt cavity/cavern is highly stable with favorable geological tightness. Currently, salt caverns are the preferred choice due to their proven capacity for hydrogen storage, while the potential of aquifers and depleted reservoirs for this purpose is still being studied[106]. This is largely due to the inert feature of the salts in the presence of hydrogen. As an emerging technology, there are only a few salt caverns being operated to store hydrogen, such as Clemens (USA), Spindletop (USA), Kiel (Germany), and Teeside (UK)[101]. But, as indicated earlier, this is an emerging field, and more research related to the feasibility of storing energy underground is underway. There are several comprehensive review papers available that discuss in detail the types, distribution, and transmission levels of various geological storage sites, including the one by Matos et al. and Tarkowski[101,107].

The remaining part of this section will focus on underground hydrogen storage, as this is closely related to the overall scope of this book as the impact of microbiology is essential for its success. Having both high energy density and versatility, hydrogen is at the forefront of the renewable energy landscape. Due to its chemical characteristics, storing hydrogen in the gaseous form is proven to be a key challenge for above-ground systems, that is, metal embrittlement due to hydrogen ingress [101,108,109]. High infrastructure costs related to maintenance and safety are required for large-scale above-ground facilities, which makes underground systems much more favorable in this context. However, a key challenge with storing hydrogen underground is contamination. Within depleted hydrocarbon reservoirs, there are risks associated with cross-contamination with previous petroleum operations, that is, hydrocarbons and chemicals. More importantly, the biological factor cannot be overlooked within underground formations[106]. There are an abundant number of microorganisms within the subsurface, with many remaining to be discovered[106]. As these microorganisms have been surviving for billions of years in an environment depleted of oxygen, hydrogen has been a highly favorable source of electrons for them[106]. In fact, hydrogen can be classified as a universal electron donor, driving many metabolic reactions (Table 10.6) including sulfate-reduction, methanogenesis, nitrate reduction, and iron reduction[106]. This proves to be problematic as such microbial reactions not only decrease the storage capacity of hydrogen but also convert it into undesirable by-products such as hydrogen sulfide and methane[106,110], within underground formations[106]. There are an abundant number of microorganisms within the subsurface, with many remaining to be discovered[106]. As these microorganisms have been surviving for billions of years in an environment depleted of oxygen, hydrogen has been a highly favorable source of electrons for them[106]. In fact, hydrogen can be classified as a universal electron donor, driving many metabolic reactions (Table 10.7), including sulfate-reduction, methanogenesis, nitrate reduction, and iron reduction[106]. This proves to be problematic as such microbial reactions not only decrease the storage capacity of hydrogen but also convert it into undesirable by-products such as hydrogen sulfide and methane[106,110]. Production of hydrogen sulfide is especially dangerous as it is a highly toxic gas. Activities of hydrogen-utilizing sulfate-reducing microorganisms convert the sulfate from minerals (i.e., gypsum) into sulfide using

TABLE 10.6
Advantages and Disadvantages of Geothermal Energy

Advantage	Drawbacks
• Renewable – Geothermal energy is continuously generated and transferred to the Earth's surface • Baseload – It offers a predictable and stable energy source, unaffected by external conditions • Domestic – Developing these resources can reduce reliance on imported fossil fuels • Small Footprint – Compared to solar or wind power plants, geothermal power plants occupy less space for equivalent electricity output • Low carbon emissions – production of electricity from geothermal resources generates less than 100 grams CO2e/kWh	• High Capital Intensity – Geothermal projects often face high capital costs and lower returns on investment. Public subsidies may be necessary, introducing additional political risks • Geographical Constraints – Most of the current geothermal power generation capacity is limited to regions with high heat flow • Geological Uncertainties – Due to geological unpredictability, not every drilled well will be productive for geothermal energy extraction • Environmental Risks – Operating geothermal plants might pose risks such as induced seismicity, water contamination, and land subsidence

TABLE 10.7
Overview of Key H_2 Microbial Consumption Processes

H_2-consuming Processes	Reaction	Free Energy $\Delta G^{0\prime}(kJ*mol^{-1}H_2)$
Methanogenesis (MA)	$\frac{1}{4} HCO_3^- + H_2 + \frac{1}{4} H^+ \rightarrow \frac{1}{4} CH_4 + \frac{3}{4} H_2O$	−33.9
Acetogenesis (APB)	$\frac{1}{2} HCO_3^- + H_2 + \frac{1}{4} H^+ \rightarrow \frac{1}{4} CH_3COO^- + 2 H_2O$	−26.1
Sulfate-reduction (SRB)	$\frac{1}{4} SO_4^{2-} + H_2 + \frac{1}{4} H^+ \rightarrow \frac{1}{4} HS^- + H_2O$	−38.0
Iron reduction (IRB)	$2FeOOH + H_2 + 4 H^+ \rightarrow 2 Fe^{2+} + 4 H_2O$	−228.3
Nitrate-reduction (NRB)	$\frac{2}{5} NO_3^- + H_2 + \frac{2}{5} H^+ \rightarrow \frac{1}{5} N_2 + 1\frac{1}{5} H_2O$	−240.1
H_2 oxidation	$H_2 + \frac{1}{2} O_2 \rightarrow H_2O$	−237

Adapted from Dopffel et al.[106].

hydrogen as an electron donor. Notably, this process was observed even at high salinity of 27% w/w[106,111]. Interestingly, such a process increases the pH of the system significantly since this is a proton-consuming process[111]. Several microbial processes are limited by high pH, including sulfate-reduction, which raises questions regarding the potential limiting factors[106,111].

There have been several model-based and experimental studies presented to illustrate the potential of microbial activities within underground gas storage[111–117]. Many of the studies have concluded that microorganisms can use hydrogen for their metabolic reactions despite the harsh conditions (i.e., high salinity) and high pressure. However, huge knowledge gaps remain regarding the impact of microorganisms within hydrogen storage sites, specifically:

1. Are there any specialized metabolic machineries used by microorganisms to convert hydrogen under various environmental stresses? Will such adaptation be common or site specific?
2. How would the site geochemistry effect the microbial reactions? Would the by-products of such reactions impact the geochemical conditions and storage capacity?
3. What is the most effective storage cycle or time to minimize microbial reactions?
4. How to monitor and predict microbial-induced risks within underground hydrogen storage to achieve effective long-term modeling and control?
5. Would mitigation strategies developed for other reservoirs be effective for underground hydrogen storage?

10.8   CONCLUDING REMARKS: THE ROLE OF MICROBIOLOGY IN THE ENERGY TRANSITION

From petroleum to renewables, microorganisms plan an integral role in our society. The degree of their impact is evaluated based on several factors, including the environment, geological conditions, temperature, nutrient availability, and microbial community structure. Microorganisms can radically influence the energy sector, both positively and negatively. On the one hand, microbes are potential energy "powerhouses" through their abilities to capture and convert CO_2 into other energy carriers, such as sugars and hydrogen[118]. For example, the photosynthetic cyanobacterium *Synechococcus elongatus* utilizes solar energy to directly convert CO_2 into chemicals such as omega-3 fatty acids and alcohols[118,119]. Scaling-up microbial-based solar capture into a form of energy "refinery" has been under development across several nations with the hope of one day providing fuel on a commercial level[118]. Similar technological advances have also been made on microbial fuel cell[121,123–125] and biofuel production[121,123–125], to further increase the fraction of bio-generated fuels in our current economy. For example, microbial fuel cells are seen as a highly promising alternative technology for self-sustaining wastewater treatment, that theoretically ends with a net energy positive reaction while recovering heavy metals such as cobalt, cadmium, and chromium[126].

However, constructing a large-scale microbial fuel cell is challenged with operation-related energy losses, as well as the high costs associated with the infrastructure, that is, electrode materials and platinum-based catalysts[126]. In addition, the shifts in the microbial community due to changes in the environment, that is, electron acceptor and donor availability, pH, and temperature will nonetheless impact the performance of the microbial fuel cell. This is further complicated by the production of negative metabolic products such as hydrogen sulfide due to the activities of sulfate-reducing prokaryotes (SRPs). In fact, production of hydrogen sulfide by the SRPs is a universal issue that is observed across several energy sectors, ranging from petroleum to renewables[106,111,127–129]. The earliest recorded observation on microorganisms-induced oilfield reservoir souring was in 1926[127], and since then several groups of SRPs have been isolated and studied with the ultimate goal to find an effective mitigation strategy[127]. In addition to the SRPs, characterization of microorganisms from other metabolic groups has expanded our knowledge on MIC, microbial-induced carbonate precipitation, microbially enhanced oil recovery, etc.[106,109,127,130,131], and since then several groups of SRPs have been isolated and studied with the ultimate goal of finding an effective mitigation strategy[127]. In addition to the SRPs, characterization of microorganisms from other metabolic groups has expanded our knowledge on MIC, microbial-induced carbonate precipitation, microbially enhanced oil recovery, etc.[106,109,127,130,131]. Interestingly, the same concerns studied within the petroleum sector is being raised again within the renewable storage systems, particularly for hydrogen storage, CO_2 storage, and geothermal energy system[89,100,106,113,129,132]. As our energy system undergoes rapid transitions with an increased number of pilot studies conducted on new underground energy technologies, we open new possibilities to expand our current understanding of microbial activity and impact. For example, the impact of high hydrogen partial pressure within underground hydrogen storage on microbial growth is currently unclear. Whether such high partial pressure will decrease microbial activity by inhibiting hydrogen-oxidizing enzymes, such as hydrogenases, is largely unknown[106,133]. In addition, many of these underground storage systems are considered hostile to microorganisms due to high salinity and low nutrient availability, yet microorganisms specifically adapted to such environments have been isolated[134–136], with many of them capable of producing sulfide, methane, acids, and other undesirable metabolic products[137,138].

A key aspect the authors wanted to highlight with this chapter and the whole book is the importance of knowledge transfer from petroleum microbiology to renewables. Many of the microbial techniques developed within the oil and gas sector, such as functional gene detection to target MIC microorganisms[139], can be applied to other energy systems, including hydrogen storage. Furthermore, the use of depleted oil reservoirs for hydrogen storage offers new potential for microbially triggered effects and risks due to the supply of oil organics for microbial growth[140,141]. As learned previously from petroleum microbiology, this is a field that requires multidisciplinary research between microbiologists, geochemists, physicists, reservoir engineers, and operators. Such collaboration will need to continue with the addition of modeling research to make accurate predictions on microbial risks and ensure a smooth and safe energy transition for our society.

REFERENCES

1. REN21, *Renewables 2019 Global Status Report*, R. Secretariat, Editor. 2019, REN21: Paris.
2. Gönül, Ö., et al., An assessment of wind energy status, incentive mechanisms and market in Turkey. *Engineering Science and Technology, an International Journal*, 2021. **24**(6): p. 1383–1395.
3. Zeman, M., *Introduction to Photovoltaic Solar Energy*. 2011, Delft University of Technology.
4. Wagner, A., *Photovoltaik Engineering-Die Methode der Effektiven Solarzellen-Kennlinie*. 1 ed. VDI-Buch. 1999, Springer: Berlin, Heidelberg.

5. Agency, I.-I.R.E., *How Falling Costs Make Renewables a Cost-effective Investment.* 2020.
6. IRENA, *Renewable Energy Statistics 2019.* IRENA-The International Renewable Energy Agency: Abu Dhabi.
7. Al-Ashouri, A., et al., Monolithic perovskite/silicon tandem solar cell with >29% efficiency by enhanced hole extraction. *Science,* 2020. **370**(6522): p. 1300–1309.
8. Fontecave, M., Energy for a sustainable world. From the oil age to a sun-powered future. By Nicola Armaroli and Vincenzo Balzani. *Angewandte Chemie International Edition,* 2011. **50**(30): p. 6704–6705.
9. Lee, J. and F. Zhao, *GWEC Global Wind Report 2021.* 2021, Global Wind Energy Council: Brussels, Belgium.
10. IEA, *Offshore Wind Outlook 2019.* 2019, IEA: Paris.
11. Fraunhofer, I.W.E.S., Windenergie report Deutschland 2012. *Wind Energy Report for Germany.* 2013.
12. Oliver, K. and H. Seitz, *Ressourceneffizienz von Windenergieanlagen.* 2014, VDI Zentrum Ressourceneffizienz GmbH: Berlin, Germany.
13. Makarichev, Y.A., A.S. Anufriev, Y.N. Ivannikov, N. Didenko, and A. Gazizulina. Low — Power wind generator, in *2018 International Conference on Information Networking (ICOIN).* 2018.
14. Adedipe, O., F. Brennan, and A. Kolios, Review of corrosion fatigue in offshore structures: Present status and challenges in the offshore wind sector. *Renew Sust Energy Rev,* 2016. **61**: p. 141–154.
15. Deborde, J., A.-M. Grolleau, P. Refait, C. Caplat, B. Olivier, M.-L. Mahaut, S. Glatin, C. Brach-Papa, G. Jean-Louis, P. Honore, and S. Pineau. Heavy metal inputs from anodic dissolution of Al-Zn-In galvanic anodes to the marine environment: TALINE Project, in *EuroCorr 2014.* 2014, Pisa, Italy.
16. Dinh, H.T., et al., Iron corrosion by novel anaerobic microorganisms. *Nature,* 2004. **427**(6977): p. 829–832.
17. Standard), D.I.F.N.E.V.G.N., *DIN 81249-2 Corrosion of metals in sea water and sea atmosphere - Part 2: Free corrosion in sea water.* 2013.
18. Momber, A.W. and T. Marquardt, Protective coatings for offshore wind energy devices (OWEAs): a review. *J Coat Technol Res,* 2017. **15**(1): p. 13–40.
19. Negro, V., et al., Monopiles in offshore wind: Preliminary estimate of main dimensions. *Ocean Eng,* 2017. **133**: p. 253–261.
20. ISO, *DIN EN ISO 9223:2012-05 Corrosion of metals and alloys - Corrosivity of atmospheres - Classification, determination and estimation.* 2012.
21. ISO, *DIN EN ISO 12944-2:2018-04 Paints and varnishes - Corrosion protection of steel structures by protective paint systems - Part 2: Classification of environments.* 2017.
22. Gibson, L., E.N. Wilman, and W.F. Laurance, How green is 'Green' energy? *Trend Ecol Evol,* 2017. **32**(12): p. 922–935.
23. Sovacool, B., The avian benefits of wind energy: A 2009 update. *Renew Energy,* 2013. **49**: p. 19–24.
24. May, R., et al., Mitigating wind-turbine induced avian mortality: Sensory, aerodynamic and cognitive constraints and options. *Renew Sust Energy Rev,* 2015. **42**: p. 170–181.
25. Lumbroso, D., A. Hurford, J. Winpenny, and S. Wade, Harnessing hydropower: Literature review. 2014.
26. Reis, M.D.P., et al., Microbial composition of a hydropower cooling water system reveals thermophilic bacteria with a possible role in primary biofilm formation. *Biofouling,* 2021. **37**(2): p. 246–256.

27. Gu, J.-D., Chapter 14 - Microbial biofilms, fouling, corrosion, and biodeterioration of materials, in *Handbook of Environmental Degradation of Materials* (Third Edition), M. Kutz, Editor. 2018, William Andrew Publishing. p. 273–298.

28. Marangoni, P.R.D., D. Robl, M.A.C. Berton, C.M. Garcia, A. Bozza, M.V. Porsani, P.D.R. Dalzoto, V.A. Vicente, and I.C. Pimentel, Occurrence of sulphate reducing bacteria (SRB) associated with biocorrosion on metallic surfaces in a hydroelectric power station in Ibirama (SC) - Brazil. *Braz Arch Biol Technol*, 2013. **56**.

29. Beech, I.B. and C.C. Gaylarde, Recent advances in the study of biocorrosion: An overview. *Revista de Microbiologia*, 1999. **30**.

30. Turnpenny, A.W.H., J. Coughlan, B. Ng, P. Crews, R. Bamber, and P. Rowles., *Cooling Water Options for the New Generation of Nuclear Power Stations in the UK*. 2017, Environment Agency: Bristol.

31. Bayoumi, A. and M.A. Meguid, Wildlife and safety of earthen structures: A review. *J Fail Anal Prev*, 2011. **11**(4): p. 295–319.

32. Legros, G., et al., *The Energy Access Situation in Developing Countries – A review focusing on the least developed countries and Sub-Saharan Africa*. 2009, UNDP and World Health Organization: New York, USA.

33. (APO), A.P.O., Solid waste management: Issues and challenges in Asia, in *Report of the APO Survey on Solid-Waste Management 2004–05*, M. Environmental Management Centre, India, Editor. 2007, Asian Productivity Organization: Tokyo, Japan.

35. Voegeli, Y. and C. Zurbrugg Decentralized anaerobic digestion of kitchen and market waste in developing countries-State-of-the-art in South India, in *Second International Symposium on Energy from Biomass and Waste*. 2008, Venice, Italy: CISA, Environmental Sanitary Engineering Centre.

36. Agoramoorthy, G. and M. Hsu, Biogas plants ease ecological stress in India's remote villages. *Hum Ecol*, 2008. **36**: p. 435–441.

37. Technology, A.C.o.S.a., *1st ASEAN Seminar-Workshop on Biogas Technology*. 1981, Working Group on Food Waste Materials: Manila, Philipines.

38. IEA, *Global Energy and Climate Model*. 2022, IEA: Paris.

39. BP, *BP Statistical Review of World Energy*. 2019, B.S.R.o.W. Energy, Editor. 2019.

40. Braun, R., Anaerobic digestion: A multi-faceted process for energy, environmental management and rural development, in *Improvement of Crop Plants for Industrial End Uses*. 2007, Springer: Dordrecht. p. 335–416.

41. Okada, K., S. Fujiwara, and M. Tsuzuki, Energy conservation in photosynthetic microorganisms. *J Gen Appl Microbiol*, 2020. **66**(2): p. 59–65.

42. Singh, V., et al., *Microbiological Aspects of Bioenergy Production: Recent Update and Future Directions*, in *Bioenergy Research: Revisiting Latest Development*, N.S. Manish Srivastava, Rajeev Singh, Editor. 2021. p. 29–52.

43. Jian, T., et al., Highly stable and tunable peptoid/hemin enzymatic mimetics with natural peroxidase-like activities. *Nat Commun*, 2022. **13**(1): p. 3025.

44. Zhou, J., et al., Analysis of the oxidative degradation of biodiesel blends using FTIR, UV–Vis, TGA and TD-DES Methods. *Fuel*, 2017. **202**: p. 23–28.

45. Li, J. and Z. Guo, Structure evolution of synthetic amino acids-derived basic ionic liquids for catalytic production of biodiesel. *ACS Sust Chem Eng*, 2017. **5**(1): p. 1237–1247.

46. Kriger, O., E. Budenkova, O. Babich, S. Suhih, N. Patyukov, Y. Masyutin, V. Dolganuk, and E. Chupakhin, The process of producing bioethanol from delignified cellulose isolated from plants of the miscanthus genus. *Bioengineering (Basel)*, 2020. **7**(2). doi:10.3390/bioengineering7020061

47. Lee, Y.-C., K. Lee, and Y.-K. Oh, Recent nanoparticle engineering advances in micro-algal cultivation and harvesting processes of biodiesel production: A review. *Bioresour Technol*, 2015. **184**: p. 63–72.
48. Bhatia, L. and S. Johri, Biovalorization potential of peels of Ananas cosmosus (L.) Merr. for ethanol production by Pichia stipitis NCIM 3498 & Pachysolen tannophilus MTCC 1077. *Indian J Exp Biol*, 2015. **53**(12): p. 819–827.
49. Singh, V., et al., Characteristics of cold adapted enzyme and its comparison with meso-philic and thermophilic counterpart. *Cell Mol Biol*, 2016. **62**.
50. Robak, K. and M. Balcerek, Review of second generation bioethanol production from residual biomass. *Food Technol Biotechnol*, 2018. **56**(2): p. 174–187.
51. Alam, F., S. Mobin, and H. Chowdhury, Third generation biofuel from algae. *Procedia Eng*, 2015. **105**: p. 763–768.
52. Scharf, M.E., Termites as targets and models for biotechnology. *Annu Rev Entomol*, 2015. **60**: p. 77–102.
53. (FNR), F.N.R.e.V., *Biofuels*, Fachagentur Nachwachsende Rohstoffe e. V. (FNR), Editor. 2016.
54. Gavrilescu, M. and Y. Chisti, Biotechnology-a sustainable alternative for chemical industry. *Biotechnol Adv*, 2005. **23**(7-8): p. 471–499.
55. Kapdan, I.K. and F. Kargi, Bio-hydrogen production from waste materials. *Enzyme Microb Technol*, 2006. **38**(5): p. 569–582.
56. Into, P., et al., Yeast diversity associated with the phylloplane of corn plants cultivated in Thailand. *Microorganisms*, 2020. **8**(1): p. 80.
57. Abdullah, B., et al., Fourth generation biofuel: A review on risks and mitigation strate-gies. *Renew Sust Energy Rev*, 2019. **107**: p. 37–50.
58. Kleinová, A., et al., Biofuels from algae. *Procedia Eng*, 2012. **42**: p. 231–238.
59. Chisti, Y., Biodiesel from microalgae. *Biotechnol Adv*, 2007. **25**(3): p. 294–306.
60. Soccol, C., et al., Brazilian biofuel program: An overview. *J Sci Ind Res*, 2005. **64**: p. 897–904.
61. (FAO), F.A.O., *Biofuels and the Sustainability Challenge: A Global Assessment of Sustainability Issues, Trends and Policies for Biofuels and Related Feedstocks*. 2013, Food and Agriculture Organization of the United Nations: Rome, Italy.
62. IRENA, *Innovation Outlook: Advanced Liquid Biofuels*. 2016, International Renewable Energy Agency: Abu Dhabi, United Arab Emirates.
63. IEA, *Renewables 2019*. 2019, IEA: Paris, France.
64. Jeswani, H.K., A. Chilvers, and A. Azapagic, Environmental sustainability of biofuels: A review. *Proc Royal Soc A Math Phys Eng Sci*, 2020. **476**(2243): p. 20200351.
65. European, C., et al., *The Use of Woody Biomass for Energy Production in the EU*. 2021: Publications Office.
66. Davies, J.H. and D.R. Davies, Earth's surface heat flux. *Solid Earth*, 2010. **1**(1): p. 5–24.
67. (EI), E.I., *Statistical Review of World Energy*. 2023, Energy Institute (EI).
68. Lund, J.W., *29. Historical Impacts of Geothermal Resources on the People of North America*. Stories from a Heated Earth: Our Geothermal Heritage, 1999. **19**: p. 451.
69. Redko, A., O. Redko, and R. DiPippo, 5 - Heating with geothermal systems, in *Low-Temperature Energy Systems with Applications of Renewable Energy*, A. Redko, O. Redko, and R. DiPippo, Editors. 2020, Academic Press. p. 177–224.
70. Batini, F., et al., Geological features of Larderello-Travale and Mt. Amiata geothermal areas (southern Tuscany, Italy). *Episodes J Int Geosci*, 2003. **26**(3): p. 239–244.
71. Lund, J.W. and A.N. Toth, Direct utilization of geothermal energy 2020 worldwide review. *Geothermics*, 2021. **90**: p. 101915.

72. IGA, I.A., *Global Geothermal Market and Technology Assessment*. 2023, International Renewable Energy Agency: Abu Dhabi.

73. Stefansson, V. World geothermal assessment, in *Proceedings World Geothermal Congress*. 2005.

74. Augustine, C., S. Fisher, J. Ho, I. Warren, and E. Witter, *Enhanced Geothermal Shot Analysis for the Geothermal Technologies Office*. 2023, National Renewable Energy Lab.(NREL): Golden, CO, USA.

75. Norbeck, A., et al., Age and referral route impact the access to diagnosis for women with advanced ovarian cancer. *J Multidiscip Healthc*, 2023. **16**: p. 1239–1248.

76. Richter, A. *ThinkGeoEnergy's Top 10 Geothermal Countries 2022 – Power Generation Capacity (MW)*. 2023; Available from: https://www.thinkgeoenergy.com/thinkgeoenergys-top-10-geothermal-countries-2022-power-generation-capacity-mw/

77. Boden, D.R., *Geologic Fundamentals of Geothermal Energy*. 2016: CRC Press.

78. Glassley, W.E., *Geothermal Energy: Renewable Energy and the Environment*. 2015: CRC Press.

79. Tyler, R.H., W.G. Henning, and C.W. Hamilton, Tidal heating in a magma ocean within Jupiter's moon Io. *Astrophys J Suppl Ser*, 2015. **218**(2): p. 22.

80. Gando, A., et al., Partial radiogenic heat model for Earth revealed by geoneutrino measurements. *Nat Geosci*, 2011. **4**: p. 647–651.

81. Favorito, J.E., et al., Heat flux from a vapor-dominated hydrothermal field beneath Yellowstone Lake. *J Geophys Res: Solid Earth*, 2021. **126**(5): p. e2020JB021098.

82. Lucazeau, F., Analysis and mapping of an updated terrestrial heat flow data set. *Geochem Geophys Geosyst*, 2019. **20**(8): p. 4001–4024.

83. Barbier, E., Geothermal energy technology and current status: An overview. *Renew Sust Energy Rev*, 2002. **6**(1–2): p. 3–65.

84. Ingrid Stober, K.B., *Geothermal Energy-From Theoretical Models to Exploration and Development*. 2021, Switzerland: Springer Cham.

85. Jaudin, F., M. Le Brun, V. Bouchot, and C. Dezaye, Atlases of geothermal waters and energy resources in Poland. *Environ Earth Sci*, 2014. **74**(12): p. 7487–7495.

86. Jaudin, F., M. Le Brun, V. Bouchot, and C. Dezaye, *French Geothermal Resources Survey BRGM Contribution to the Market Study in the LOW-BIN Project*. 2009: France. p. 39.

87. Bertani, R., Geothermal power generation in the world 2010–2014 update report. *Geothermics*, 2016. **60**: p. 31–43.

88. Duffield, W.A. and J.H. Sass, *Geothermal Energy: Clean Power from the Earth's Heat*. Vol. 1249. 2003: US Geological Survey.

89. Gniese, C., et al., Relevance of deep-subsurface microbiology for underground gas storage and geothermal energy production. *Adv Biochem Eng Biotechnol*, 2014. **142**: p. 95–121.

90. Davidova, I.A., et al., Involvement of thermophilic archaea in the biocorrosion of oil pipelines. *Environ Microbiol*, 2012. **14**(7): p. 1762–1771.

91. Lerm, S., et al., Thermal effects on microbial composition and microbiologically induced corrosion and mineral precipitation affecting operation of a geothermal plant in a deep saline aquifer. *Extremophiles*, 2013. **17**(2): p. 311–327.

92. Madirisha, M., R. Hack, and F. van der Meer, Simulated microbial corrosion in oil, gas and non-volcanic geothermal energy installations: The role of biofilm on pipeline corrosion. *Energy Rep*, 2022. **8**: p. 2964–2975.

93. Power, J.F., et al., Temporal dynamics of geothermal microbial communities in Aotearoa-New Zealand. *Front Microbiol*, 2023. **14**: p. 1094311.

94. Torres-Sanchez, R., et al., High temperature microbial corrosion in the condenser of a geothermal electric power unit. *Mater Perform*, 1997. **36**: p. 43–46.

95. Koohi-Fayegh, S. and M.A. Rosen, A review of energy storage types, applications and recent developments. *J Energy Storage*, 2020. 27. doi:10.1016/j.est.2019.101047

96. Guney, M.S. and Y. Tepe, Classification and assessment of energy storage systems. *Renew Sust Energy Rev*, 2017. **75**: p. 1187–1197.

97. Luo, X., et al., Overview of current development in electrical energy storage technologies and the application potential in power system operation. *Appl Energy*, 2015. **137**: p. 511–536.

98. Aneke, M. and M. Wang, Energy storage technologies and real life applications – A state of the art review. *Appl Energy*, 2016. **179**: p. 350–377.

99. Caglayan, D.G., et al., Technical potential of salt caverns for hydrogen storage in Europe. *Int J Hydrog Energy*, 2020. **45**(11): p. 6793–6805.

100. Portarapillo, M. and A. Di Benedetto, Risk assessment of the large-scale hydrogen storage in salt caverns. *Energies*, 2021. **14**(10). doi: 10.3390/en14102856

101. Tarkowski, R., Underground hydrogen storage: Characteristics and prospects. *Renew Sust Energy Rev*, 2019. **105**: p. 86–94.

102. Uliasz-Misiak, B., et al., Prospects for the implementation of underground hydrogen storage in the EU. *Energies*, 2022. **15**(24). doi:10.3390/en15249535

103. Amid, A., D. Mignard, and M. Wilkinson, Seasonal storage of hydrogen in a depleted natural gas reservoir. *Int J Hydrog Energy*, 2016. **41**(12): p. 5549–5558.

104. Andersson, J. and S. Grönkvist, Large-scale storage of hydrogen. *Int J Hydrog Energy*, 2019. **44**(23): p. 11901–11919.

105. Bünger, U., J. Michalski, F. Crotogino, and O. Kruck, Large-scale underground storage of hydrogen for the grid integration of renewable energy and other applications, in *Compendium of Hydrogen Energy*. 2016. p. 133–163.

106. Dopffel, N., S. Jansen, and J. Gerritse, Microbial side effects of underground hydrogen storage – Knowledge gaps, risks and opportunities for successful implementation. *Int J Hydrog Energy*, 2021. **46**(12): p. 8594–8606.

107. Matos, C.R., J.F. Carneiro, and P.P. Silva, Overview of large-scale underground energy storage technologies for integration of renewable energies and criteria for reservoir identification. *J Energy Storage*, 2019. **21**: p. 241–258.

108. Enning, D. and J. Garrelfs, Corrosion of iron by sulfate-reducing bacteria: new views of an old problem. *Appl Environ Microbiol*, 2014. **80**(4): p. 1226–1236.

109. Knisz, J., et al., Microbiologically influenced corrosion - More than just microorganisms. *FEMS Microbiol Rev*, 2023. doi: 10.1093/femsre/fuad041

110. Thauer, R.K., et al., Hydrogenases from methanogenic archaea, nickel, a novel cofactor, and H_2 storage. *Annu Rev Biochem*, 2010. **79**: p. 507–536.

111. Dopffel, N., et al., Microbial hydrogen consumption leads to a significant pH increase under high-saline-conditions: implications for hydrogen storage in salt caverns. *Sci Rep*, 2023. **13**(1): p. 10564.

112. Hagemann, B., et al., Hydrogenization of underground storage of natural gas. *Comput Geosci*, 2015. **20**(3): p. 595–606.

113. Eddaoui, N., et al., Impact of pore clogging by bacteria on underground hydrogen storage. *Transp Porous Media*, 2021. **139**(1): p. 89–108.

114. Feldmann, F., B. Hagemann, L. Ganzer, and M. Panfilov, Numerical simulation of hydrodynamic and gas mixing processes in underground hydrogen storages. *Environ Earth Sci*, 2016. **75**(16). doi:10.1007/s12665-016-5948-z

115. Liu, N., A.R. Kovscek, M.A. Fernø, and N. Dopffel, Pore-scale study of microbial hydrogen consumption and wettability alteration during underground hydrogen storage. *Front Energy Res*, 2023. 11. doi:10.3389/fenrg.2023.1124621

116. Hemme, C. and W. van Berk, Potential risk of H 2 S generation and release in salt cavern gas storage. *J Nat Gas Sci Eng*, 2017. **47**: p. 114–123.
117. Dohrmann, A.B. and M. Krüger, Microbial H$_2$ consumption by a formation fluid from a natural gas field at high-pressure conditions relevant for underground H$_2$ storage. *Environ Sci Technol*, 2023. **57**(2): p. 1092–1102.
118. Donohue, T.J. and R.J. Cogdell, Microorganisms and clean energy. *Nat Rev Microbiol*, 2006. **4**(11): p. 800–800.
119. Santos-Merino, M., et al., Highlighting the potential of Synechococcus elongatus PCC 7942 as platform to produce alpha-linolenic acid through an updated genome-scale metabolic modeling. *Front Microbiol*, 2023. **14**: p. 1126030.
120. Deng, X. and A. Okamoto, Electrode potential dependency of single-cell activity identifies the energetics of slow microbial electron uptake process. *Front Microbiol*, 2018. **9**: p. 2744.
121. Ramamurthy, P.C., et al., Microbial biotechnological approaches: Renewable bioprocessing for the future energy systems. *Microb Cell Fact*, 2021. **20**(1): p. 55.
122. Fredrickson, J.K., et al., Towards environmental systems biology of Shewanella. *Nat Rev Microbiol*, 2008. **6**(8): p. 592–603.
123. Zabranska, J. and D. Pokorna, Bioconversion of carbon dioxide to methane using hydrogen and hydrogenotrophic methanogens. *Biotechnol Adv*, 2018. **36**(3): p. 707–720.
124. Timilsina, G.R., Biofuels in the long-run global energy supply mix for transportation. *Philos Trans A Math Phys Eng Sci*, 2014. **372**(2006): p. 20120323.
125. Nakayama, C.R., E.D. Penteado, R.T.D. Duarte, A.J. Giachini, and F.T. Saia, Improved methanogenic communities for biogas production, in *Improving Biogas Production*. 2019. p. 69–98.
126. Boas, J.V., et al., Review on microbial fuel cells applications, developments and costs. *J Environ Manage*, 2022. **307**: p. 114525.
127. Gieg, L.M., T.R. Jack, and J.M. Foght, Biological souring and mitigation in oil reservoirs. *Appl Microbiol Biotechnol*, 2011. **92**(2): p. 263–282.
128. An, B.A., Y. Shen, and G. Voordouw, Control of sulfide production in high salinity Bakken shale oil reservoirs by halophilic bacteria reducing nitrate to nitrite. *Front Microbiol*, 2017. **8**: p. 1164.
129. Hashemi, L., M. Boon, W. Glerum, R. Farajzadeh, and H. Hajibeygi, A comparative study for H2–CH4 mixture wettability in sandstone porous rocks relevant to underground hydrogen storage. *Adv Wat Resour*, 2022. **163**. doi:10.1016/j.advwatres.2022.104165
130. Foght, J.M., L.M. Gieg, and T. Siddique, The microbiology of oil sands tailings: Past, present, future. *FEMS Microbiol Ecol*, 2017. **93**(5). doi:10.1093/femsec/fix034
131. da Silva, M.L., et al., Effects of nitrate injection on microbial enhanced oil recovery and oilfield reservoir souring. *Appl Biochem Biotechnol*, 2014. **174**(5): p. 1810–1821.
132. Małachowska, A., N. Łukasik, J. Mioduska, and J. Gębicki, Hydrogen storage in geological formations—The potential of salt caverns. *Energies*, 2022. **15**(14). doi:10.3390/en15145038
133. Greening, C. and E. Boyd, Editorial: Microbial hydrogen metabolism. *Front Microbiol*, 2020. **11**: p. 56.
134. Rodriguez-Valera, F., G. Juez, and D.J. Kushner, Halobacterium mediterranei spec, nov., a new carbohydrate-utilizing extreme halophile. *Syst Appl Microbiol*, 1983. **4**(3): p. 369–381.
135. Oren, A., Life at high salt concentrations, intracellular KCl concentrations, and acidic proteomes. *Front Microbiol*, 2013. **4**: p. 315.
136. Oren, A., Microbial life at high salt concentrations: Phylogenetic and metabolic diversity. *Saline Syst*, 2008. **4**: p. 2.

137. Oren, A., Halophilic microbial communities and their environments. *Curr Opin Biotechnol*, 2015. **33**: p. 119–124.
138. Oren, A., Thermodynamic limits to microbial life at high salt concentrations. *Environ Microbiol*, 2011. **13**(8): p. 1908–1923.
139. Lahme, S., J. Mand, J. Longwell, R. Smith, and D. Enning, Severe corrosion of carbon steel in oil field produced water can be linked to methanogenic archaea containing a special type of [NiFe] hydrogenase. *Appl Environ Microbiol*, 2021. **87**(3). doi:10.1128/AEM.01819-20
140. Muhammed, N.S., et al., A review on underground hydrogen storage: Insight into geological sites, influencing factors and future outlook. *Energy Rep*, 2022. **8**: p. 461–499.
141. Muhammed, N.S., et al., Hydrogen storage in depleted gas reservoirs: A comprehensive review. *Fuel*, 2023. **337**: p. 127032.

Index

Pages in *italics* refer to figures and pages in **bold** refer to tables.

For Product Safety Concerns and Information please contact our EU
representative GPSR@taylorandfrancis.com
Taylor & Francis Verlag GmbH, Kaufingerstraße 24, 80331 München, Germany

www.ingramcontent.com/pod-product-compliance
Lightning Source LLC
Chambersburg PA
CBHW060356220326
41598CB00023B/2937

9 781032 269566